Active Strike-Slip and Collisional Tectonics of the Northern Caribbean Plate Boundary Zone

Edited by

James F. Dolan
Department of Earth Sciences
University of Southern California
Los Angeles, California 90089

and

Paul Mann
Institute for Geophysics
University of Texas
8701 North Mopac Boulevard
Austin, Texas 78759-8345

THE GEOLOGICAL SOCIETY OF AMERICA
·1888·

SPECIAL PAPER

326

1998

Published by The Geological Society of America, Inc.
3300 Penrose Place, P.O. Box 9140, Boulder, Colorado 80301

Printed in U.S.A.

GSA Books Science Editor Abhijit Basu

Library of Congress Cataloging-in-Publication Data

Active strike-slip and collisional tectonics of the Northern Caribbean plate boundary zone / edited by James F. Dolan and Paul Mann.
p. cm. -- (Special paper ; 326)
Includes bibliographical references and index.
ISBN 0-8137-2326-4
1. Plate tectonics--Caribbean Area. 2. Plate tectonics--South America. 3. Plate tectonics--Hispaniola. 4. Geology--Caribbean Area. 5. Geology--South America. 6. Geology--Hispaniola. I. Dolan, James F., 1959- . II. Mann, Paul, 1956- . III. Series: Special papers (Geological Society of America) ; 326.
QE511.4.A25 1998
551.1′36′09729--dc21 98-28431
CIP

Cover: Onshore topography and offshore bathymetry of the north-central and northeastern Caribbean-North America plate boundary zone (Hispaniola is the island in the center of the map; the eastern edges of the islands of Cuba and Jamaica are on the left side of the map; and Puerto Rico and the Virgin Islands are shown on the right side of the map). The distance across the map is ~1,500 km. The offshore bathymetric map combines the predicted seafloor topography from satellite altimetry with ship depth soundings (Smith and Sandwell, 1997). Onshore topography is based on the GTOPO3 database compiled by the U.S. Geological Survey's EROS data center. The image was created by Lisa Gahagan at the University of Texas at Austin, Institute for Geophysics, using the GMT software system (Wessel and Smith, 1991).

Smith, W. H. F., and Sandwell, D. T., 1997, Global sea floor topography from satellite altimetry and ship depth soundings: Science, v. 277, p. 1956–1962.

Wessel, P., and Smith, W. H. F., 1991, Free software helps map and display data: EOS (Transactions, American Geophysical Union), v. 72, p. 445–446.

10 9 8 7 6 5 4 3 2 1

Contents

Preface

TECTONIC SETTING

The purpose of this book is to present four well-integrated chapters that focus on the complex active tectonics of the northern margin of the Caribbean plate (Fig. 1). Geologic and earthquake studies indicate that the Caribbean plate is moving eastward relative to North and South America and that this movement is accommodated by left-lateral faults along its northern boundary within continental, island arc, and oceanic lithosphere bounding the North America plate, and right-lateral faults along its southern boundary within continental, island arc, and oceanic lithosphere bounding the South America plate. Oceanic lithosphere is consumed along the eastern edge of the Caribbean at the Lesser Antilles subduction zone, and oceanic lithosphere of the Cocos plate is consumed along the western edge of the plate at the Middle America subduction zone (cf. Mann et al., 1990, for a detailed review of neotectonic features of the Caribbean plate).

Most of the chapters in this volume focus on the complex, late Neogene to present-day tectonic interaction of the northeastern margin of the Caribbean plate and the southeastern extension of the Bahama platform north of Hispaniola (Fig. 1). Studies by Pindell and Barrett (1990) and others have shown that the Caribbean plate was Pacific derived and entered the space between North and South America during Late Cretaceous time

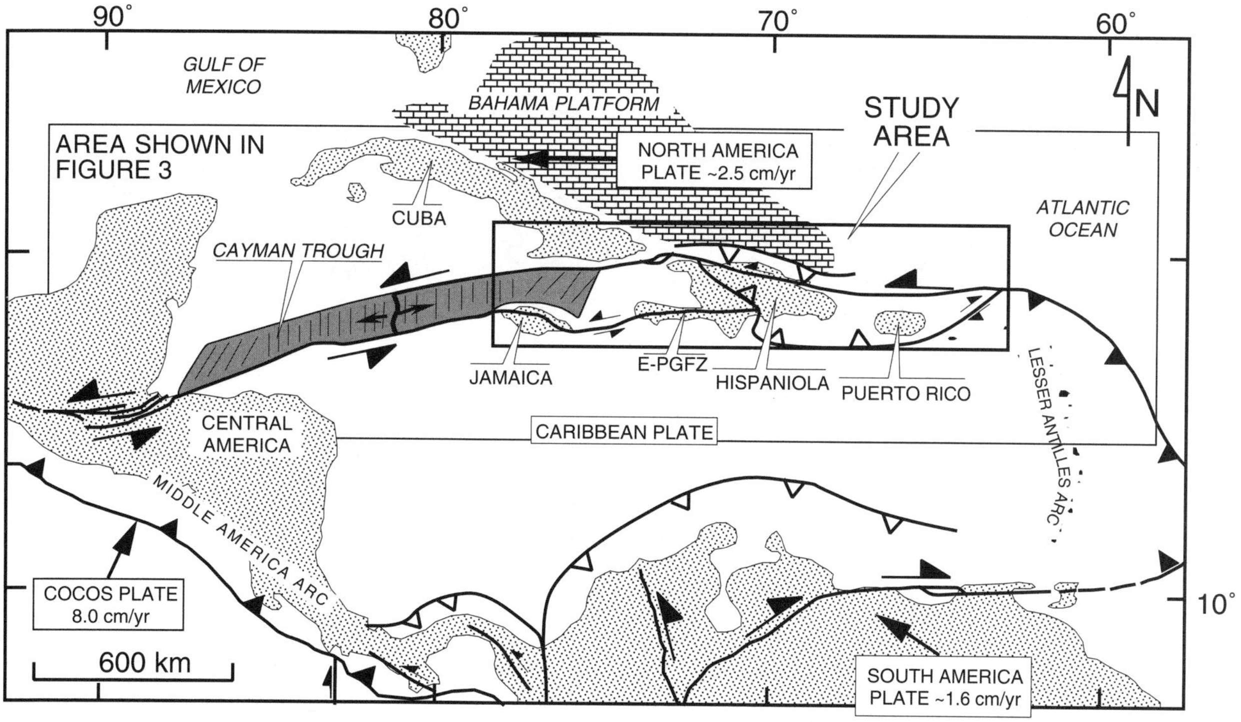

Figure 1. Present-day plate boundary faults of the Caribbean plate modified from Mann et al. (1991b) and rates of motion relative to the Caribbean plate from DeMets et al. (1990, 1996). EPGFZ is an abbreviation for Enriquillo-Plantain Garden fault zone. Boxed area of northern Caribbean plate boundary indicates area of maps shown in Figure 3.

(Fig. 2A). Overall motion of the Caribbean plate during Cenozoic time was from west to east and was accommodated by oblique collision and strike-slip faulting along the northern and southern margins of the plate (Fig. 2B–F).

In the northern Caribbean region, late Paleocene–early Eocene convergence of the Caribbean plate against North America proceeded in a clockwise direction from an early phase of northward to northeastward convergence with outliers of the Bahama platform along the edge of the North America plate in western Cuba (Bralower and Iturralde-Vinent, 1997; Gordon et al., 1997) (Fig. 2B). Subsequent convergence migrated to the southeast with east-northeastward overthrusting of arc rocks of the Caribbean plate onto the Bahama platform of central and eastern Cuba during the middle Eocene (Fig. 2C). A Neogene phase of eastward

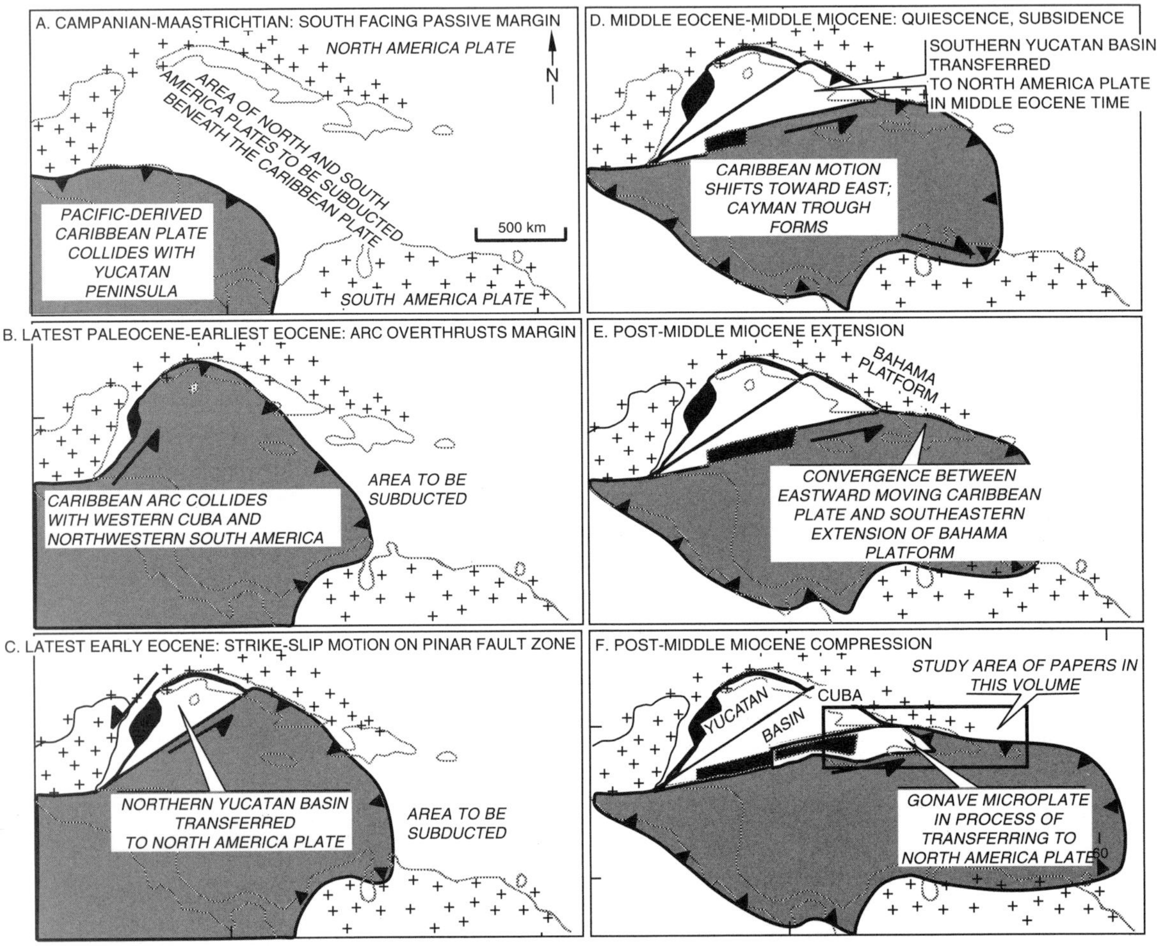

Figure 2. Tectonic setting of the Caribbean plate from Maastrichtian to Holocene time (modified from Pindell and Barrett, 1990, and Mann et al., 1995). (A) Initial collision of a Pacific-derived, eastward-moving Caribbean plate and North and South America plates occurred in Central America and western South America during Maastrichtian time. (B) In latest Paleocene/earliest Eocene time, continued movement of the arc and trailing Caribbean plate leads to collision with passive margins in western Cuba and northwestern South America. (C) In latest early Eocene time, migration of collision from western to central Cuba induces clockwise rotation of the direction of collision as the forward progress of the arc is stopped by the entry of thicker crust of the Bahama platform into the subduction zone. As a result, the northern Yucatan basin is transferred from the Caribbean to the North America plates. (D) Middle Eocene/early Miocene collision can advance no farther to the north-northeast above the Bahama platform, and the present-day Cayman trough forms as the Caribbean plate is rotated clockwise and is twisted in a more easterly direction. (E) In post-early Miocene time, oblique Miocene collision between the Caribbean plate and southeastern Bahama platform produces tectonic transpression and uplift in Hispaniola. In northern South America, oblique collision continues to migrate along the passive margin. (F) In post-early Miocene time, transpression between the Caribbean plate and southeastern Bahama platform has detached the Gonave microplate along the Enriquillo-Plantain Garden fault zone. Less transpression is observed in South America because the direction of plate motion is roughly parallel to the passive margin.

oblique convergence between the Caribbean plate in Hispaniola and the southeastern extension of the Bahama platform occurs today and forms the main subject of this book (Fig. 2D, E, F).

NORTHERN CARIBBEAN PLATE BOUNDARY ZONE

The North America–Caribbean plate boundary zone consists of a 100- to 250-km-wide seismogenic zone of mainly left-lateral strike-slip deformation extending over 2,000 km along the northern edge of the Caribbean Sea (Fig. 3A). The dominant structural element in the central plate boundary zone is the Cayman trough, a submarine pull-apart basin formed by at least 1,100 km of oceanic spreading at the mid-Cayman spreading center, a 100-km-long jog between left-lateral faults of the plate boundary (CAYTROUGH, 1979). The spreading center has been active since middle Eocene time and is currently spreading at a rate of ~15 mm/yr (Rosencrantz, et al., 1988). In the western Cayman trough, the plate boundary appears to be a single, active strike-slip fault exhibiting locally complex restraining bends (Mann et al., 1991a).

To the east of the Cayman trough in Jamaica, Hispaniola, Puerto Rico, and the Virgin Islands, the plate boundary is especially wide, with a seismogenic zone up to 250 km wide at the longitude of the island of Hispaniola (Fig. 3A). The width of the plate boundary in this area is reflected in:

(1) *An active, multibranched zone of both left-lateral strike-slip and contractional faulting*. Two distinct strike-slip fault systems are present in onland areas (Fig. 3A). In general, convergent faults and accretionary prisms form the break between abyssal depths of the Caribbean Sea and the Atlantic Ocean, and strike-slip faults occupy an inboard position on the higher standing land and shallow marine areas (Calais and Mercier de Lépinay, 1995; Dolan et al., this volume). The northern strand of the strike-slip zone is known as the Septentrional-Oriente fault zone (Mann et al., this volume), and the southern strand is known as the Enriquillo-Plantain Garden fault zone (Mann et al., 1995).

(2) *A dual zone of shallow active seismicity west of 71°W* (Sykes et al., 1982; Calais et al., 1992, this volume). The vague, southern belt of seismicity appears most intense around the island of Jamaica although some of this intensity reflects a denser distribution of seismographs on Jamaica (Fig. 1A). The location of events in other areas (e.g., southern Hispaniola) is poor because most stations are in an east-west line and because Haiti (which occupies the western third of Hispaniola) and the Dominican Republic (which occupies the eastern two-thirds of the island) do not have operational seismographic networks. Local events recorded by the seismograph network in southern Cuba are not reported to the International Seismological Centre (ISC), and these local network data remain unpublished (E. Calais, personal communication, 1993). The northern zone of seismicity is parallel to the Septentrional-Oriente fault zone and to the offshore, south-dipping collisional contact between Hispaniola and the Bahama platform (Dolan and Wald, this volume); the southern zone of seismicity is parallel to the Enriquillo-Plantain Garden fault zone west of central Hispaniola (Mann et al., 1995), and to the Los Muertos deformed belt to the east of central Hispaniola (Masson and Scanlon, 1991).

(3) *Dual rupture zones of large, historical earthquakes.* A northern zone of historical seismicity passes through northern Hispaniola and southern Cuba, and a southern zone passes through southern Hispaniola and Jamaica (Kelleher et al., 1973) (Fig. 1B). Many areas of the Greater Antilles were settled by Spanish colonists by the first quarter of the 16th century and, therefore, this region has the longest and most complete record of destructive, historical earthquakes in the Western Hemisphere. The northern zone of historical seismicity is parallel to the Septentrional-Oriente fault zone, the contractional structures of the northern Hispaniola deformed belt, and the Santiago deformed belt off southern Cuba (Mann et al., this volume; Dolan et al., this volume; Calais et al., this volume). The southern zone of historical seismicity is parallel to the Enriquillo-Plantain Garden fault zone from central Hispaniola westward and the Los Muertos deformed belt to the east of central Hispaniola. Marine surveys have shown that the northern zone of strike-slip faulting extends offshore eastward of Hispaniola and along the northern coast of Puerto Rico (Masson and Scanlon, 1991; Grindlay et al., 1997; Dolan et al., this volume).

(4) *The widest and most elevated zone of uplifted late Quaternary marine terraces* in the North America–Caribbean plate boundary zone is centered on western Hispaniola (Horsfield, 1975) (Fig. 3C). The area of maximum uplift straddles most of the western part of Hispaniola, which is also characterized by the widest onland segment of the northeastern plate boundary zone and the highest topography in the northern Caribbean (maximum elevation in central Hispaniola = 3,081 m). Plots of the mean annual change in land level at tide gauges in Jamaica and Hispaniola suggest variable vertical movements (Emery and Aubrey, 1991). According to these authors, the three tide gauges in Hispaniola have the two fastest sinking stations (Port-au-Prince, Haiti: –12.3 mm/yr; Puerto Plata, Dominican Republic: –4.4 mm/yr) and the only rising station in the Caribbean region (Barahona, Dominican Republic: +2.3 mm/yr). Mann et al. (1995) have compiled much slower uplift rates for the island of Hispaniola based on isotopic dating of raised late Quaternary coral reef terraces. These data indicate that the area of maximum uplift in Hispaniola is along the west coast of the island in Haiti where uplift rates range from 0.19 to 0.37 mm/yr. Uplift of the eastern part of the island in the Dominican Republic has been slow or stable over the past 125 ka. Data derived from the elevations of late Quaternary coral terraces along the northwestern coast of Puerto Rico also indicate slow rates of uplift ranging from 0.03 mm/yr and 0.05 mm/yr (Taggert and Joyce, 1990).

MODEL FOR MICROPLATE TECTONICS IN THE NORTHERN CARIBBEAN REGION

Mann et al. (1995) proposed that these four characteristics of the northern margin of the Caribbean plate may represent the continuation of the much longer term tectonic process of oblique collision, microplate detachment, and eastward motion of a smaller Caribbean plate as shown in Figure 2A–F. In this hypoth-

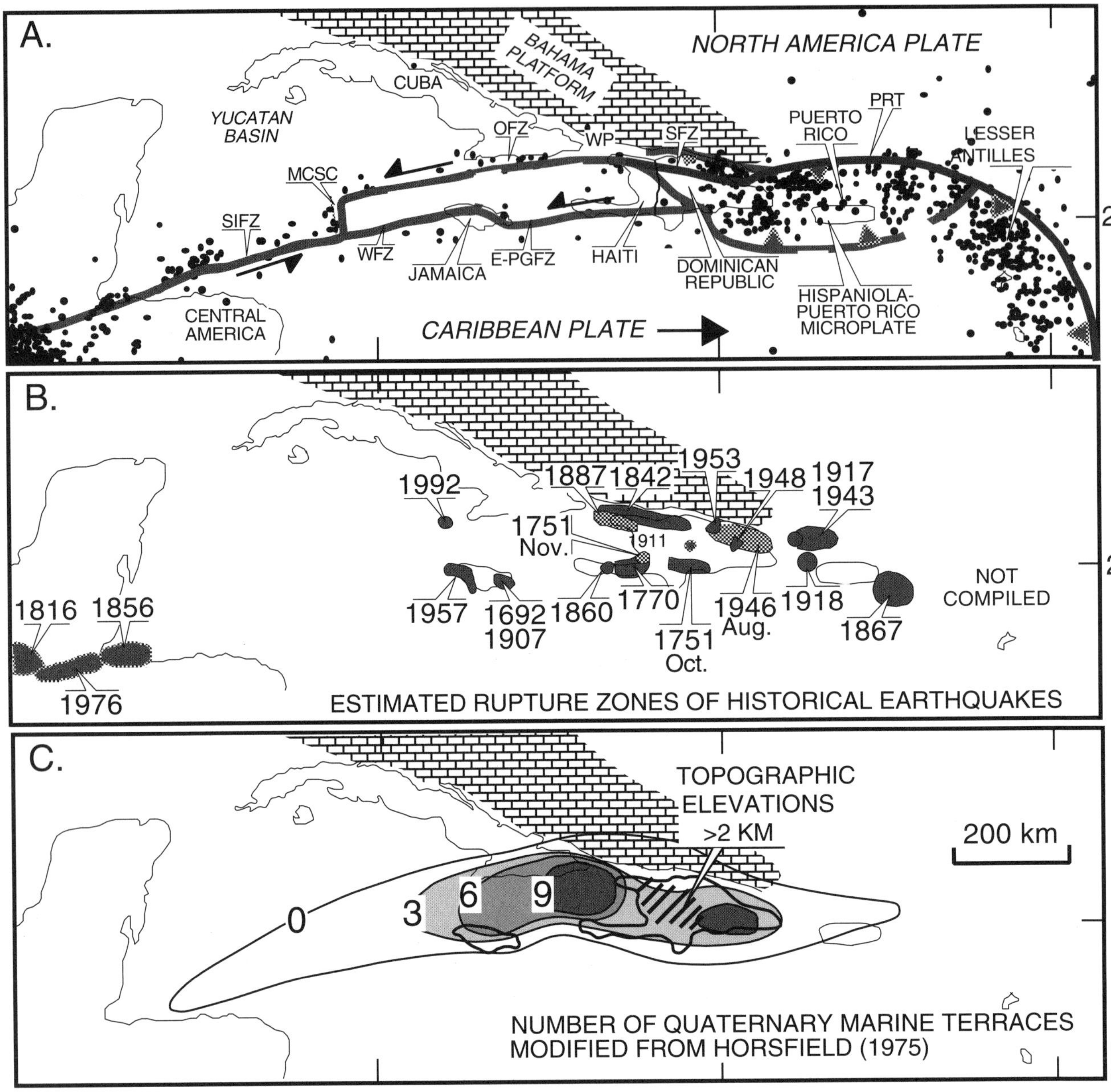

Figure 3. (A) Tectonic setting and seismicity (1963–1992, M > 4.5, source: International Seismological Centre) of the North America–Caribbean plate boundary zone (modified from Mann et al., 1995). Heavy gray lines indicate major plate boundary faults. Key to abbreviations: SIFZ = Swan Islands fault zone; WFZ = Walton fault zone; MCSC = Mid-Cayman spreading center; OFZ = Oriente fault zone; E-PGFZ = Enriquillo-Plantain Garden fault zone; SFZ = Septentrional fault zone; JP = Jamaica Passage, WP = Windward Passage; MT = Muertos trench; PRT = Puerto Rico trench; AP = Anegada Passage. (B) Estimated rupture zones of northern Caribbean historical earthquakes modified from Robinson et al. (1958), Kelleher et al. (1973), Plafker (1976), Osieki (1981), and White (1985). Note approximate parallelism of earthquake rupture zones to faults shown in A. Rupture zones were determined prior to the discovery of the continuity of the Enriquillo-Plantain Garden fault zone. (C) Number of Quaternary marine terraces modified from Horsfield (1975) and area of topographic elevation >2 km on the island of Hispaniola (Haiti and Dominican Republic).

esis, the Septentrional-Oriente fault zones and the Enriquillo-Plantain Garden fault zones bound the Gonave microplate, a 190,000 km^2 area of the northeastern Caribbean plate that is in the process of shearing off the Caribbean plate and accreting to the North America plate. Late Cenozoic transpression between the southeastern Bahama platform and the Caribbean plate in Hispaniola has inhibited the eastward motion of the northeastern corner of the plate. Transpression is manifested in western and central Hispaniola by the formation of regional-scale folds that correspond to present-day, anticlinal topographic mountain chains continuous with offshore anticlinal ridges. Areas of most rapid Quaternary uplift determined from onland coral reefs 125 ka and younger coincide with the axial traces of these folds (Fig. 3C). Formation of the 1,200-km-long Enriquillo-Plantain Garden-Walton fault zone as a "bypass" strike-slip fault has isolated the southern edge of the Gonave microplate and is allowing continued, unimpeded eastward motion of a smaller Caribbean plate past the zone of late Neogene convergence and Quaternary uplift of coral reefs in Hispaniola. A similar history of oblique collision and microplate detachment has been proposed for the Paleogene deformation of western Cuba (Gordon et al., 1997).

This book was motivated in part by the need to better understand the complex tectonics of the broad plate boundary zone in the Hispaniola region and to investigate the nature of the active contact zone between Caribbean plate rocks in the Hispaniola area and underthrusting Bahama platform rocks offshore. Such information is critical for testing the regional tectonic hypothesis shown in Figure 2A–F and gaining a better understanding of the mechanism for destructive earthquakes in the region.

SEISMIC HAZARDS AND STRAIN PARTITIONING IN THE NORTH-CENTRAL CARIBBEAN

The chapters in this book delineate several major seismic sources within the northern Caribbean plate boundary zone, including the left-lateral Septentrional-Oriente fault zone (Mann et al., this volume; Calais et al., this volume), and the south-dipping, plate-interface thrust fault offshore northern Hispaniola and Puerto Rico (Dolan et al., this volume; Dolan and Wald, this volume). Dolan et al. (this volume) show that these two major fault systems represent an excellent example of a phenomenon commonly known as strain partitioning. In a strain-partitioned system, obliquely convergent (or transpressional) deformation is split, or partitioned, onto two parallel fault strands, a strike-slip strand that accommodates predominantly margin-parallel motion, and a thrust strand that accommodates predominantly margin-perpendicular motion (Fig. 4) (e.g., Platt, 1993).

Early reconstructions of the plate boundary in the northern Caribbean assumed that plate convergence was entirely accommodated by thrust faulting and subduction (e.g., reconstructions of Sykes et al., 1982) or by strike-slip faulting with limited thrusting at strike-slip fault bends (Heubeck and Mann, 1991). These early, rigid plate kinematic models did not consider the possibility of the combined paleogeographic and strain effects of simultaneous strike-slip and subduction on roughly parallel fault zones. Such fault geometries and the strain partitioning process they implied were pointed out when marine side-scan sonar mapping of the plate boundary zone was completed offshore southern Cuba (Calais et al., 1989) in the Windward Passage area (Calais and Mercier de Lépinay, 1995), offshore northwestern (Dillon et al., 1992) and north-central Hispaniola (Dolan et al., this volume), and south of the Puerto Rico trench (Masson and Scanlon, 1991; Grindlay et al., 1997). These data suggest that the northern margin of the Caribbean plate cannot be thought of as a simple boundary between two rigid plates (i.e., the Caribbean and North America plates). Rather, the northern Caribbean plate boundary zone (NCPBZ) is characterized by numerous active fault systems that bound several major blocks, which move relative to one another within the plate boundary zone (Mann et al., 1984, 1995; Byrne et al., 1985; Calais and Mercier de Lépinay, 1993). This view of a wide, diffuse plate boundary zone, which is based primarily on geologic and seismicity data, is now being confirmed by Global Positioning System (GPS)-based geodetic measurements that show motion occurring over the 250 m width of the plate boundary zone (DeMets et al., 1996).

The two major fault systems of the northern deformation belt of the NCPBZ represent significant seismic hazards to the densely populated islands of Hispaniola and Puerto Rico. In particular, the proximity of these faults to major population centers in northern Hispaniola and Puerto Rico (e.g., Santiago, Dominican Republic, population >700,000; and San Juan, Puerto Rico, population >2,000,000), coupled with the commonly poor construction practices in much of the region, provides an ample impetus for developing a better understanding of the risks associated with major earthquakes generated by these faults. Both fault systems have generated major, destructive earthquakes. For example, Mann et al. (1984, this volume) propose that the great 1842 northwestern Hispaniola earthquake may be related to slip along the Septentrional fault, and Dolan et al. (this volume) show that the 1943 Ms 7.8 northern Mona Passage earthquake and the 1946 Ms 8.1 northeastern Hispaniola earthquake were generated by the offshore, plate-interface thrust fault.

In addition to the hazards posed by strong ground shaking, the many low-lying coastal cities of Hispaniola and Puerto Rico are subject to tsunamis caused by sea-floor deformation during major thrust earthquakes. Further compounding the problem, evidence from marine geophysical surveys for widespread, large-scale slumping and mass wasting of the northern slopes of Puerto Rico and Hispaniola (Mullins, 1991, 1992; Schwab et al., 1991; Dolan et al., this volume) suggests that strong ground shaking during major earthquakes on either the offshore thrust or the Septentrional fault could trigger large underwater landslides that could in turn generate major tsunamis, even in the absence of major tectonic displacements of the sea floor.

Liquefaction of water-rich sediments onshore Hispaniola and Puerto Rico also represents a major potential hazard to the region. Tuttle et al. (1997) have documented recent evidence for multiple liquefaction events affecting large alluvial areas of the eastern and western Cibao Valleys, and Prentice et al. (1993) documented liq-

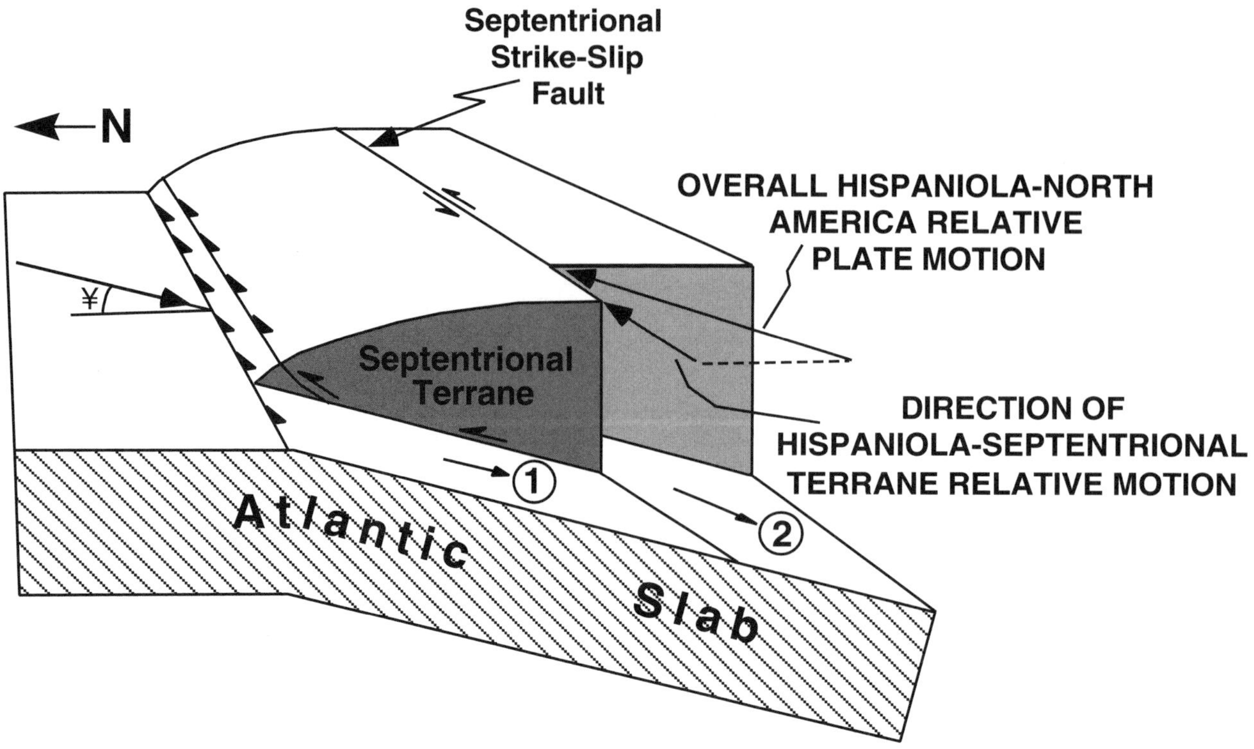

Figure 4. Schematic block diagram showing proposed mechanical relationships of parallel strike-slip and thrust faults within the obliquely convergent, strain-partitioned northern Caribbean plate boundary zone in the northern Hispaniola region (modified from Platt, 1993, to illustrate the model of Dolan et al., this volume). Number 1 shows relative motion direction between the Septentrional terrane and the underthrust Atlantic slab along the south-dipping plate interface thrust fault. Number 2 shows the relative motion direction between the main part of Hispaniola (south of the left-lateral Septentrional fault) and the underthrust slab. Note that 2 is parallel to the overall motion direction vector between Hispaniola and the North America (Atlantic) plate. Also note that 2 is subparallel to the plate boundary, whereas 1 is at a higher angle to the plate boundary due to partitioning of much of the margin-parallel slip onto the left-lateral Septentrional strike-slip fault. ¥ denotes angle between local motion vector and normal to margin at deformation front of the northern Hispaniola deformed belt.

uefaction features associated with the most recent earthquake on the central part of the Septentrional fault zone ~800 years ago.

GOALS OF THIS BOOK

The specific goals of this book include the following:

(1) Integrate offshore, marine geophysical observations of predominantly contractional features mapped from the marine geophysical survey of the northern margin of Hispaniola with geologic observations of predominantly strike-slip features recognized in the Cibao Valley of northern Hispaniola (Dominican Republic).

(2) Provide a set of base maps of the oblique collision area both onland and offshore that can be used for future, more detailed and focused studies.

(3) Integrate both offshore and onland data sets with earthquake data from the region. Are geologic, geophysical, and earthquake data sets consistent with the same tectonic processes?

(4) Use these combined data sets to better understand the complex tectonic process of "strain partitioning," where oblique plate motion is manifested in offshore thrust faulting and onland strike-slip faulting.

(5) Establish the oblique collision between the northeastern margin of the Caribbean plate and the Bahama platform as an important area to study the tectonic processes of oblique collision at all scales.

ORGANIZATION

We have organized the four chapters of this book into two sections. Part 1 consists of two chapters that describe the geologic and tectonic setting of the plate boundary, including detailed analysis of the major plate-boundary fault systems. Chapter 1 describes the marine geology and active structures of the offshore part of the northern plate boundary from central Hispaniola to northwestern Puerto Rico. This region includes the active collision zone between the overriding Paleogene arc rocks of Hispaniola and carbonate banks of the obliquely underthrusting Bahama platform. Original data include SeaMARC II side-scan bathymetric and sea-floor reflectivity maps along with single-channel seismic-reflection data. The collision zone and several major offshore fault systems are described for the first time.

Chapter 2 describes evidence for strike-slip faulting along the length of the Septentrional fault zone in late Holocene sediments of the Cibao Valley of the Dominican Republic. Original data include regional maps showing the extent of the late Holocene fault scarp over the 200-km-length of the Cibao Valley, aerial photographs, and field observations documenting the character of the fault scarp and the results of fault excavations across the fault. The fault is described in great detail for two reasons: (1) it has never been described before and was only proposed to be an active plate-boundary fault by Mann et al. (1984), and (2) the rapid rate of development and the growth of population in the Dominican Republic is destroying many of the fault zone localities. Radiocarbon dating keyed to logs of trench excavations allows us to show that the most recent movement on the fault occurred ~800 yr ago.

Part 2 also comprises two chapters, which describe and interpret the recent seismicity and major earthquakes of the plate boundary zone. Chapter 3 describes marine geophysical data and recent seismicity from the northern part of the plate boundary zone between Cuba and northwestern Hispaniola. This chapter includes a quantitative analysis of the recent seismicity data that is used to calculate a seismic slip rate for the major plate-boundary strike-slip fault in the north-central Caribbean, thus providing an independent check on previous estimates of the plate-boundary slip rate.

Chapter 4 describes a series of major thrust earthquakes that occurred along the oblique Hispaniola-Bahama collision zone offshore northeastern Hispaniola between 1943 and 1953. The database for these earthquake studies includes both analog seismograms recorded during the great earthquakes of the 1940s and 1950s and digital teleseismic data archived by the International Seismological Centre and Harvard focal mechanism library. These data, including new focal mechanisms for the six largest events in the 1943–1953 sequence, provide insights into both the seismic hazards facing the rapidly growing and densely populated northern Caribbean region, as well as the relative motion direction between the North America plate and the Hispaniola and Puerto Rico blocks of the northern Caribbean plate boundary zone.

The combined chapters allow us to illustrate an obliquely convergent boundary characterized by predominantly contractional, late Neogene structures in the offshore area and predominantly strike-slip late Holocene faults in the onshore area of the Cibao Valley of Hispaniola. The mechanical and temporal relations in this zone of strain partitioning are a recurrent theme discussed in several of the chapters.

POTENTIAL TOPICS OF FUTURE RESEARCH EFFORTS

Several important aspects of the active tectonics of the northern Caribbean margin remain relatively poorly understood and would benefit greatly from future study. The following list, while by no means comprehensive, represents what we consider to be some of the more important and interesting questions in Caribbean tectonics that are not yet completely resolved.

What is the deep geometry of the plate boundary beneath northwestern Hispaniola, west of the western limit of the seismically defined underthrust slab? Is there a south-dipping slab beneath northwestern Hispaniola? If so, is this the source of thrust events at 20–30 km depths beneath the Cibao Valley reported by Kafka (1980)?

Dolan and Wald (this volume) suggest that collisional parts of the south-dipping plate interface are strongly coupled and capable of producing great thrust earthquakes (e.g., 1946 northeastern Hispaniola earthquake). What is the degree of interplate mechanical coupling along parts of the plate-interface thrust fault that are not currently undergoing collision?

Is the north-dipping plate-interface thrust fault beneath the Los Muertos deformed belt along the southern edge of the NCPBZ strongly coupled, as suggested by the moderate (Ms 6.7) 1984 San Pedro basin earthquake (Byrne et al., 1985)? If so, what are the seismic hazards posed by this very large potential seismic source to southeastern Hispaniola and southern Puerto Rico?

What is the deep geometry of the Septentrional fault beneath Hispaniola? Does it merge into a gently south-dipping decollement along the top of underthrust North America (Atlantic) lithosphere, as proposed by Dolan et al. (this volume)? Or does it exhibit some more complex relationship with the underthrust slab?

What are the true slip rates for the major onshore (Septentrional fault) and offshore (thrust faults of the northern Hispaniola deformed belt) structures within the northern Caribbean plate boundary zone? More detailed and longer term Global Positioning System geodetic measurements may be able to answer this question in coming years. Continued onshore trench studies may also help better define the slip rate of the Septentrional fault onshore Hispaniola (e.g., Mann et al., this volume).

What are the seismic hazards posed to the urban centers of northern Hispaniola and Puerto Rico by the faults of the Septentrional-Oriente fault system and offshore thrust faults of the northern Hispaniola deformed belt?

How will the deep sedimentary fill of the Cibao Valley of northern Hispaniola affect strong ground motions in future earthquakes generated by these faults?

What are the tsunami hazards facing the densely populated, low-lying coastal regions of southern Cuba, Hispaniola, Puerto Rico, and the Virgin Islands?

What role, if any, does the Camu fault play in accommodating active plate motion in the NCPBZ? Is the Camu fault even an active structure? Detailed geomorphologic and paleoseismological studies of the fault may resolve these questions.

Why are young coral reefs along most of the coast of northeastern Hispaniola not being uplifted (Mann et al., 1995) despite the transpressional nature of the plate boundary?

Do major earthquake sequences occur only on faults within either the northern or southern deformation belts of the NCPBZ at any given time, as suggested by the past four centuries of historical data (Kelleher et al., 1973; Dolan et al., this volume; Mann et al., this volume)? Or is this on/off behavior merely a coincidence or an artifact of an incomplete historical record? More detailed studies of historical earthquake data are needed to better answer this question.

Did the 1842 northwestern Hispaniola earthquake occur on the Septentrional fault, as suggested by Mann et al. (1984, this volume)? Or could it have occurred on offshore thrust faults within the northern Hispaniola deformed belt?

What were the sources of the poorly understood 1917 and 1918 Mona Passage earthquakes? Did they occur on normal faults within the Mona Passage? Or were they generated by rupture of the offshore subduction thrust or the eastern Septentrional fault (Dolan et al., this volume)? What is the hazard to Puerto Rico posed by the Mona Passage normal faults?

What is the geometry of the plate boundary beneath Windward Passage between Hispaniola and Cuba? Specifically, how is the reversal in dip directions of margin-parallel thrust faults between the northern Hispaniola and Santiago deformed belts (south-dipping north of Hispaniola versus north-dipping south of Cuba) (Calais et al., 1989, 1992, this volume; Dillon et al., 1992; Dolan et al., this volume) accommodated at depth?

Does the North America (Atlantic) slab "break up" into several pieces as it is underthrust beneath eastern Hispaniola and Puerto Rico, as proposed by Dillon et al. (1994)? Or is it a relatively continuous feature at depth, with only local disruption beneath the Mona Passage between Hispaniola and Puerto Rico, as proposed by Dolan et al. (this volume)? Alternatively, is the Atlantic slab beneath Hispaniola detached from the shallower parts of the North America plate by deep-seated strike-slip faults, as proposed by Calais et al. (1992)?

Is Puerto Rico actively separating from Hispaniola, accompanied by active normal faulting in the Mona Passage between the islands? Or are these normal faults relict from an earlier phase of deformation?

Is Puerto Rico actively rotating, as proposed by Masson and Scanlon (1991)? Or has the island stopped rotating, as suggested by Dolan et al. (this volume)?

Has Hispaniola undergone any significant rotations in Neogene time? Is the island, or any part of the island, currently undergoing rotation? Paleomagnetic studies of young rocks in northern Hispaniola may resolve these questions.

What is the slip rate of the Enriquillo-Plantain Garden fault zone in southern Hispaniola? Has this rate varied through Neogene time? If so, do these variations support the model of Mann et al. (1995) that this fault accommodates transfer of the Gonave microplate from the Caribbean to the North America plate (Fig. 2F)?

Is motion along the strike-slip faults bounding the northeast-striking Anegada Passage through the Virgin Islands right lateral, as suggested by Jany et al. (1990), despite the overall left-lateral nature of the plate boundary?

Did the 1867 Anegada Passage earthquake (Reid and Taber, 1920) occur on a strike-slip fault? Or was it generated by a normal fault within the passage?

What are the geometry and mechanics of the interaction between the Anegada Passage normal faults and the northern edge of the NCPBZ northeast of Puerto Rico?

CATALOG OF ACTIVE TECTONIC THESIS STUDIES IN THE REGION OF SOUTHERN CUBA, HISPANIOLA, PUERTO RICO, AND THE VIRGIN ISLANDS

Past geologic exploration in this region has been driven by economic geology, especially the search for minerals and hydrocarbons. Academic studies have traditionally focused on the Cretaceous to Eocene island-arc basement of the islands (cf. preface of Mann et al., 1991b, for a review of the history of academic and economic geologic investigations in Hispaniola).

To place these chapters in a better historical context, we catalog master's and Ph.D. theses addressing the active tectonics of this region over the past several decades (Table 1). As can be judged from the titles, some of the studies listed are only peripherally related to active tectonics but nonetheless were important for recognizing many of the active structures discussed in this book.

The reader can see that active tectonic studies are rather limited, as is true for most parts of the circum-Caribbean. We hope this compilation will point out fruitful areas of future research that will in turn stimulate further work on active tectonics.

ACKNOWLEDGMENTS

Our field-based work in the Dominican Republic began in the mid-1980s with two grants to Mann and J. C. Moore and Dolan from the Petroleum Research Fund of the American Chemical Society. These grants supported Ph.D.-level graduate work by Dolan at the University of California at Santa Cruz and M.A.-level graduate work by R. de Zoeten and C. Heubeck at the University of Texas at Austin. Dolan continued offshore work north of Hispaniola with a grant from the Ocean Sciences Division of the National Science Foundation (NSF) to Dolan, N. Breen, and H. Mullins. Mann, C. Prentice, L. Peña, F. Taylor, and G. Burr began onland fault mapping and excavation of the Septentrional fault zone with a grant from the Earth Sciences Division of the NSF in 1991 and continued this work through 1996 with grants to Mann and Prentice from the National Earthquake Hazards Reduction Program of the U.S. Geological Survey. We thank all of the funding agencies involved for the opportunity to complete this work. Field and offshore work in the Hispaniola area was greatly facilitated by the cooperation of government geologists in the Dominican Republic, Haiti, and Puerto Rico. Key long-term collaborators in the Dominican Republic include I. Tavares, L. Peña, D. Sanchez, N. Gil, S. Brouwer, J. Espaillat, R. Llinas, and C. Herridge de Guerrero. We would like to thank the following colleagues for volunteering their time to review the papers presented in this book: E. Calais, J. W. Dewey, W. Dillon, G. Draper, W. McCann, P. McCrory, C. Prentice, and T. Tuttle.

TABLE 1. LIST OF UNPUBLISHED THESES CONCERNING THE ACTIVE TECTONICS OF THE NORTHERN CARIBBEAN PLATE BOUNDARY ZONE*

1. Amilcar, H., 1997, Etude géologique de l'Ouest du Massif de La Hotte (Haïti, Grandes Antilles) dans son cadre terrestre et marin. Implications géodynamiques [Ph.D. thesis (Thèse de doctorat)]: Toulouse, France, Université Paul-Sabatier (Toulouse-III), 295 p. (biostratigraphy and geological mapping in southern Haiti).

2. Amilcar-Cenatus, H., 1997, Etude sédimentaire et tectonique des terrasses marines plio-quaternaires et de leur substratum éocène a miocene dans la Presqu'ile du Nord-Ouest de l'ile de La Tortue, Haiti, Grandes Antilles [Ph.D. thesis (Thèses de doctorat)]: Paris, Université de Paris Sud (Orsay), 301 p. (sedimentology of recent terraces in northern Haiti).

3. Antonini, G. A., 1968, Processes and patterns of landscape change in the Linea Nordeste, Dominican Republic [Ph.D. thesis]: New York, Columbia University, 196 p. (landscape study of the south flank of the Cibao Valley, Dominican Republic).

4. Bien Aimé-Momplaisir, R., 1986, Contribution à l'étude géologique de la partie orientale du massif de la Hotte (Presqu'île du Sud d'Haïti). Synthèse structurale des marges de la presqu'île à partir de données sismiques [Thèse de 3ème cycle]: Paris, Université Pierre-et-Marie-Curie (Paris-VI), 199 p. (geological mapping and cross-sections in the southern peninsula of Haiti).

5. Boisseau, M., 1987, Le flanc nord-est de la Cordillère Centrale dominicaine (Hispaniola, Grandes Antilles): Un édifice de nappes Crétacé polyphasé [doctoral thesis 3rd cycle]: Paris, Université Pierre-et-Marie-Curie (Paris-VI), 173 p. (used radar imagery to delineate young faults in northern Hispaniola).

6. Boisson, D., 1987, Etude géologique du massif du nord d'Haiti (Hispaniola-Grandes Antilles) [Ph.D. thesis]: Paris, Université Pierre-et-Marie-Curie (Paris-VI), 232 p. (geological mapping and construction of cross-sections in northern Haiti).

7. Bourdon, L., 1985, La cordillère orientale dominicaine (Hispaniola, Grandes Antilles): un arc insulaire crétacé polystructuré [Thèse de 3ème cycle]: Paris, Université Pierre-et-Marie-Curie (Paris-VI), 149 p. (geological mapping, structural analysis in the Eastern Cordillera of the Dominican Republic).

8. Calais, E., 1990, Relations cinématique/déformation le long des limites de plaques en coulissage: l'exemple de la limite de plaques nord caraïbe de Cuba à Porto Rico [Ph.D. thesis]: Sophia Antipolis, Nice, France, Université de Nice–Sophia Antipolis, 296 p. (study of marine geophysical data, onshore structural analysis, and seismological data; includes a deformation model for the northern Caribbean plate boundary zone).

9. Calmus, T., 1983, Contribution à l'étude géologique du Massif du Macaya (sud-ouest d'Haiti, Grandes Antilles): Sa place dans l'évolution de l'orogène Nord-Caraïbe [Ph.D. thesis]: Paris, Université Pierre-et-Marie-Curie, 163 p. (geologic map of part of the Enriquillo-Plantain Garden fault zone).

10. Cooey, J. C., 1978, Structure and stratigraphy of the offshore margin of the southern Dominican Republic [M.S. thesis]: Hattiesburg, Mississippi, University of Southern Mississippi, 53 p. (mapping study of active features associated with the Muertos trench and Beata Ridge).

11. Cooper, J. C., 1982, Geology of the Fondo Negro region, Dominican Republic [M.S. thesis]: Albany, New York, State University of New York at Albany, 145 p. (mapping study of active faults in the eastern Enriquillo Valley, Dominican Republic).

12. Desreumaux, C., 1987, Contribution à l'étude géologique des régions centrales et méridionales d'Haïti (Grandes Antilles) du Crétacé à l'Actuel [Ph.D. thesis]: Bordeaux, France, Université de Bordeaux I, n 904, 424 p. (mapping of late Neogene faults and deformed sedimentary rocks in central Hispaniola).

13. De Zoeten, R., 1988, Structure and stratigraphy of the central Cordillera Septentrional, Dominican Republic [M.A. thesis]: Austin, Texas, University of Texas, 299 p. (mapping study included part of the central Septentrional fault zone, Dominican Republic).

14. **Dolan, J. F.**, 1988, Paleogene sedimentary basin development in the eastern Greater Antilles: Three studies in active-margin sedimentology [Ph.D. thesis]: Santa Cruz, California, University of California, 236 p. (mapping study included possibly active faults affecting Paleogene rocks, south-central Dominican Republic).

15. Dubreuilh, P., 1982, Contribution à l'étude du bassin néogène du Plateau central (Haïti) [Thèse de 3ème cycle]: Bordeaux, France, Université de Bordeaux, 200 p. (mapping of late Neogene faults and deformed sedimentary rocks in central Hispaniola).

16. Evans, C. C., 1986, Facies evolution in a Neogene transpressional basin: Cibao Valley, Dominican Republic [Ph.D. thesis]: Miami, Florida, University of Miami, 103 p. (mapping study of fault controlled sedimentation, Cibao basin, Dominican Republic).

17. Erikson, J. P., 1993, Neogene sedimentation and tectonics in the Cibao basin and northern Hispaniola [Ph.D. thesis]: Hanover, New Hampshire, Dartmouth College (study included mapping of young faults in Puerto Rico and gravity modeling of the Septentrional fault in northern Hispaniola).

18. Feliciano, J., 1985, Stratigraphic and structural analysis of the Middle to Late Tertiary sequence on the southwest coast of Puerto Rico [Undergraduate senior thesis]: Mayaguez, Puerto Rico, University of Puerto Rico, 12 p. plus appendix (mapping of late Neogene faults in the youngest exposed section in southern Puerto Rico).

19. Feliciano, J., 1993, Paleomagnetic study of mid-Tertiary rocks in the northeastern Caribbean: Investigations of suspected tectonic block rotations. [M.S. thesis]: Syracuse, New York, Syracuse University, 61 p. (paleomagnetic study of Neogene rocks on Antigua showing 25° of counterclockwise rotation since Oligocene time).

20. Fernandez, S., 1988, Recent faulting on southeastern Puerto Rico [Undergraduate senior thesis]: Mayaguez, Puerto Rico, University of Puerto Rico–Mayaguez, 20 p. (mapping of young faults in bedrock and alluvium in southwestern Puerto Rico).

21. Fischer, K., M., 1984, Velocity modeling and earthquake relocation in the northeast Caribbean [Ph.D. thesis]: New York, Columbia University (study of the seismotectonics of the northeastern Caribbean region).

TABLE 1. LIST OF UNPUBLISHED THESES CONCERNING THE ACTIVE TECTONICS OF THE NORTHERN CARIBBEAN PLATE BOUNDARY ZONE (continued, p. 2)

22. Fox, P. J., 1972, The geology of some Atlantic fracture zones, Caribbean escarpments, and the nature of the oceanic basement and crust [Ph.D. thesis]: New York, Columbia University, 359 p. (marine geological study including dredge results from numerous sites along the northern Caribbean margin).

23. Frankel, A., 1982, Earthquake source parameters and seismic attenuation in the northeast Caribbean [Ph.D. thesis]: New York, Columbia University, 159 p. (seismological study of source parameters, scaling and attenuation relationships, precursors, and a composite focal mechanism of small earthquakes in the Virgin Islands area).

24. Gill, I. P., 1989, The evolution of Tertiary St. Croix [Ph.D. thesis]: Baton Rouge, Louisiana, Louisiana State University, 320 p. (biostratigraphic study of Neogene rocks on St. Croix documenting fault movements along Anegada Passage fault zone and transtensional separation of St. Croix and Puerto Rico).

25. Goreau, P. D. E., 1983, The tectonic evolution of the north central Caribbean plate margin [Ph.D. thesis]: Woods Hole, Massachusetts, Woods Hole Oceanographic Institute, 244 p. (pioneer marine mapping study of offshore margins of southern Cuba and Hispaniola).

26. Gugliemo, G., Jr., 1986, Push-up structure in a transpressional environment, northeastern Hispaniola [M.A. thesis]: New York, City College of New York, 52 p. (mapping study of active push-up structure related to the Septentrional fault zone, Dominican Republic).

27. Harms, F. J., 1989, Konglomeratisches Jungetertiär im Valle de San Juan (südwestliche Dominikanische Republik, Grosse Antillen): Zusammensetzung, Herkunft und Alter der Gerölle Fazies [Ph.D. dissertation]: Hannover, Federal Republic of Germany, University of Hanover, 200 p. (mapping study that included active faults bounding the southern flank of the Cordillera Central, Dominican Republic).

28. Heubeck, C. E., 1988, Geology of the southeastern termination of the Cordillera Central, south-central Hispaniola, Greater Antilles [M.A. thesis]: Austin, Texas, University of Texas, 333 p. (mapping study that included active faults bounding the Cordillera Central and within the Beata Ridge fault system).

29. Houlgatte, E., 1983, Etude d'une Partie de la Frontière Nord-est de la Plaque Caraïbe: Brest, Université de Bretagne Occidentale [M.A. thesis]: 69 p. (marine geophysical study of active faulting in the Anegada Passage using seismic reflection profiles).

30. Huerfano, V., 1996, Modelo de Corteza y Mecanica del Bloque de Puerto Rico (Crustal model and Stress Regime of the Puerto Rico block) [M.S. thesis]: Mayaguez, Puerto Rico, University of Puerto Rico–Mayaguez (Physics Dept.), 90 p. (mechanical modeling of the Puerto Rico microplate).

31. Idris, F. M., 1975, Reflection Seismic Measurements in the Old Bahama Channel North of Cuba [M.S. thesis]: Miami, Florida, University of Miami, 93 p. (early seismic reflection measurements showing some evidence for Neogene faulting and diapirism).

32. Jany, I., 1989, Néotectonique au sud des Grandes Antilles, collision (ride de Beata, Presquîle de Bahoruco); subduction (fosse de Muertos); transtension (passage d'Anegada) [Ph.D. thesis]: Paris, Université Pierre-et-Marie-Curie (Paris-VI), 306 p. (marine geophysical mapping of active faulting at the northern end of the Beata Ridge, the Muertos Trench and the Anegada Passage).

33. Joyce, J. J., 1985, High pressure–low temperature metamorphism and tectonic evolution of the Samana Peninsula, Dominican Republic (Greater Antilles) [Ph.D. thesis]: Evanston, Illinois, Northwestern University, 246 p. (mapping study included the eastern part of the Septentrional fault zone, Dominican Republic).

34. Kafka, A. L., 1980, Caribbean tectonic processes, seismic surface wave source and path property analysis [Ph.D. thesis]: Stonybrook, New York, State University of New York at Stonybrook, 121 p. (early study of earthquake focal mechanisms in northern Caribbean region; includes compilation of focal mechanisms in the Hispaniola, Puerto Rico and central Caribbean areas).

35. Ladd, J., 1974, South Atlantic seafloor spreading and Caribbean tectonics [Ph.D. thesis]: New York, Columbia University, 251 p. (includes marine geophysical data that showed active accretionary prism development in the Los Muertos Deformed Belt south of Hispaniola and Puerto Rico, as well as other major tectonic features along the southern and northern margins of the Caribbean basin).

36. Lamar, M. E., 1990, Structure and Stratigraphy of Pliocene Evaporites, Enriquillo Valley, Dominican Republic [unpublished M.A. thesis]: Austin, Texas, University of Texas at Austin, 181 p. (stratigraphic and structual study of late Neogene evaporitic diapirism, faulting and folding in the Enriquillo Valley).

37. Leroy, S., 1995, Structure et Origine de la plaque Caraïbe. Implications Géodynamiques [Ph.D. thesis (Doctorat d'Université)]: Paris, Université Pierre-et-Marie-Curie, 250 p. (marine geophysical mapping of active, intraplate faults along the Beata Ridge, Cayman Trough, and adjacent areas in the Colombian and Venezuelan basins).

38. Llinas, R., 1971, Geología del area Polo-Duvergé, Cuenca de Enriquillo, República Dominicana [Tesis profesional]: Mexico City, Universidad Nacional Autónoma de México (Facultad de Ingeniera), 83 p. (mapping study included active faults bounding the southern edge of the Enriquillo Valley, Dominican Republic).

39. **Mann, P.**, 1983, Cenozoic tectonics of the northern Caribbean: Structural and stratigraphic studies in Jamaica and Hispaniola [Ph.D. thesis]: Albany, New York, State University of New York at Albany, 688 p. (mapping study that included the Enriquillo–Plantain Garden fault zone in Hispaniola and Jamaica).

40. Matthews, J. E., 1970, Geology of the northeastern Caribbean [M.A. thesis]: Salt Lake City, Utah, University of Utah (pioneer study of the marine geology of the Beata Ridge and Los Muertos trench).

41. McCann, W., 1980, Large and moderate-size earthquakes: Their relationships to the tectonics of subduction [Ph.D. thesis]: New York, Columbia University, 189 p. (seismological and marine geophysical study that included analysis of earthquakes and active faults of the northern and eastern margins of the Caribbean plate).

42. McLaughlin, P. P., Jr., 1989, Neogene basin evolution in the southwestern Dominican Republic: A foraminiferal study [Ph.D. thesis]: Baton Rouge, Louisiana, Louisiana State University, 318 p. (mapping study that included late Neogene-fault controlled sedimentation in the Enriquillo Valley, Dominican Republic).

TABLE 1. LIST OF UNPUBLISHED THESES CONCERNING THE ACTIVE TECTONICS OF THE NORTHERN CARIBBEAN PLATE BOUNDARY ZONE (continued, p. 3)

43. **Mercier-de-Lépinay, B.**, 1987, Evolution géologique de la bordure nord-caraïbe: l'exemple de la transversale de l'île d'Hispaniola, (Grandes Antilles) [Ph.D. thesis]: Paris, Université Pierre-et-Marie-Curie (Paris-VI), 341 p. (geological mapping and structural study including active faults in Hispaniola).

44. Michael, R. C., 1978, Geology of the south-central flank of the Cordillera Central and the adjacent portions of the San Juan Valley between Rio San Juan and Rio Yacahueque, Dominican Republic [M.S. thesis]: Washington, D.C., The George Washington University, 162 p. (mapping study that included active faults along the southern flank of the Cordillera Central, Dominican Republic).

45. Nadai, A. C., 1987, Tectonostratigraphy of the San Francisco ridge area in the northeastern Cibao Valley, Dominican Republic [M.A. thesis]: New York, City College of New York, 34 p. (mapping study of fault-controlled late Neogene sedimentation, Cibao Valley, Dominican Republic).

46. Nagle, F., 1966, Geology of the Puerto Plata area, Dominican Republic [Ph.D. thesis]: Princeton, New Jersey, Princeton University, 171 p. (pioneer mapping of the Camu fault zone, northern Dominican Republic).

47. Pindell, J. L., 1985, Plate tectonic evolution of the Gulf of Mexico and Caribbean region [Ph.D. thesis]: Durham, United Kingdom, University of Durham, 287 p. (mapping study included part of the Camu fault zone, northern Dominican Republic).

48. Pisot, N., 1989, Synthèse lithostratigraphique et tectonique de l'île de la Jamaïque et de sa marge septentrionale [Ph.D. thesis]: Sophia Antipolis, Nice, France, Université de Nice–Sophia Antipolis, 366 p. (geological mapping study in Jamaica).

49. Reblin, M. T., 1973, Regional gravity survey of the Dominican Republic [M.S. thesis]: Salt Lake City, Utah, University of Utah, 129 p. (map showing regional gravity data with some modeling of crustal structure).

50. Redmond, B., 1982, Sedimentary processes and products: An amber-bearing turbidite complex from the northern Dominican Republic [Ph.D. thesis]: Troy, New York, Rensselaer Polytechnic Institute, 195 p. (mapping study that included identification of active faults, northern Dominican Republic).

51. Reid, J. A., 1992, Paleomagnetism and rock magnetics of the middle Tertiary carbonate sequence, North Coast, Puerto Rico, with tectonic implications [M.S. thesis]: Syracuse, New York, Syracuse University, 117 p., with five appendices (paleomagnetic study of Neogene limestones that revealed 24.5° of counter-clockwise rotation of Puerto Rico during late Miocene-early Pliocene time).

52. Roemer, L. B., 1973, Geology and geophysics of the Beata Ridge, Caribbean Sea [M.S. thesis]: College Station, Texas, Texas A&M University, 93 p. (sseismic reflection study showing late Neogene faults of the Beata Ridge area with some mechanical modeling of the region).

53. Ruth, M. D., 1989, Cenozoic geology of the western San Juan Valley, Dominican Republic [M.S. thesis]: Washington, D.C., The George Washington University, 252 p. (mapping and remote sensing study of active faults in the San Juan Valley, Dominican Republic).

54. Taggart, B. E., 1991, Radiometric dating of marine terraces and coral reefs on northwestern Puerto Rico [Ph.D. thesis]: Mayaguez, Puerto Rico, University of Puerto Rico–Mayaguez, 252 p. (mapping and radiometric dating study of young coral reef terraces in Puerto Rico).

55. Thom, M., 1976, Geomorphic and structural studies in Hispaniola using LANDSAT and low altitude aerial photographs [M.S. thesis]: Washington, D.C., The George Washington University, 53 p. (pioneer study of LANDSAT imagery showing major lineaments in Hispaniola, including the Septentrional and Enriquillo-Plantain Garden fault zones).

56. Thomas, S., 1995, Zur Driftgeschichte Karibischer Terranes: Paläomagnetische Untersuchungen auf Hispaniola (Grosse Antillen) [Ph.D. dissertation]: Universität zu Köln, 148 p. (Paleomagnetic analysis of rotations of fault-bounded blocks in Hispaniola).

57. Treadgold, G. E., 1975, Modeling and interpretation of the oceanic crustal structure north of the Puerto Rico trench [M.A. thesis]: Austin, Texas, The University of Texas at Austin, 190 p. (seismic reflection and refraction study of normal faults along the outer bulge of the subducting North America plate adjacent to the Puerto Rico trench).

58. van den Berghe, B., 1983, Evolution sédimentaire et structurale depuis le Paléocène du secteur du massif de la Selle (Haïti), Bahoruco (République Dominicaine) au Nord de la ride de Beata, dans l'orogène nord-caraïbe (Hispaniola, Grandes Antilles) [Ph.D. thesis]: Paris, Université Pierre-et-Marie-Curie, 205 p. (microtectonic study that includes late Neogene sedimentary rocks of southern Hispaniola).

59. van der Hilst, R., 1990, Tomography with P, PP, and pP delay-time data and the three-dimensional mantle structure below the Caribbean region [Ph.D. dissertation]: Utrecht, The Netherlands, University of Utrecht, 250 p. (tomographic study of the subducted Atlantic slab beneath the Lesser Antilles and Caribbean plate).

60. Velez, R., 1985, A survey of seismic reflection data for Puerto Rico and its environs [Undergraduate senior thesis]: Mayaguez, Puerto Rico, University of Puerto Rico–Mayaguez, 25 p. (compilation of petroleum industry seismic reflection profiles and maps of late Neogene structures around Puerto Rico).

61. Vespucci, R. M., 1976, Petrology and geochemistry of the Late Cenozoic volcanic rocks of the Dominican Republic [Ph.D. dissertation]: Washington, D.C., The George Washington University, 286 p. (study of Plio-Pleistocene volcanism of the central Dominican Republic).

62. Wirtz, W. K., 1985, The petrochemistry and genesis of Late Cenozoic shoshonitic basalts, Dominican Republic, and their tectonic significance [M.S. thesis]: Gainesville, Florida, University of Florida, 113 p. (study of Plio-Pleistocene volcanism of the central Dominican Republic).

* Note: Authors contributing to this volume are in bold type.

REFERENCES CITED

Bralower, T. J., and Iturralde, M. A., 1997, Micropaleontological dating of the collision between the North American plate and Greater Antilles arc in western Cuba: Palaios, v. 12, p. 133–150.

Byrne, D. B., Suarez, G., and McCann, W. R., 1985, Muertos Trough subduction—Microplate tectonics in the northern Caribbean: Nature, v. 317, p. 420–421.

Calais, E., and Mercier de Lépinay, B., 1993, Semiquantitative modeling of strain and kinematics along the Caribbean/North America strike-slip plate boundary zone: Journal of Geophysical Research, v. 98, p. 8293–8308.

Calais, E., and Mercier de Lépinay, B., 1995, Strike-slip tectonic processes in the northern Caribbean between Cuba and Hispaniola (Windward Passage): Marine Geophysical Research Letters, v. 17, p. 63–95.

Calais, E., Mercier de Lépinay, B., Renard, V., and Tardy, M., 1989, Géométrie et régime tectonique le long d'une limite de plaques en coulissage: la frontière nord-Caraibe de Cuba à Hispaniola, Grandes Antilles (Geometry and tectonic regime along a transcurrent plate boundary: the northern Caribbean border from Cuba to Hispaniola, Greater Antilles): Comptes Rendu de l'Académie des Sciences, v. 308, p. 131–135.

Calais, E., Bethoux, N., and Mercier de Lépinay, B., 1992, From transcurrent faulting to frontal subduction: A seismotectonic study of the northern Caribbean plate boundary from Cuba to Puerto Rico: Tectonics, v. 11, p. 114–123.

CAYTROUGH, 1979, Geological and geophysical investigation of the mid-Cayman rise spreading center: Initial results and observations, *in* Talwani, M., Pitman, W. C., and Hayes, D. E., eds., Deep drilling results in the Atlantic Ocean, ocean crust: Washington, D.C., American Geophysical Union, Maurice Ewing Series, v. 2, p. 66–93.

DeMets, C., Gordon, R. G., Argus, D. F., and Stein, S., 1990, Current plate motions: Geophysical Journal International, v. 101, p. 425–478.

DeMets, C., Dixon, T., Farina, F., Calais, E., Jansma, P., and Mann, P., 1996, GPS-derived velocities in the northeastern Caribbean: Implications for Caribbean–North America motion and the NUVEL–1A model [abs.]: Eos, (Transactions of the American Geophysical Union), v. 77, p. F143.

Dillon, W., Austin, J. A., Scanlon, K. M., Edgar, N. T., and Parson, L. M., 1992, Structure and development of the insular margin north of western Hispaniola: A tectonic accretionary wedge on the Caribbean plate boundary: Marine and Petroleum Geology, v. 9, p. 70–88.

Dillon, W. P., Edgar, N. T., Scanlon, K. M. and Coleman, D. F., 1994, A review of the tectonic problems of the strike-slip northern boundary of the Caribbean Plate and examination by GLORIA, *in* Gardner, J. V., Field, M. E., and Twichell, D. C., eds., Geology of the United States' Seafloor: The view from Gloria: United Kingdom, Cambridge University Press, p. 135–164.

Emery, K. O., and Aubrey, D. G., 1991, Sea levels, land levels, and tide gauges: Springer Verlag, New York, New York, 237 p.

Gordon, M. B., Mann, P., Cáceres, D., and Flores, R., 1997, Cenozoic tectonic history of the North America–Caribbean plate boundary zone in Cuba: Journal of Geophysical Research, v. 102, p. 10055–10082.

Grindlay, N., Mann, P., and Dolan, J., 1997, Researchers investigate submarine faults north of Puerto Rico: Eos, (Transactions of the American Geophysical Union), v. 78, p. 404.

Heubeck, C., and Mann, P., 1991, Geologic evaluation of plate kinematic models for the North America–Caribbean plate boundary zone: Tectonophysics: v. 191, p. 1–26.

Horsfield, W. T., 1975, Quaternary vertical movements in the Greater Antilles: Geological Society of America Bulletin, v. 86, p. 933–938.

Jany, I., Scanlon, K. M., and Mauffret, A., 1990, Geological interpretation of combined SEABEAM, GLORIA, and seismic data from Anegada Passage (Virgin Islands, North Caribbean): Marine Geophysical Research, v. 12, p. 173–196.

Kafka, A. L., 1980, Caribbean tectonic processes, seismic surface wave source, and path property analysis [Ph.D. thesis]: Stonybrook, New York, State University of New York at Stonybrook, 121 p.

Kelleher, J. L., Sykes, L. R., and Oliver, J., 1973, Possible criteria for predicting earthquake locations and their application to major plate boundaries of the Pacific and Caribbean: Journal of Geophysical Research, v. 78, p. 2547–2585.

Mann, P., Burke, K., and Matumoto, T., 1984, Neotectonics of Hispaniola: Plate motion, sedimentation, and seismicity at a restraining bend: Earth and Planetary Science Letters, v . 70, p. 311–324.

Mann, P., Schubert, C., and Burke, K., 1990, Review of Caribbean neotectonics, *in* Dengo, G., and Case, J. E., eds., The Caribbean region: Boulder, Colorado, Geological Society of America, The Geology of North America, v. H., p. 307–338.

Mann, P., Tyburski, S. A., and Rosencrantz, E., 1991a, Neogene development of the Swan Islands restraining bend complex, Caribbean Sea: Geology, v. 19, p. 823–826.

Mann, P., Draper, G., and Lewis, J. F., 1991b, An overview of the geologic and tectonic development of Hispaniola, *in* Mann, P., Draper, G., and Lewis, J. F., eds., Geologic and tectonic development of the North America–Caribbean plate boundary in Hispaniola: Boulder, Colorado, Geological Society of America Special Paper 262, p. 1–28.

Mann, P., Taylor, F., Edwards, R. L., and Ku, T., 1995, Actively evolving microplate formation by oblique collision and sideways motion along strike-slip faults: An example from the northeastern margin of the Caribbean plate: Tectonophysics, v. 246, p. 1–69.

Masson, D. G., and Scanlon, K. M., 1991, The neotectonic setting of Puerto Rico: Geological Society of America Bulletin, v. 103, p. 144–154.

Mullins, H. T., Dolan, J., Breen, N., Andersen, B., Gaylord, M., Petruccione, J. L., Wellner, R. W., Melillo, A. J., and Jurgens, A. D., 1991, Retreat of carbonate platforms: Response to tectonic processes: Geology, v. 19, p. 1089–1092.

Mullins, H. T., Breen, N., Dolan, J., Wellner, R. W., Petruccione, J. L., Gaylord, M., Anderses, B., Melillo, A. J., Jurgens, A. D., and Orange, D., 1992, Carbonate platforms along the southeast Bahamas-Hispaniola collision zone: Marine Geology, v. 105, p. 169–209.

Osieki, P. S., 1981, Estimated intensities and probable tectonic sources of historic (pre-1898) Honduran earthquakes: Bulletin of the Seismological Society of America, v. 71, p. 865–881.

Pindell, J. L., and Barrett, S. F., 1990, Geologic evolution of the Caribbean region; a plate-tectonic perspective, *in* Dengo, G., and Case, J. E., eds., The Caribbean region: Boulder, Colorado, Geological Society of America, The Geology of North America, v. H., p. 405–432.

Plafker, G., 1976, Tectonic aspects of the Guatemala earthquake of 4 February, 1976: Science, v. 193, p. 1201–1208.

Platt, J. P., 1993, Mechanics of oblique convergence: Journal of Geophysical Research, v. 98, p. 16239–16256.

Prentice, C. S., Mann, P, Taylor, F. W., Burr, G., and Valastro, S., 1993, Paleoseismicity of the North American–Caribbean plate boundary (Septentrional fault), Dominican Republic: Geology, v. 21, p. 49–52.

Reid, H., and Taber, S., 1920, The Virgin Islands earthquakes of 1867–1868: Bulletin of the Seismological Society of America, v. 10, p. 9–30.

Robinson, E., Versey, H. R., and Williams, J. B., 1958, The Jamaica earthquake of March 1, 1957: Transactions of the 2nd Caribbean Geological Conference, Mayaguez, Puerto Rico, p. 50–57.

Schwab, W. C., Danforth, W. W., Scanlon, K. M., and Masson, D. G., 1991, A giant submarine slope failure on the northern insular slope of Puerto Rico: Marine Geology, v. 96, p. 237–246.

Sykes, L. R., and McCann, W. R., and Kafka, A. L., 1982, Motion of the Caribbean plate during the last 7 million years and implications for earlier Cenozoic movements: Journal of Geophysical Research, v. 87, p. 10656–10676.

Taggart, B. E., and Joyce, J., 1990, Radiometrically dated marine terraces on northwestern Puerto Rico, *in* Larue, D. K., and Draper, G., eds., Transactions of the 12th Caribbean Geological Conference, St. Croix, U.S. Virgin Islands: Coral Gables, Florida, Miami Geological Society, p. 248–258.

Tuttle, M. P., Dyer-Williams, K., Prentice, C. S., and Peña, L. R., 1997, Liquefaction features contribute to paleoseismology study of the Septentrional fault, Dominican Republic: Geological Society of America Abstracts with Programs, v. 29, p. A-441.

White, R. A., 1985, The Guatemala earthquake of 1816 on the Chixoy-Polochic fault: Bulletin of the Seismological Society of America, v. 75, p. 455–473.

MANUSCRIPT ACCEPTED BY THE SOCIETY JANUARY 19, 1998

Geological Society of America
Special Paper 326
1998

Active tectonics of the north-central Caribbean: Oblique collision, strain partitioning, and opposing subducted slabs

James F. Dolan
Department of Earth Sciences, University of Southern California, Los Angeles, California 90089
Henry T. Mullins
Department of Earth Sciences, Syracuse University, Syracuse, New York 13244
David J. Wald
U.S. Geological Survey, 525 South Wilson Avenue, Pasadena, California 91106

ABSTRACT

SeaMARC II side-scan sonar and single-channel seismic reflection data collected off northern Hispaniola and northwestern Puerto Rico provide the first detailed observations of a 400-km-long segment of the offshore part of the northern Caribbean Plate boundary zone. The complex deformation zone revealed by these data is highlighted by active, oblique collision between Hispaniola and the southeasternmost Bahamas banks at the southeastern edge of a major restraining bend. Pronounced along-strike changes in structural style along the Hispaniola–Puerto Rico slope record an east-to-west transition from highly oblique, predominantly left-lateral underthrusting of Atlantic oceanic lithosphere (North America plate) along the 085° margin northwest of Puerto Rico, to oblique left-lateral contractional deformation along the 110° plate boundary off north-central Hispaniola.

Three major structural provinces are discernible along the northern Hispaniola slope, with the central collision zone flanked on the west by a small accretionary prism composed of offscraped Hispaniola basin turbidites, and on the east by a wide zone of predominantly strike-slip faulting. The slope within the central collision zone has failed extensively by slumping yet exhibits little direct evidence of compression, save the presence of the underthrust carbonate bank itself. On the Rio San Juan Peninsula onshore Hispaniola, however, collisional underthrusting of a prominent spur of high-standing carbonate basement has uplifted a spectacular flight of at least 15 concentric marine terraces. The slope in the southeastern survey area exhibits few discernible contractional structures, although the gently south-dipping Atlantic slab underthrusts the slope. Several east- to east-northeast–trending strike-slip faults extend along the slope from the collision zone eastward into the southeastern survey area. The most prominent of these strike-slip faults is the offshore, eastward extension of the left-lateral Septentrional fault system of northern Hispaniola. The opening of two major pull-apart basins along two splays of this fault system records sinistral offset along the Septentrional fault system of as much as several tens of kilometers. These data represent the first direct evidence of large-scale, left-lateral offsets along the plate boundary, long inferred on the basis of regional geological reconstructions.

On August 4, 1946, a great earthquake ruptured a 195-km-long section of the plate boundary beneath northeastern Hispaniola. The focal mechanism and after-

Dolan, J. F., Mullins, H. T., and Wald, D. J., 1998, Active tectonics of the north-central Caribbean: Oblique collision, strain partitioning, and opposing subducted slabs, *in* Dolan, J. F., and Mann, P., eds., Active Strike-Slip and Collisional Tectonics of the Northern Caribbean Plate Boundary Zone: Boulder, Colorado, Geological Society of America Special Paper 326.

shock distribution of the event indicate that it ruptured a gently south-dipping zone of strong coupling between Hispaniola and underthrust, buoyant Bahamas carbonate basement. The absence of active tectonic structures within the colliding Bahamas province, except along their extreme southern edges, suggests that the 5.5-km-thick, well-lithified carbonate banks are behaving as a relatively rigid block during collisional underthrusting. The presence of the left-lateral Septentrional fault onshore Hispaniola to the south of the southwest-dipping, oblique-thrust interface indicates partitioning of the compressional and translational components of plate motion in northern Hispaniola. Thus, between the Septentrional fault and the offshore contractional deformation front, northernmost Hispaniola is a thin crustal sliver underlain by a shallowly southwest-dipping decollement.

Recent seismicity data define the three-dimensional geometry of the south-dipping Atlantic oceanic slab, as well as an opposing, north-dipping slab of Caribbean lithosphere that is being thrust beneath Hispaniola and Puerto Rico along the Los Muertos trough. The near-vertical, deep part of the Atlantic slab is continuous with the gently south-dipping shallow slab, except beneath the Mona Passage, where an absence of intermediate-depth earthquakes may reflect local detachment of the deep part of the slab. The Atlantic and Caribbean slabs appear to collide in the upper mantle at ~50 km depth beneath eastern Hispaniola and the Mona Passage. The site of the collision corresponds to the steepest part of the deep Atlantic slab and the sharpest inflection point in the dip of the slab, suggesting that the steepness of the deep Atlantic slab is a response to the collision.

Highly oblique, westward-younging collisional underthrusting of the buoyant Bahamas ridge has controlled many of the structural, seismological, and sedimentological features of the north-central Caribbean margin, including: (1) the location of the Hispaniola restraining bend; (2) early(?) to mid-Pliocene onset of uplift in northern Hispaniola; (3) initiation of the Septentrional fault system; (4) a sharp, 30° bend in the strike of the deeply subducted Atlantic lithospheric slab beneath Mona Passage; (5) erosion, collapse, and step-back of the southern edges of the colliding carbonate banks; and (6) 25° counterclockwise rotation of Puerto Rico during late Miocene–Pliocene time. In addition, the collision has retarded eastward motion of Hispaniola, which in turn has allowed Puerto Rico, which overthrust the southwestern extension of the west-northwest-trending Bahamas ridge in late Miocene–Pliocene time, to move eastward relative to Hispaniola, opening Mona Passage and Mona Canyon in its wake.

INTRODUCTION

Since middle Eocene time (~45 Ma) the Caribbean plate has moved generally eastward relative to both North and South America, with active arc volcanism occurring in the Lesser Antilles in response to west-dipping subduction of Atlantic oceanic lithosphere of the North America plate (Fig. 1; Pindell and Dewey, 1982; Burke, 1988; Pindell and Barrett, 1990). There is general agreement that this eastward relative motion results in predominantly left-lateral motion along the northern boundary of the Caribbean plate (Molnar and Sykes, 1969; Bracey and Vogt, 1970; Jordan, 1975; Minster and Jordan, 1978; Schell and Tarr, 1978; Mann and Burke, 1984; Mann et al., 1984). The exact direction of relative motion between the Caribbean and North America plates, however, has long been the subject of debate, with a range of proposed motions from pure east-west, left-lateral strike-slip (e.g., Molnar and Sykes, 1969; Mann and Burke, 1984; Mann et al., 1984; Stein et al., 1988; Mann et al., 1990; Calais et al., 1992) to northeast-southwest oblique convergence (Sykes et al., 1982; McCann and Sykes, 1984; Deng and Sykes, 1995). Unfortunately, global plate motion models (e.g., NUVEL-1) are poorly constrained for the northern Caribbean region and may not accurately record relative motion (DeMets et al., 1990; DeMets, 1993). The velocity of Caribbean–North America relative plate motions is also debated, with a range of proposed values from 11 mm/year (NUVEL-1 global plate motion model; DeMets et al., 1990; DeMets, 1993) to 37 mm/yr (Sykes et al., 1982). Most estimates suggest that the Caribbean–North America plate motion rate is ~20 to 25 mm/year (see discussion later in this chapter).

This debate results at least partly from a lack of data concerning the location, geometry, and kinematics of major plate-boundary structures along a 375-km-long section of the offshore part of the boundary between north-central Hispaniola and northwestern Puerto Rico. In this chapter we describe the results of the first detailed marine geophysical investigations of this part of the plate boundary. In addition to these side-scan sonar and seismic reflection data, we also describe the recent seismicity of the region and use this combined data set to illustrate the three-dimensional geometry of the plate boundary zone. We then discuss the implications of the geophysical and seismicity data for the structural, seismological, and sedimentological development of the plate margin and conclude by presenting a model for the tectonic evolution of the north-central region during late Cenozoic time.

REGIONAL GEOLOGIC SETTING

The boundary between the Caribbean and North America plates in the north-central Caribbean region is a complex, 250-km-wide zone known as the northern Caribbean plate boundary zone (NCPBZ) (Fig. 2; Mann et al., 1984). The NCPBZ comprises two major zones of deformation, one along the northern boundary of the islands of Puerto Rico and Hispaniola, and one along the southern edge of the islands (Mann et al., 1984; Burke, 1988; Mann et al., 1990). In the northern zone, which forms the focus of this study, much of the plate-boundary, left-lateral strike-slip is accommodated on the Septentrional fault system, a major fault system that extends from the Windward Passage between Hispaniola and Cuba, through the northern part of Hispaniola, and offshore eastward to the north of Puerto Rico (Figs. 1 and 2).

The architecture of major plate boundary structures is markedly different on either side of our study area offshore northern Hispaniola. To the east, the plate boundary exhibits a well-defined, south-dipping Wadati-Benioff zone associated with highly oblique underthrusting of Atlantic oceanic crust of the North America plate beneath Puerto Rico and the Virgin Islands; this zone extends to at least 150 km depth beneath Puerto Rico and projects to the surface at the 8-km-deep Puerto Rico trench (McCann and Sykes, 1984; McCann and Pennington, 1990). In contrast, the area to the west of our study area is dominated by strike-slip tectonics, with no seismological evidence for a deep, underthrust slab similar to that beneath Puerto Rico (McCann and Sykes, 1984; McCann and Pennington, 1990).

In the Windward Passage offshore northwestern Hispaniola, the Septentrional fault merges westward with the Oriente fault system, which in turn forms the northern margin of the Cayman trough pull-apart basin (Figs. 1, 2, and 3) (Mann et al., 1984; Rosencrantz et al., 1988; Rosencrantz and Mann, 1991; Calais

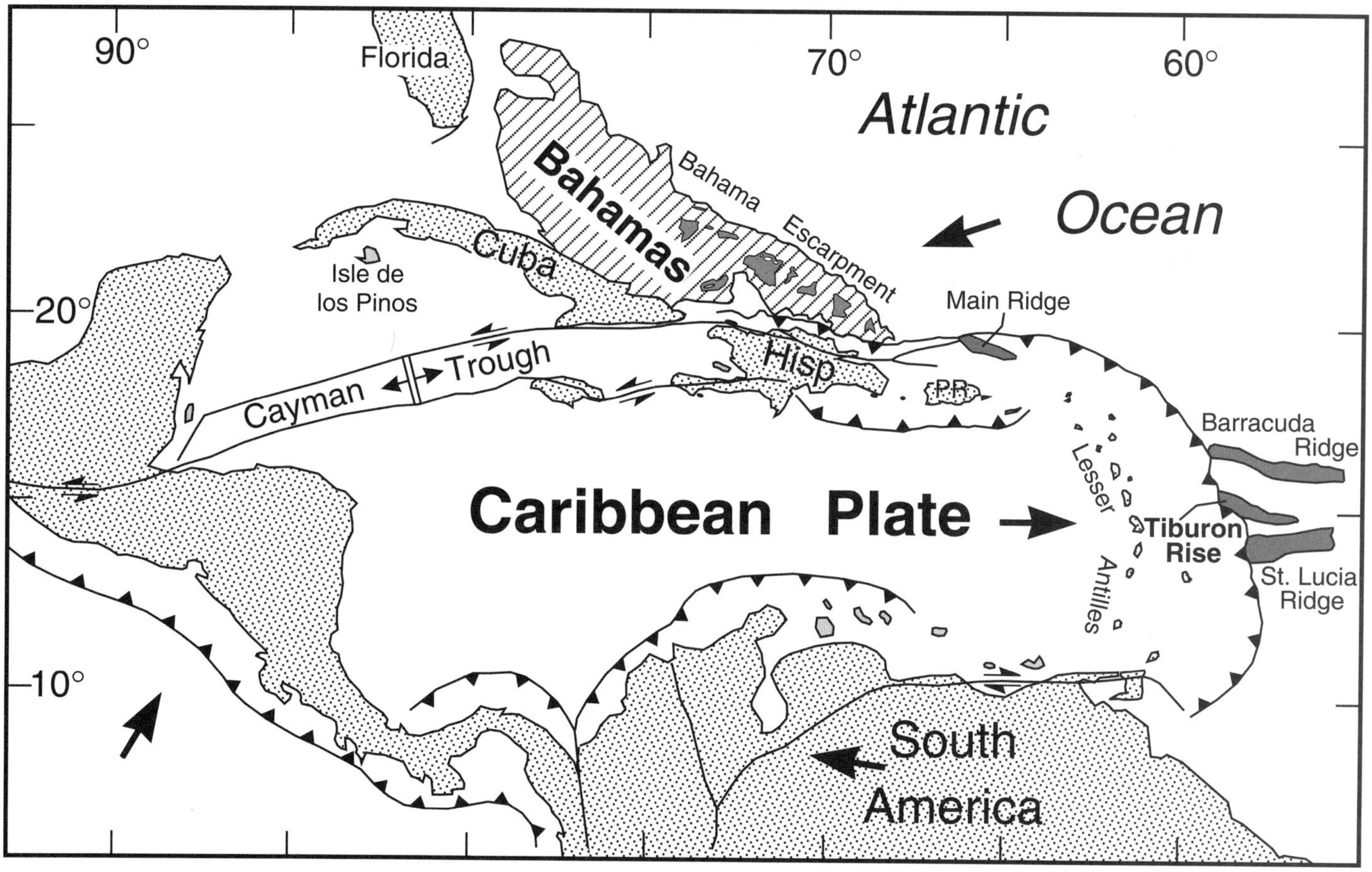

Figure 1. Location and tectonic map of the Caribbean region. Modified from Case and Holcombe (1980) and Mann et al. (1984). Large arrows represent relative plate motions vectors from Mann et al. (1984). The North America (Atlantic Ocean)–Caribbean vector is from Dolan and Wald (this volume). Hisp is Hispaniola; PR is Puerto Rico.

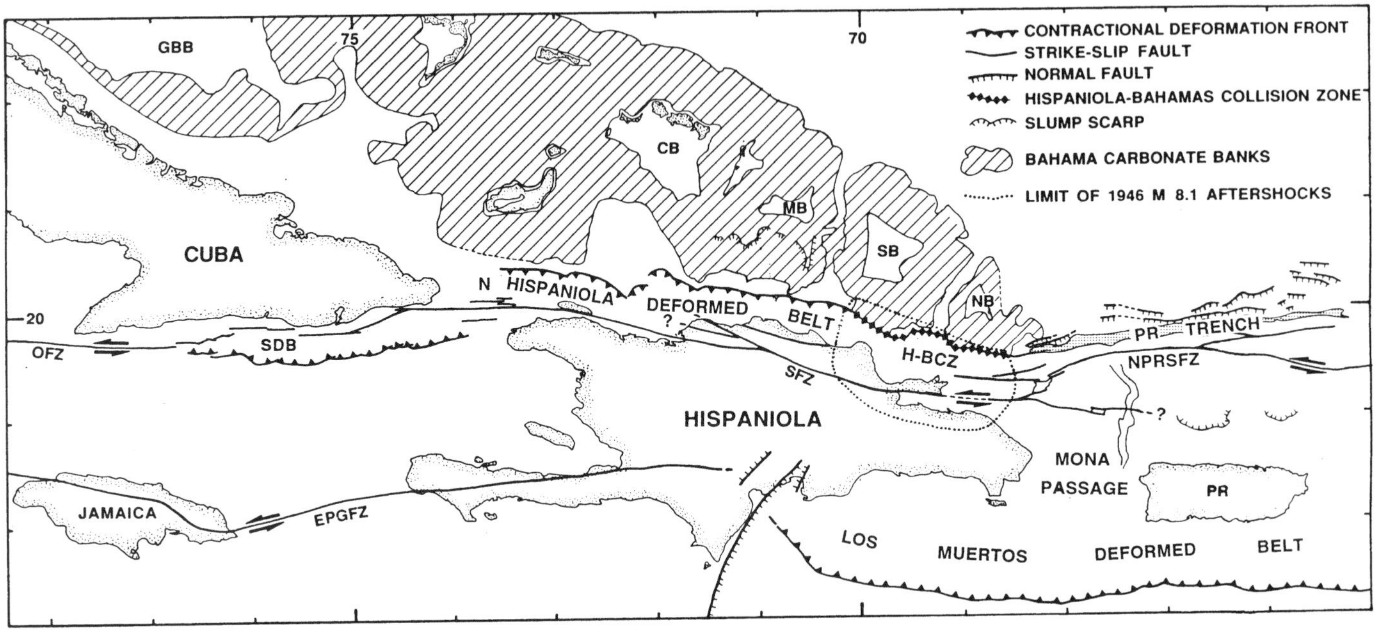

Figure 2. Summary map of major tectonic features of the north-central Caribbean region. Location of Oriente Fault Zone (OFZ) and Santiago deformed belt (SDB) from Calais et al. (1989) and Calais and Mercier de Lépinay (1995). Location of Enriquillo-Plantain Garden fault zone (EPGFZ) from Mann et al. (1984). Location of western part of northern Hispaniola accretionary prism (NHAP) from Dillon et al. (1992). Location of structures north and south of Puerto Rico from Masson and Scanlon (1990). Location of western Los Muertos trench deformation front from Case and Holcombe (1980). Major onshore faults of northern Hispaniola from de Zoeten et al. (1991). Limits of 1946 northeast Hispaniola earthquake aftershock zone from Kelleher et al. (1973) slightly modified by the event relocations of Russo and Villaseñor (1995) (see Dolan and Wald, this volume, for discussion of these data). All other structural data from this study. GBB is Great Bahama bank; CB is Caicos bank; MB is Mouchoir bank; SB is Silver bank; NB is Navidad bank; H-BCZ is Hispaniola-Bahamas collision zone (see text for discussion); PR is Puerto Rico; SFZ is Septentrional fault zone.

et al., 1989, 1992, this volume; Calais and Mercier de Lépinay, 1995). Deformation west of our study area is not limited to the Septentrional fault system, however, as marine geophysical data from offshore northwestern Hispaniola reveal a zone of south-dipping contractional deformation, highlighted by the development of a small accretionary prism (Austin, 1983; Dillon et al., 1992). This accretionary prism, together with evidence of other potentially active structures in northernmost Hispaniola, defines a zone of active deformation north of the Septentrional fault that we refer to as the northern Hispaniola deformed belt (NHDB) (Fig. 2).

MARINE GEOPHYSICAL DATA

Our survey was carried out aboard the R/V *Moana Wave* during June and July, 1989. In addition to the SeaMARC II side-scan sonar data, we collected digitally recorded, single-channel seismic reflection data as well as gravity and 3.5 kHz high-resolution seismic reflection data. The main study area encompasses a 375-km-long segment of the northern part of the NCPBZ offshore north-central and northeastern Hispaniola, as well as three shallow-water carbonate banks of the southeasternmost Bahamas (Fig. 3). The origin and evolution of these banks are discussed in two companion studies presented elsewhere (Mullins et al., 1991; 1992). In addition to our main study area, we collected side-scan data from a 40-km-wide swath that extended 190 km from the southeast corner of the main survey area eastward across northern Mona Passage (the 125-km-wide, north-south passage separating Hispaniola and Puerto Rico) and along the northern slope of Puerto Rico (Fig. 3). We refer to this ancillary mosaic as the northwestern Puerto Rico slope mosaic.

SeaMARC II produces a side-scan reflection image and bathymetry for a swath 10 km wide. Our main survey area consisted of 61 total lines, oriented at a high angle to the roughly east-west structural grain of the Hispaniola slope (Fig. 4). Line spacing averaged ~8 km, allowing overlap of adjacent lines. Initial navigation was by a combination of Global Positioning System (GPS) satellite and dead reckoning, with later adjustments made by aligning reflectors and bathymetric features on the SeaMARC II imagery.

The side-scan image was processed to remove noise and was then plotted and formed into a mosaic according to navigation. The bathymetry was likewise corrected, plotted, and formed into a mosaic (Fig. 5). The bathymetry was then redrawn by hand, interpolating between discrepancies in the overlapping swaths. Although SeaMARC II bathymetry pro-

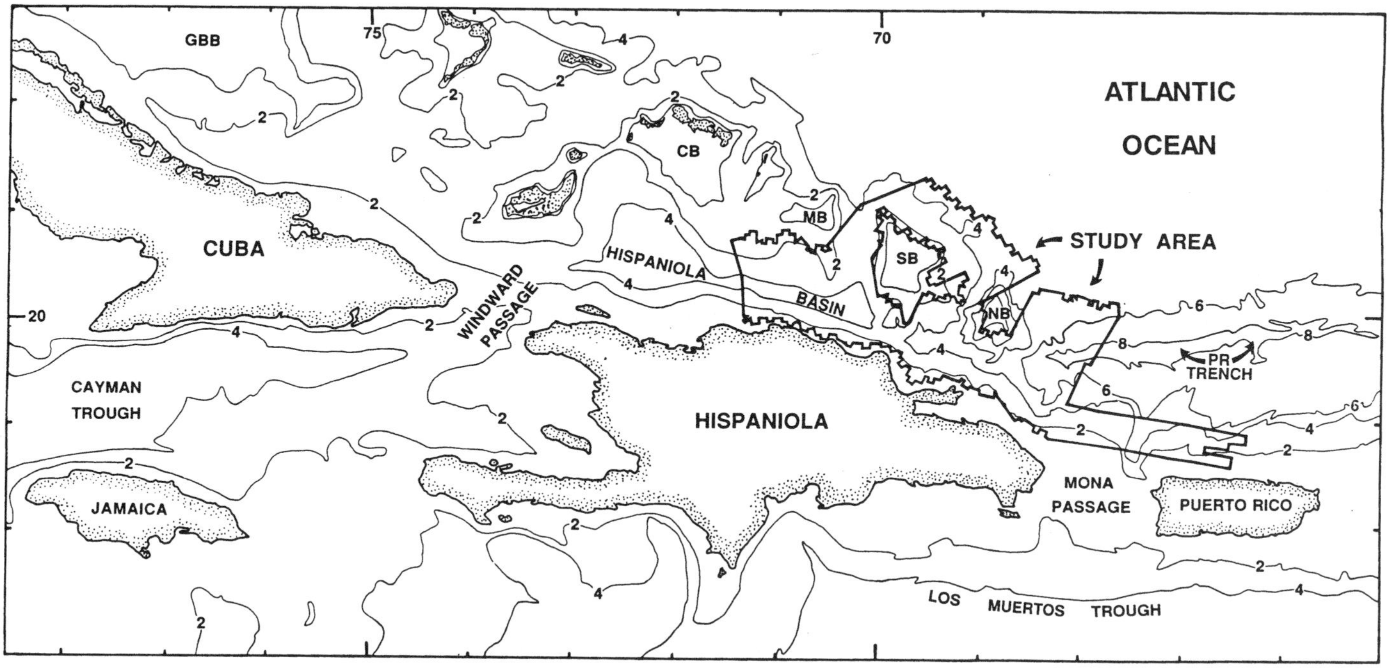

Figure 3. Location map of the SeaMARC II side-scan sonar study area. Bathymetric contours (in km) are from Case and Holcombe (1980). GBB is Great Bahama bank; CB is Caicos bank; MB is Mouchoir bank; SB is Silver bank; NB is Navidad bank; PR is Puerto Rico.

vides a 100 m contour interval, in the most detailed bathymetric maps we will present we have plotted the bathymetry at a 200 m contour interval for clarity. Some regional maps that we will present show only 500 m contours for clarity.

The energy source for the digitally recorded single-channel seismic-reflection data was generally either one or two 120-in^3 air guns, although equipment problems during the cruise resulted in some lines being shot with various combinations of 120-in^3 and 40-in^3 air guns, especially in the southeastern and far-western parts of the survey. Processing of the seismic reflection data was routine, and included deconvolution, automatic gain control (AGC), high- and low-cut filters, and a ramp filter designed to remove ship noise.

The geophysical data, in combination with GLORIA side-scan sonar and seismic reflection studies offshore northwestern Hispaniola (Edgar, 1991; Dillon et al., 1992) and Puerto Rico (EEZ SCAN 85 Staff, 1987; Masson and Scanlon 1991), multi-channel seismic reflection profiles offshore northern Hispaniola and Puerto Rico (Austin, 1983; Larue and Ryan, 1990, 1998; Larue et al., 1990), and Seabeam/seismic reflection studies offshore southeastern Cuba and northern Haiti (Calais et al., 1989, 1992, this volume; Calais and Mercier de Lépinay, 1995), provide near-complete swath-map and seismic-reflection coverage of active deformation in the northern offshore portion of the NCPBZ.

STRUCTURAL PROVINCES

Analysis of the side-scan images, seismic reflection profiles, and SeaMARC II bathymetry enables division of the main survey area into three major provinces (Fig. 6): (1) Atlantic oceanic crust in the northeast corner of the survey area; (2) steep-sided, shallow-water carbonate banks of the southeastern Bahamas in the north-central and northwestern parts of the survey area; and (3) the northern slope of Hispaniola and the northwestern slope of Puerto Rico, which can be further subdivided into a central zone of Bahamas bank collision that is flanked by structurally distinct western and eastern subprovinces; the eastern Hispaniola slope exhibits many similarities to the slope off northwestern Puerto Rico, which is described separately. These three offshore structural subprovinces, together with onshore active structures north of the Septentrional fault, constitute the northern Hispaniola deformed belt (Fig. 2).

NORTHERN HISPANIOLA SLOPE AND NORTHWESTERN PUERTO RICO SLOPE PROVINCE

Western Hispaniola slope subprovince (71° 20′W to 70° 08′W)

In the western part of our survey area the 4-km-deep Hispaniola basin separates Hispaniola from Mouchoir carbonate bank to the north (Fig. 7). This flat-floored, turbidite-filled basin, which extends for ~400 km along the northern margin of Hispaniola, terminates abruptly against the western edge of the Bahamas-Hispaniola collision zone at 70°W (Fig. 7A). Strong, continuous reflectors underlying the Hispaniola basin can be traced northward into carbonate slope deposits of southern Mouchoir bank (Fig. 8A). These deposits exhibit marked lateral

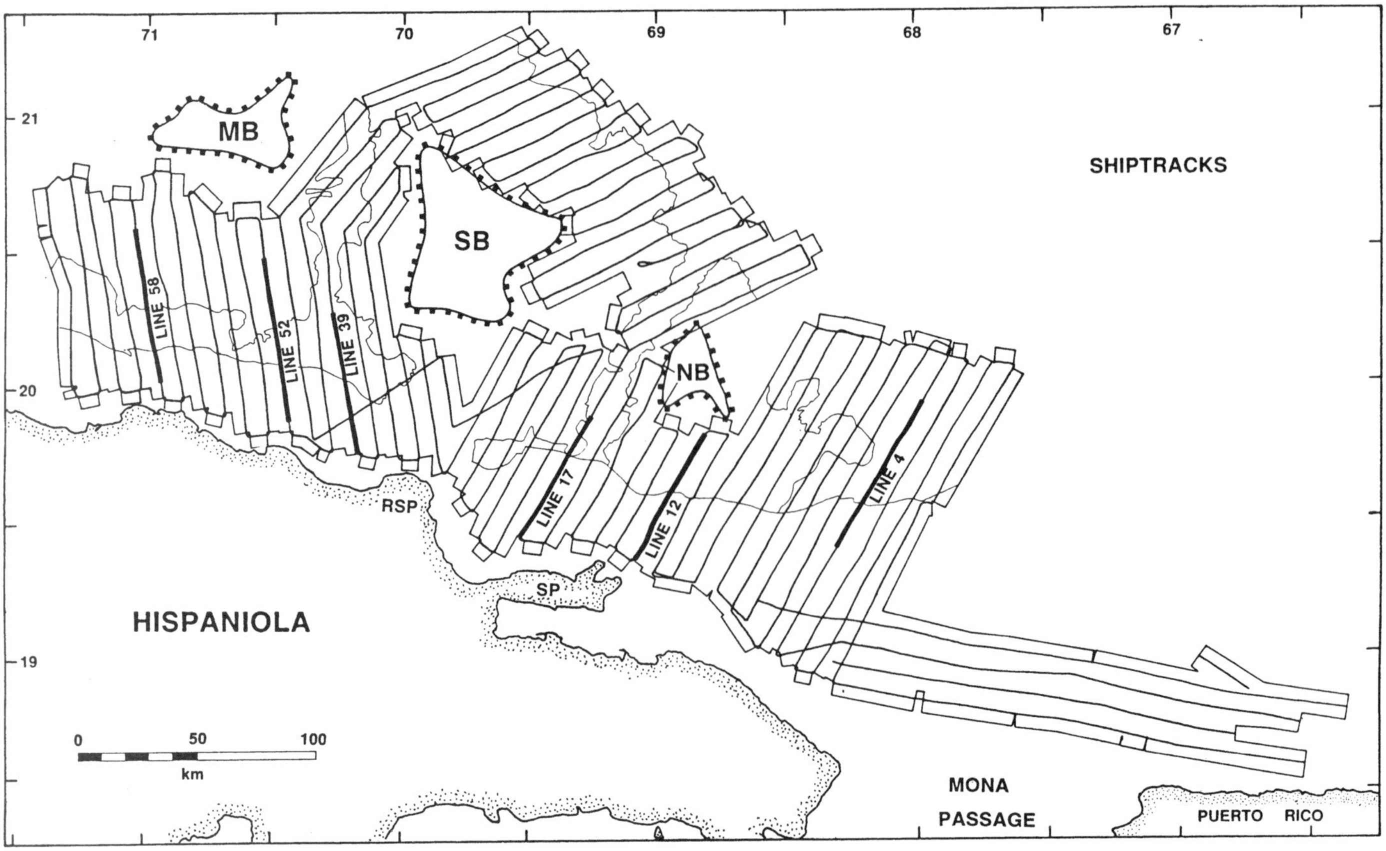

Figure 4. Ship tracks for SeaMARC II cruise MW 8909. Thicker parts of tracklines denote portions of seismic-reflection profiles shown in Figures 8, 13, and 17. MB is Mouchoir bank; SB is Silver bank; NB is Navidad bank; RSP is Rio San Juan Peninsula; SP is Samana Peninsula.

changes in thickness, suggestive of deposition as isolated submarine fans or aprons (Mullins and Cook, 1986). Recent basin fill consists of a mixture of carbonate and terrigenous turbidites, reflecting sediment supply from both the Bahamas carbonate banks to the north and Hispaniola to the south (Ditty et al., 1977). Thickness of the basinal sediments typically ranges from ~900 to 1,350 m (1–1.5 sec two-way travel time [twtt], assuming a 1.8 km/sec interval velocity), with moderate basement relief. The basinal strata are remarkably flat lying and exhibit little to no southward fanning of bedding dips (Fig. 8). Along the northern margin of the basin, basinal turbidites onlap the shallowly south-dipping carbonate slope strata (Fig. 8A). The horizontal nature of the basinal turbidites precludes significant post-depositional tilting. The southern edge of carbonate basement is quite irregular and, at 70° 40′W, a high-standing prong of Mouchoir bank carbonate basement, which we refer to as Mouchoir Spur, extends southward almost to the Hispaniola slope (Fig. 7A).

The Hispaniola slope west of the collision zone is characterized by a small accretionary prism, herein named the northern Hispaniola accretionary prism (NHAP), composed of offscraped Hispaniola basin sediments (Fig. 7B). Fold axes and thrust faults generally trend east-west to west-northwest, although structures at the eastern end of the NHAP, just west of the collision zone, swing into a more northwesterly orientation (Fig. 7). Anticlines in the NHAP are nearly concentric and exhibit 450 to 800 m of frontal relief (Fig. 8). Major thrust faults are typically spaced 5–8 km apart. In several cases, frontal offscraping appears to involve the entire sedimentary section, including at least the uppermost carbonate slope deposits lying beneath Hispaniola basin turbidites (e.g., Fig. 8A).

The SeaMARC II mosaic reveals a generally sharp deformation front at the slope-basin contact that correlates with frontal thrust faults on seismic-reflection profiles (Fig. 9). Over most of this area the prism consists of two frontal anticlines that are traceable for ~50 km along strike. Upslope of the two frontal folds, folds become less well defined, although in at least one locality (70° 30′W) as many as four south-dipping thrust faults are discernible; at least three of these exhibit associated anticlines (Fig. 7). Much of the Hispaniola slope upslope of the frontal folds exhibits a relatively steep northward surface slope without obvious large-scale folding.

Lateral propagation of the frontal fold-thrust fault couplet occurs at two localities; east of 70° 17′W, where the frontal system is propagating eastward, and west of 70° 58′W, where the frontal thrust is propagating westward (e.g., Fig. 8A). At both localities clearly defined frontal thrust faults die out laterally, whereas their

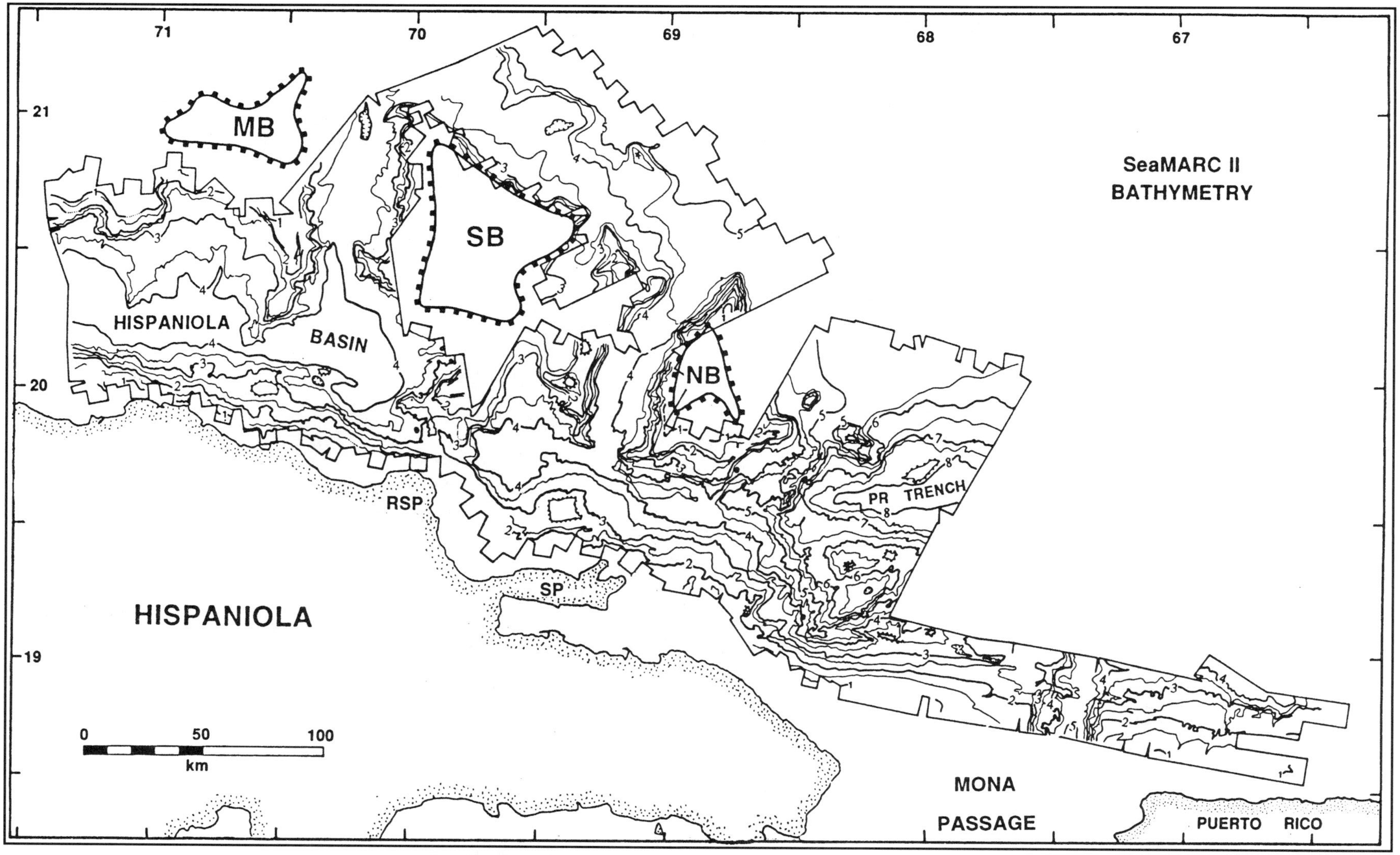

Figure 5. Simplified SeaMARC II bathymetry for study area. Contour interval shown is 500 m for clarity. Lines with small black squares denote edges of shoal-water tops of southeastern Bahamas carbonate banks. MB is Mouchoir bank; SB is Silver bank; NB is Navidad bank; PR is Puerto Rico.

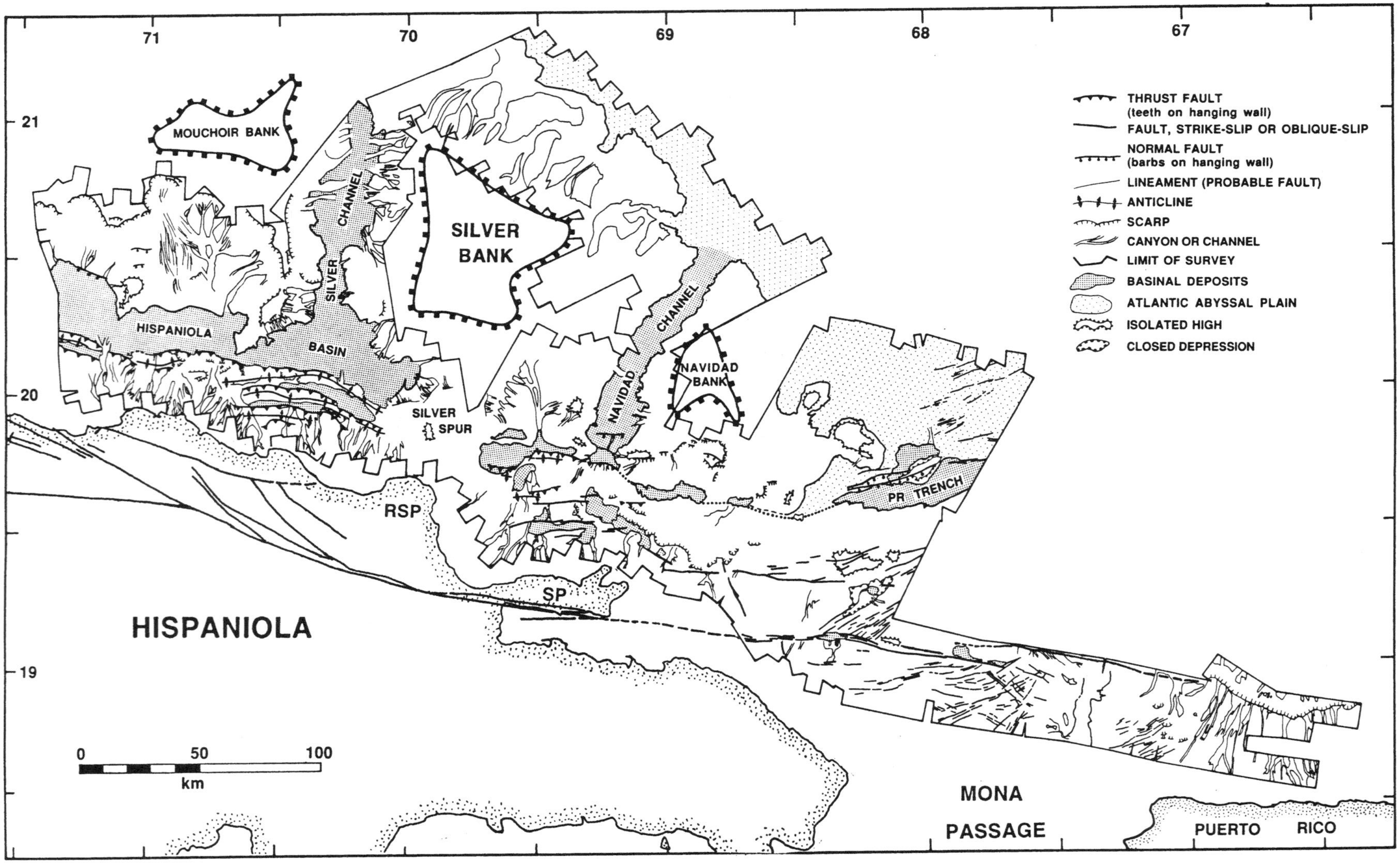

Figure 6. Summary map of side-scan sonar and seismic-reflection data showing major structural and sedimentologic features of the study area. See text for complete discussion. PR is Puerto Rico; SP is Samana Peninsula; RSP is Rio San Juan Peninsula.

associated frontal anticlines continue. At the eastern edge of the NHAP the frontal fold continues 25 km east of the easternmost expression of the surficial frontal thrust fault. There the fold gradually broadens eastward; at its easternmost extent (70° 05′W) the frontal deformation zone is represented by a very broad (12 km), gentle upwarp (Fig. 8C). These data indicate that folding precedes propagation of frontal thrust fault to the sea floor.

Much of the NHAP exhibits extensive erosion via channels and slumping (Fig. 10). Small gullies highlight the limbs of most folds in the NHAP, and in several cases the frontal folds have been truncated by large slumps (e.g., Fig. 8B). Slumping is particularly widespread between 70° 45′W and 70° 25′W, just southeast of Mouchoir Spur (Fig. 10). Analysis of the mechanics associated with the underthrusting of bathymetrically high-standing blocks (e.g., seamounts) shows that upon initial underthrusting, the overriding margin is uplifted, followed by collapse of the overriding margin in response to oversteepening of the slope (Lallemand and Le Pinchon, 1987; Collot and Fisher, 1991; Hagen and Moberly, 1994). We suggest that the slumps and intense gullying of the NHAP southeast of Mouchoir Spur resulted from a similar process, with erosion and collapse of the accretionary prism occurring in response to uplift and oversteepening during oblique underthrusting of Mouchoir Spur.

On the side-scan mosaic a number of these gullies exhibit small, highly reflective bright spots, which we interpret as vent localities associated with dewatering of the accretionary prism ("fluid vents" in Fig. 10B). Submersible dives on other plate margins have shown that similar bright spots observed on side-scar sonar data are caused by carbonate crusts and organic buildups associated with fluid vents (e.g., Carson et al., 1991, 1994, and references discussed therein). These deposits, which have a much higher acoustic impedance than the surrounding accretionary prism sediments, backscatter the sonar energy more effectively than the surrounding sea floor, and appear as highly reflective patches, or "bright spots," on the side-scan image. The occurrence of most of these bright spots in gully bottoms within the NHAP indicates that gullying of the frontal folds has apparently reached deep enough to intersect sedimentary aquifers within the folds. These bright spots also occur along the thalwegs of several channels cut into the offscraped sediments. These data have been used by others to illustrate the hydrologic and structural controls on deep-marine seeps (Orange and Breen, 1992).

Western continuation of the northern Hispaniola deformed belt. West of our survey area, offshore of northwestern Hispaniola, seismic reflection and GLORIA side-scan data show that active compression and accretionary prism development extends westward within the northern Hispaniola deformed belt to ~73° 30′W (Dillon et al., 1992). The accretionary prism in this area is similar to that offshore of central Hispaniola, with compression accommodated by large-amplitude folds forming above south-dipping thrust faults (Dillon et al., 1992), and we suggest that the name northern Hispaniola accretionary prism be extended westward to include the accretionary prism offshore western Hispaniola. The western part of the left-lateral Septentrional fault, which trends offshore at ~71° 45′W, extends along the Hispaniola shelf and upper slope ~25–50 km south of the contractional deformation front (Case and Holcombe, 1980; Edgar, 1991; Calais et al., 1992; Mann et al., this volume).

The NHAP dies out west of 73° 30′W in the Windward Passage (Figs. 2 and 3; Dillon et al., 1992). Farther west, south of Cuba, active contractional deformation within the Santiago deformed belt occurs predominantly on north-dipping structures (Fig. 2; Calais et al., 1989, 1992; Calais and Mercier de Lépinay, 1995), indicating a switch in the dip of partitioned contractional structures across the Windward Passage. The Septentrional fault changes name in the Windward Passage, becoming the Oriente fault west of ~73° 30′W. Along the south coast of Cuba the left-lateral Oriente fault is located ~10 to 25 km north of the contractional deformation front (Calais et al., 1989; Calais and Mercier de Lépinay, 1995).

Central Hispaniola slope subprovince (70°08′W to 68°30′W)

The SeaMARC II mosaic reveals that in the central part of the survey area the southern margins of both Silver and Navidad banks are actively colliding with the northern Hispaniola slope (Figs. 6 and 11). The most pronounced example of this collision occurs between 70° 05′W and 69° 42′W, where a bathymetrically high-standing ridge of Silver bank carbonate basement, herein named Silver Spur, impinges on the Hispaniola slope and projects beneath the Rio San Juan Peninsula of Hispaniola (Fig. 12). SeaMARC II bathymetry indicates that the crest of Silver Spur ranges in depth from 1,800 to 2,200 m; basinal sediments to the west lie at ~4,000 m, those to the east at 4,410 m (Fig. 11A). Seismic reflection data from within the collision zone are of generally poor quality, and thus the three-dimensional geometry of the south-dipping collisional interface is difficult to discern. Nevertheless, the deformation front within the collision zone, which we interpret to be at the southern edge of reflective carbonate basement, is generally well defined on the side-scan sonar image (Fig. 12). For example, the highly reflective, lithified carbonate of Silver Spur extends south to at least 19° 45′N along the eastern edge of the spur (Fig. 11B). Thus, the presumed deformation front within the Silver Spur part of the collision is approximately along trend with the well-defined deformation front both to the west and east (Fig. 12B).

East of Silver Spur the collisional deformation front is characterized by a series of discontinuous, flat-floored basins (Fig. 11). These small, east-west elongate basins descend in step-like fashion from west to east toward the Puerto Rico trench; depth increases from 4,410 m for the basin just east of Silver Spur (Navidad basin of Sieglie et al., 1976) to 5,300 m for the basin at the eastern edge of the collision zone. As in the Hispaniola basin to the west of the collision zone, the basin fill is a mixture of carbonate and terrigenous turbidites derived from Silver and Navidad banks to the north, as well as Hispaniola to the south (Sieglie et al., 1976). Seismic reflection profiles reveal that the basinal deposits are generally underlain by shallowly southward-

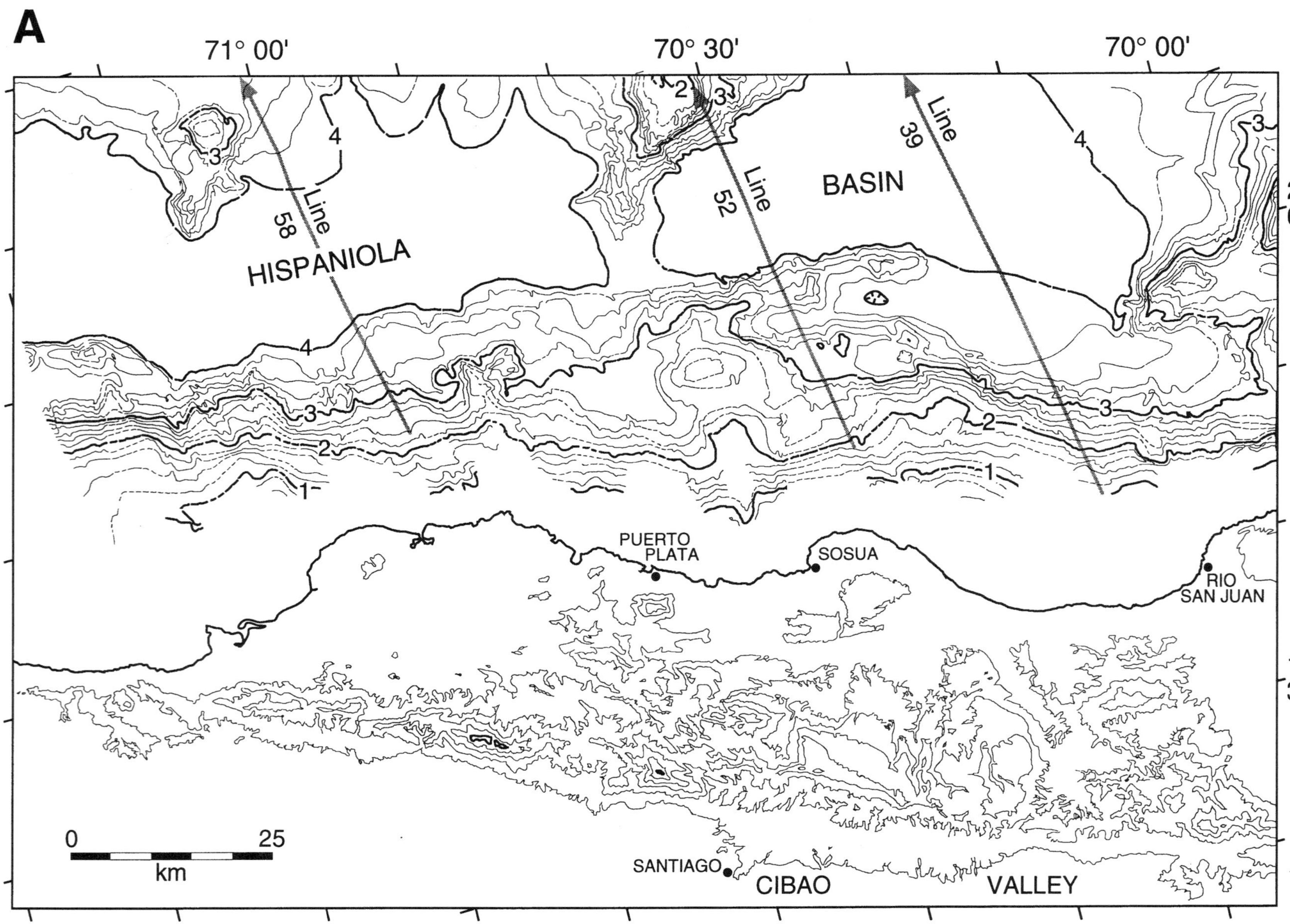

Figure 7 (this and facing page). Data and interpretation from the western survey area. (A) SeaMARC II bathymetry of the western Hispaniola slope. Contour interval 200 m. Onland topography from compilation by de Zoeten et al. (1991). Gray lines with arrows show locations of seismic-reflection profiles shown in Figure 8. Note all three profiles continue north off edge of figure (line 58 continues for 14 km beyond north edge of figure, line 52 continues for 16 km, and line 39 continues for 1 km). See Figure 4 for location of entire length of profiles shown in Figure 8. (B) Interpretation of major structures in the western part of survey area. Dark gray shading shows outcrop of Miocene-Pliocene Villa Trina Formation limestones. Pale gray shading denotes flat-floored, turbidite-filled basins. Onshore data from de Zoeten and Mann (1991) and de Zoeten et al. (1991). Epicenter (star) of 1953 Sosua earthquake from International Seismological Summary of 1953 (1961) and Sykes and Ewing (1965). Symbols as in Figure 6.

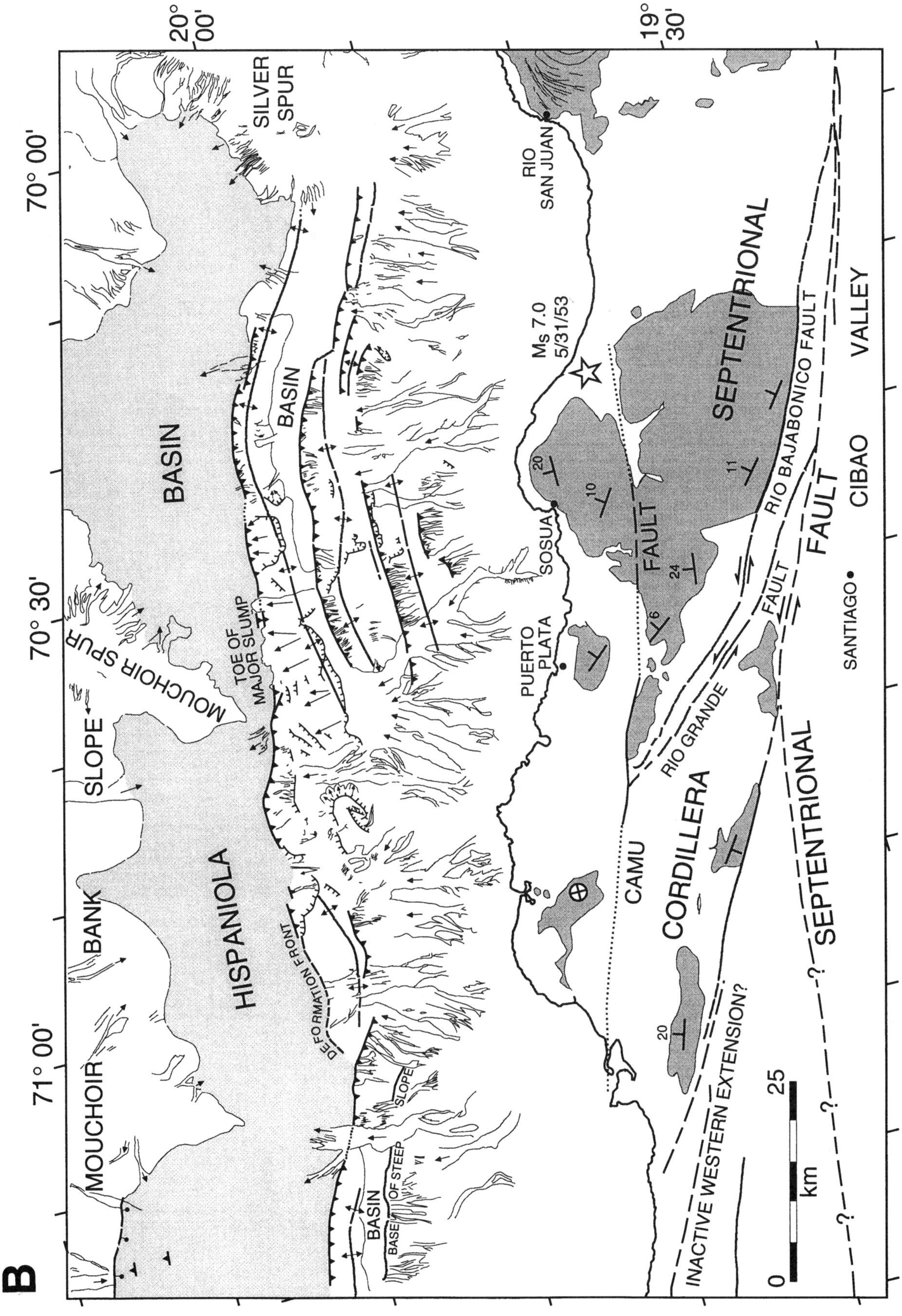
B
20° 00'
19° 30'
71° 00'
70° 30'
70° 00'
MOUCHOIR BANK
MOUCHOIR SLOPE
HISPANIOLA BASIN
SILVER SPUR
MOUCHOIR SPUR
TOE OF MAJOR SLUMP
DEFORMATION FRONT
BASIN
BASE OF STEEP SLOPE
RIO SAN JUAN
Ms 7.0 5/31/53
SOSUA
PUERTO PLATA
CAMU
CORDILLERA SEPTENTRIONAL
SEPTENTRIONAL FAULT
FAULT
RIO GRANDE FAULT
RIO BAJABONICO FAULT
CIBAO VALLEY
SANTIAGO
INACTIVE WESTERN EXTENSION?
20
10
11
24
6
20
0
25
km

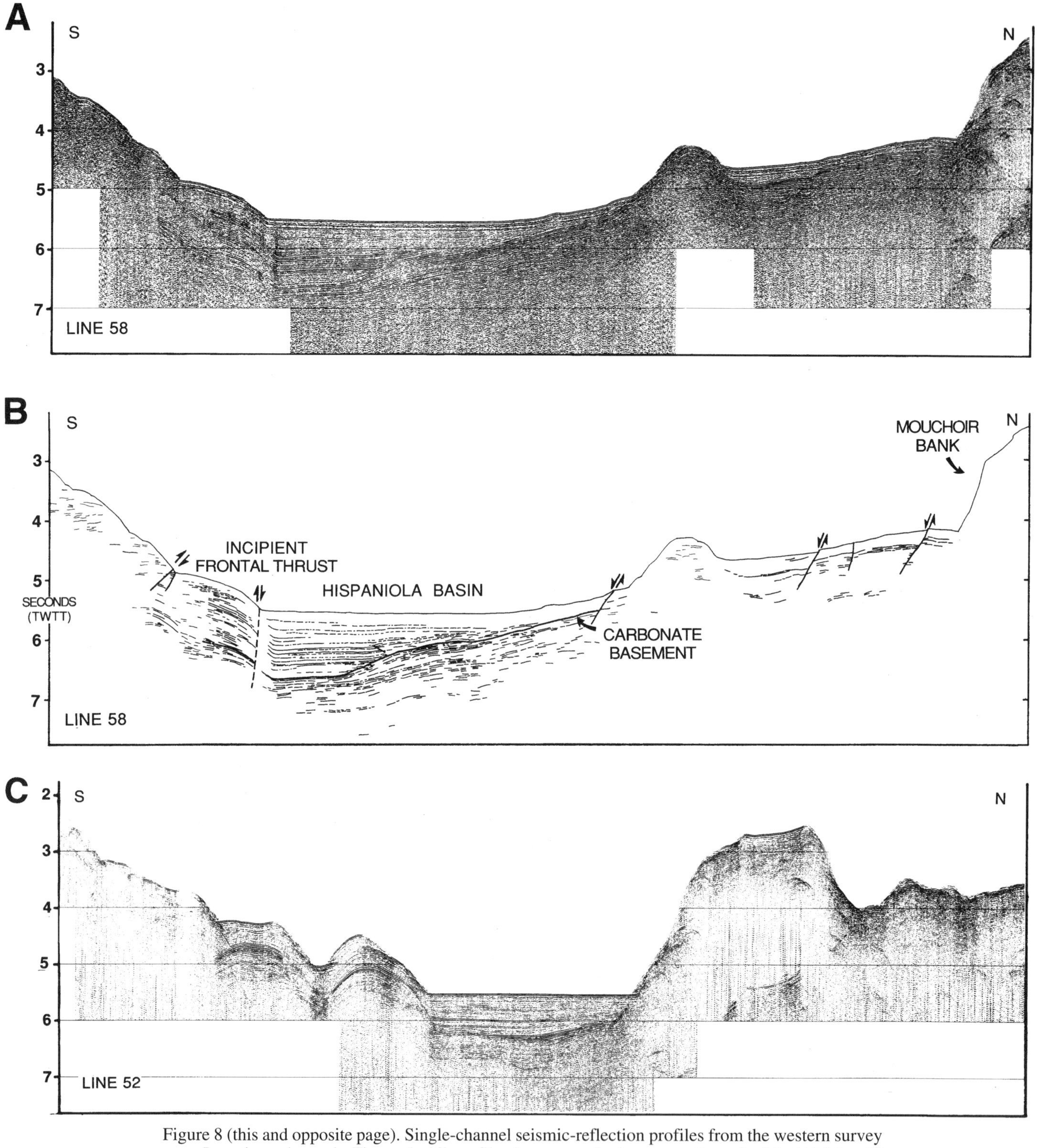

Figure 8 (this and opposite page). Single-channel seismic-reflection profiles from the western survey area. Locations shown in Figure 4. All profiles have a vertical exaggeration of ~4:1. (A) Line 58 from the western part of northern Hispaniola accretionary prism (NHAP) showing offscraping of thick Hispaniola basin sedimentary sequence including upper part of Mouchoir bank carbonate slope deposits. (B) Line-drawing interpretation of Line 58. (C) Line 52 from central part of NHAP showing extensive erosion of frontal fold. (D) Line-drawing interpretation of Line 52. (E) Line 39 from easternmost part of NHAP, just west of western edge of Silver Spur collision. Note the development of a broad upwarp well seaward of the thrust front. (F) Line-drawing interpretation of Line 39.

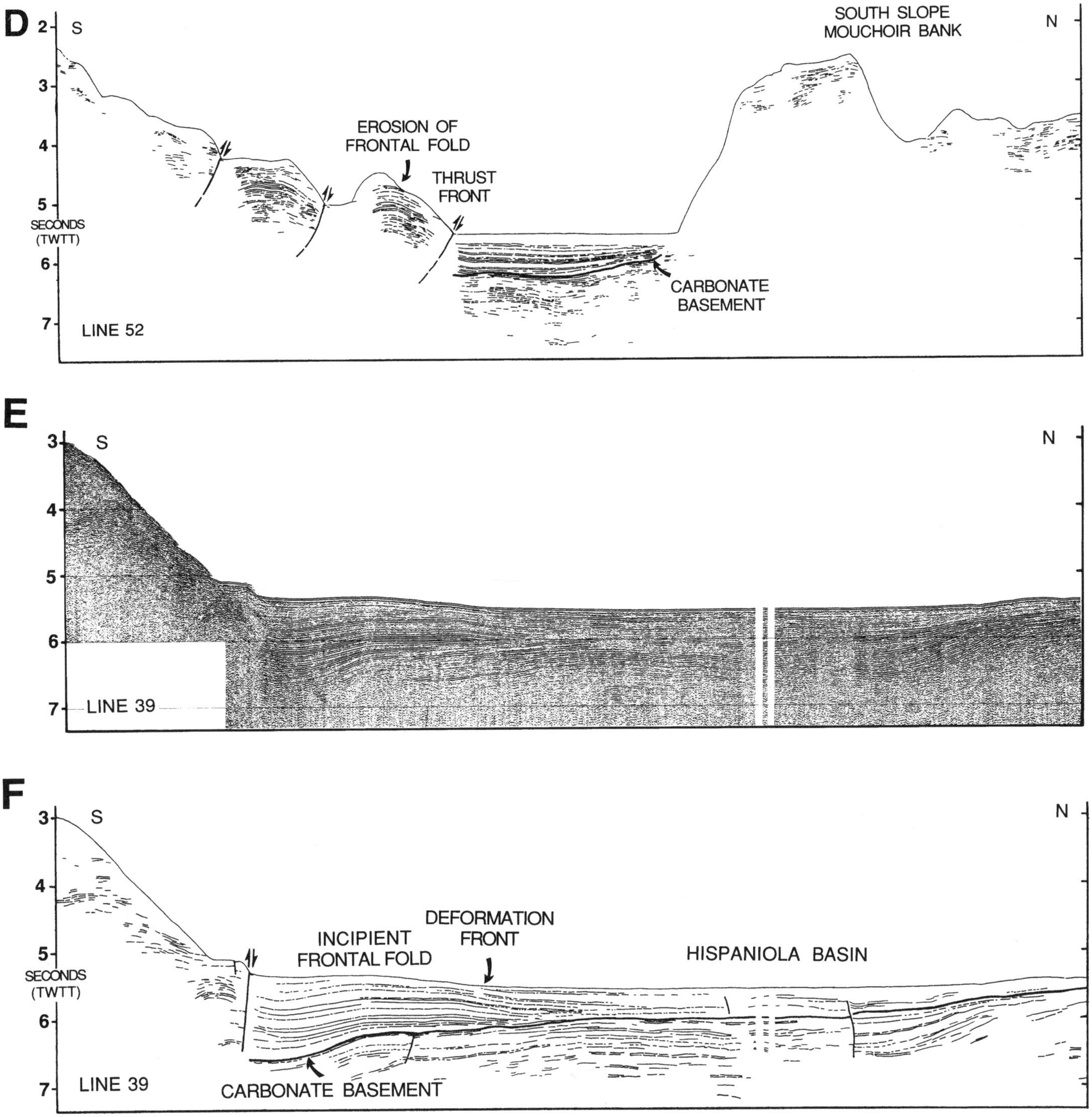

dipping carbonate basement. Between the discontinuous basins highly reflective carbonate bank deposits directly abut lower reflectivity Hispaniola slope deposits.

Between the eastern edge of Silver Spur (69° 42′W) and the southwestern edge of the Navidad bank collision (69° 12′W) seismic-reflection profiles and side-scan sonar data reveal a well-defined deformation front characterized by small-scale folding and faulted basinal strata (Fig. 13). These structures consistently lie along the southern margins of the discontinuous turbidite-filled basins. Just east of Silver Spur the deformation front exhibits a 10-km-long, northeast-trending left step, which we interpret as a left-lateral tear fault in the deformation-front thrust system (Fig. 11B). Within the Navidad bank collision, east of 69° 12′W, deformation-front contractional structures are either poorly defined or absent. The location of the deformation front in this region is inferred from the location of the discontinuous basins in

A

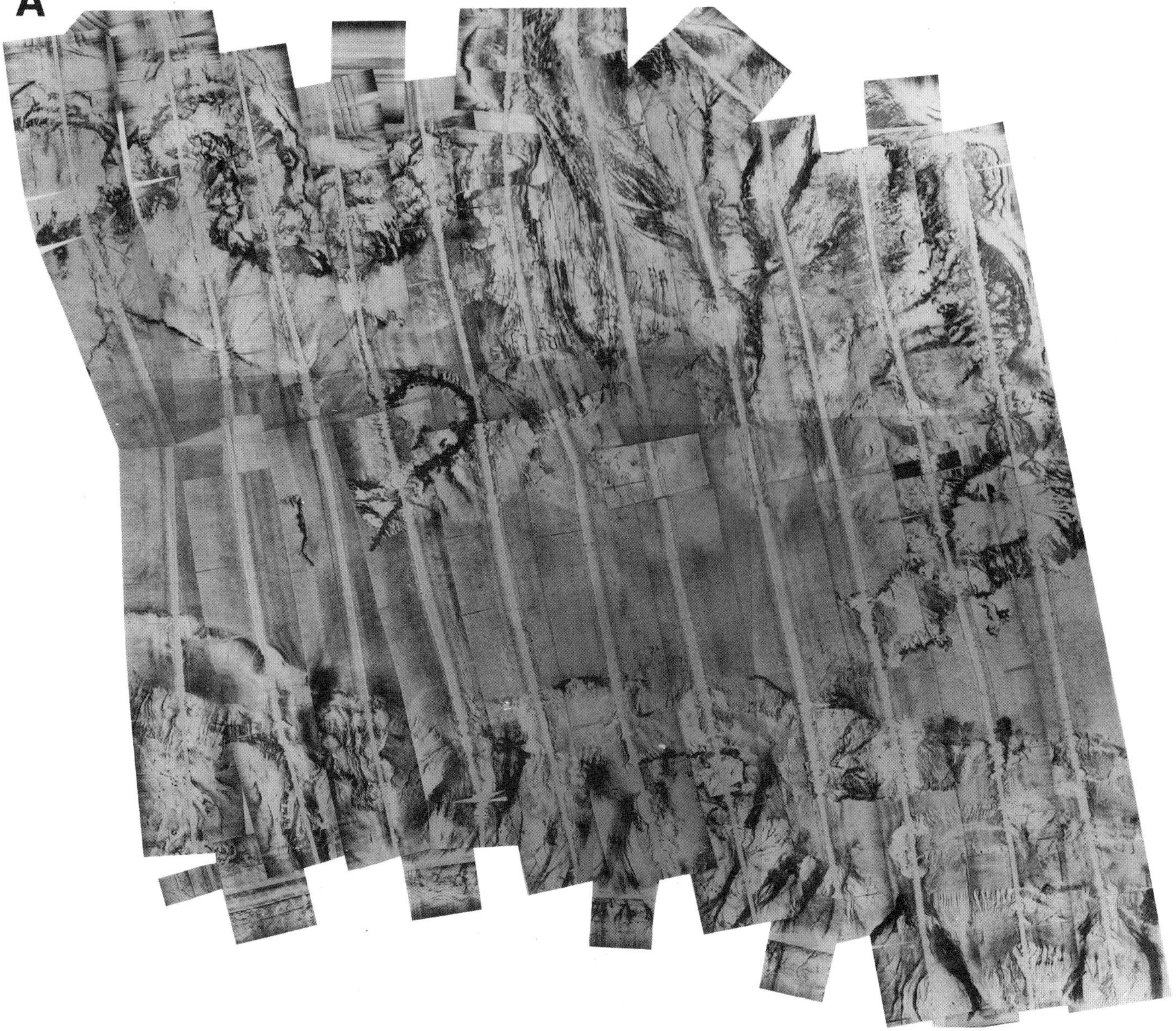

Figure 9A. SeaMARC II side-scan sonar mosaic of western survey area. Note the spectacularly imaged, 50-km-long avalanche chute extending southward from the southern margin of the shoal-water crest of Mouchoir bank. Also note the amphitheater-shaped, erosional "scallops" that characterize the bank edge.

combination with the sharp contrast in reflectivity between the bank and the Hispaniola slope (shown by dotted line in Fig. 11B).

One particularly interesting aspect of the Navidad bank collision is an anomalous zone of highly reflective outcrop on the Hispaniola slope that extends from 68° 48′W to 69° 08′W. This reflective zone, which lies south of the presumed deformation front, represents the northern slope of a 35-km-long bathymetric ridge that projects up to 1,400 m above the basin floor to the north (Fig. 11). In contrast to the gently north-dipping, northernmost Hispaniola slope both east and west of the highly reflective zone, the ridge exhibits a southern, south-facing slope with several hundred meters of relief (Fig. 13B). Seismic-reflection profiles across the ridge show generally flat-lying internal reflectors, suggesting that the feature is not a frontal fold above a thrust fault. The highly reflective northern slope of the ridge closely resembles the slopes of Navidad bank to the north, but it is quite unlike the generally low-reflectivity Hispaniola slope to the south. Based on this similarity and the anomalous bathymetry of this ridge, we suggest that it may represent a piece of southernmost Navidad bank that is being thrust beneath the Hispaniola slope. Thus, locally the true deformation front may actually lie along the southern

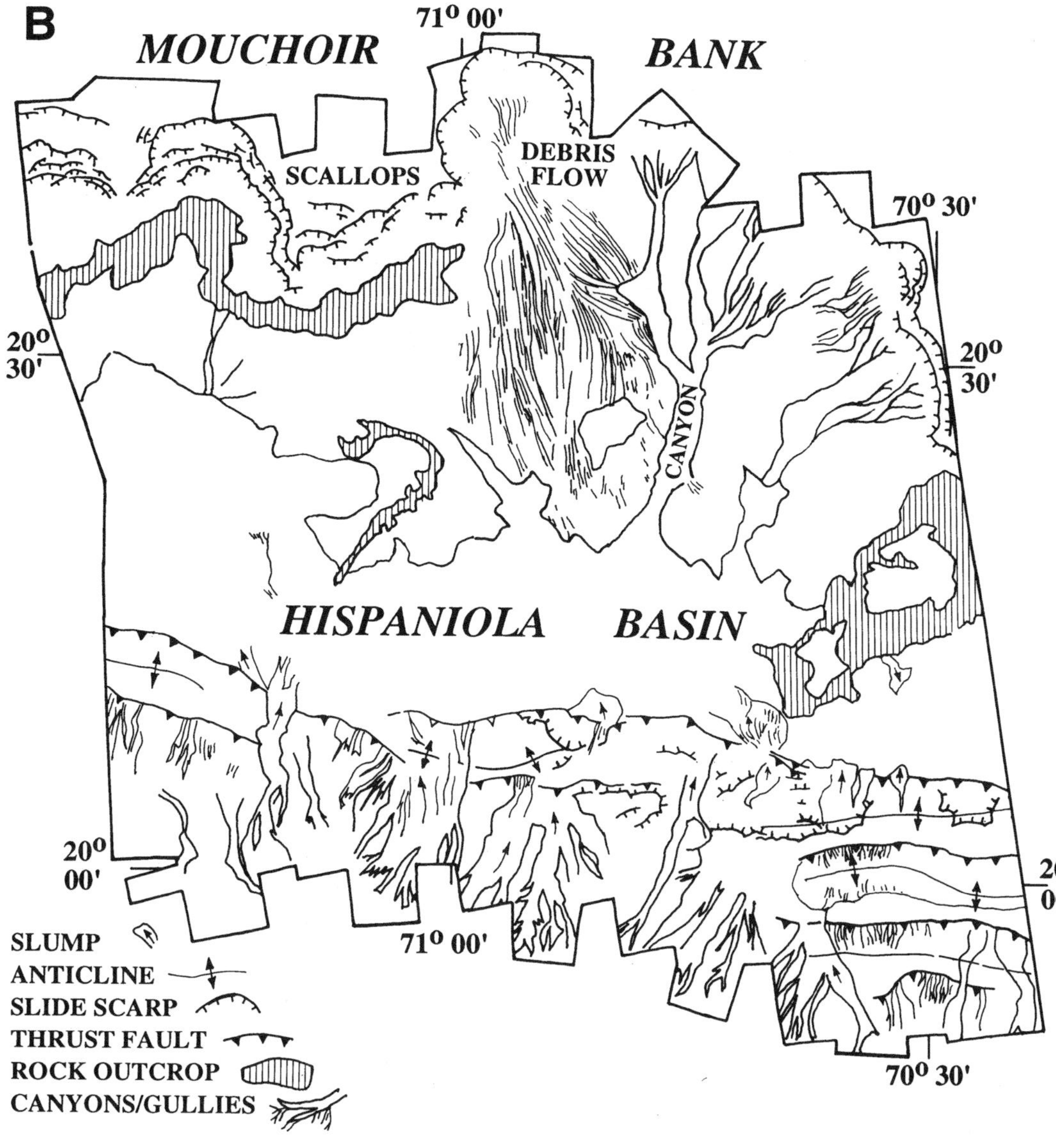

Figure 9B. Line drawing of major structural and sedimentary features observed on the side-scan sonar mosaic.

edge of this ridge, rather than along the edge of the small sedimentary basin to the north.

Structures in the Hispaniola slope trend generally east-west, and several are traceable for >40 kilometers along strike on the side-scan image. Many of these structures, particularly on the lower slope, exhibit contractional deformation features such as anticlinal folds. These structures are generally small in scale, in marked contrast to the large-scale fold-and-thrust belt structure evident in the accretionary prism to the west of the collision zone.

Unlike structures in the lower slope, structures in the upper slope do not exhibit clear-cut contractional features on seismic-reflection profiles. Furthermore, there is no consistent sense of higher topography on the north or south along the length of the side-scan lineaments. This is strikingly evident in the square bathymetric high centered at 69° 23′W (Fig. 11A). Both the north and south edges of this feature correspond to major lineaments visible on the SeaMARC II mosaic. Traced westward from the south-facing slope along the south edge of the bathymetric high, the southern lineament trends directly into a north-facing slope. Seismic-reflection profiles crossing these features indicate that the lineament to the south of the square bathymetric high locally exhibits a probable component of normal displacement, whereas the structure just to the north exhibits a probable component of reverse displacement (Fig. 13A). Based on the linear traces of these features, their lack of consistent along-strike topographic slope direction, and the laterally inconsistent indicators of vertical movement, we interpret them as strike-slip faults with subordinate vertical movements.

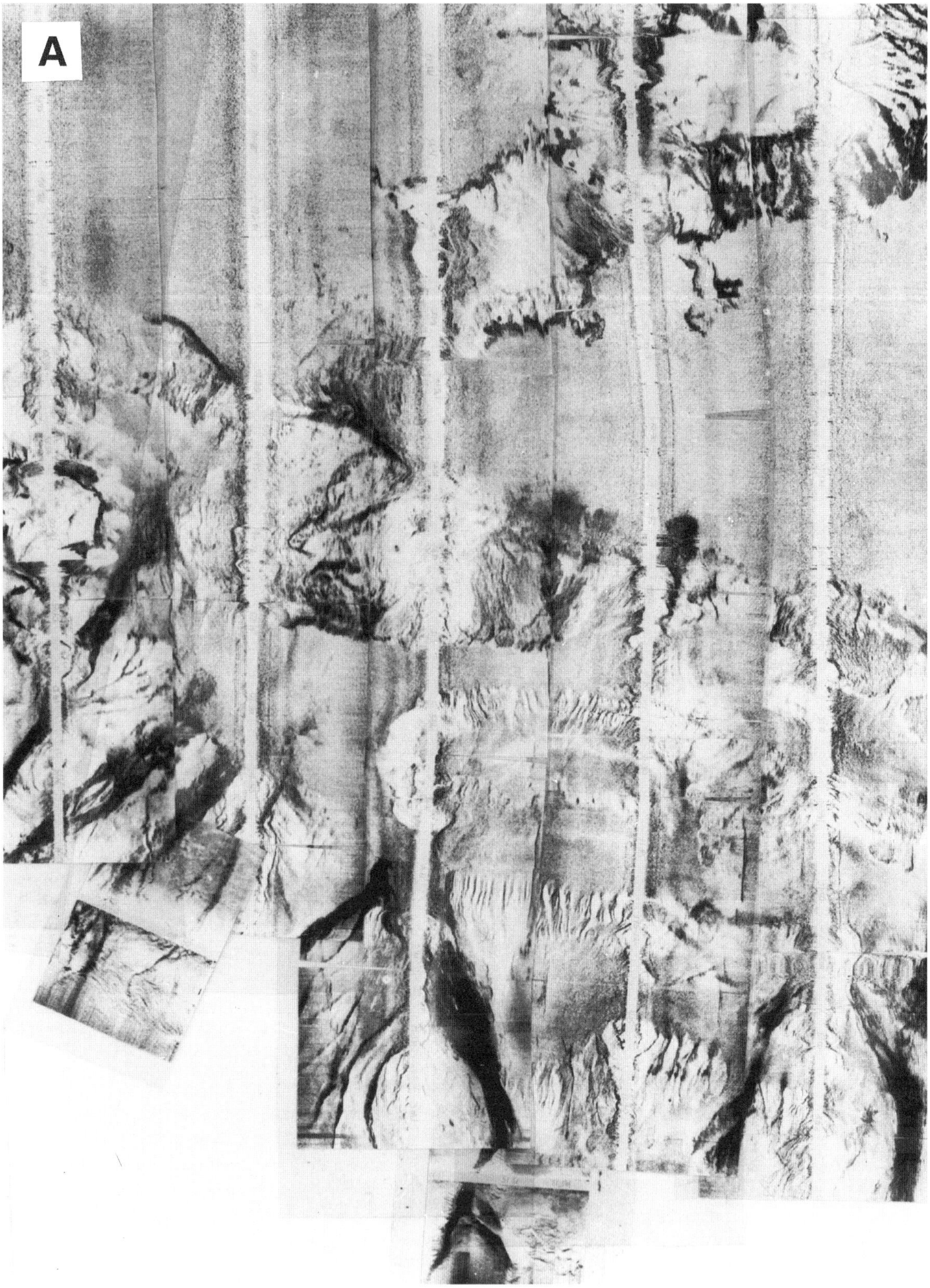

Figure 10A. Detail of SeaMARC II side-scan sonar image for the eastern half of the northern Hispaniola accretionary prism (NHAP).

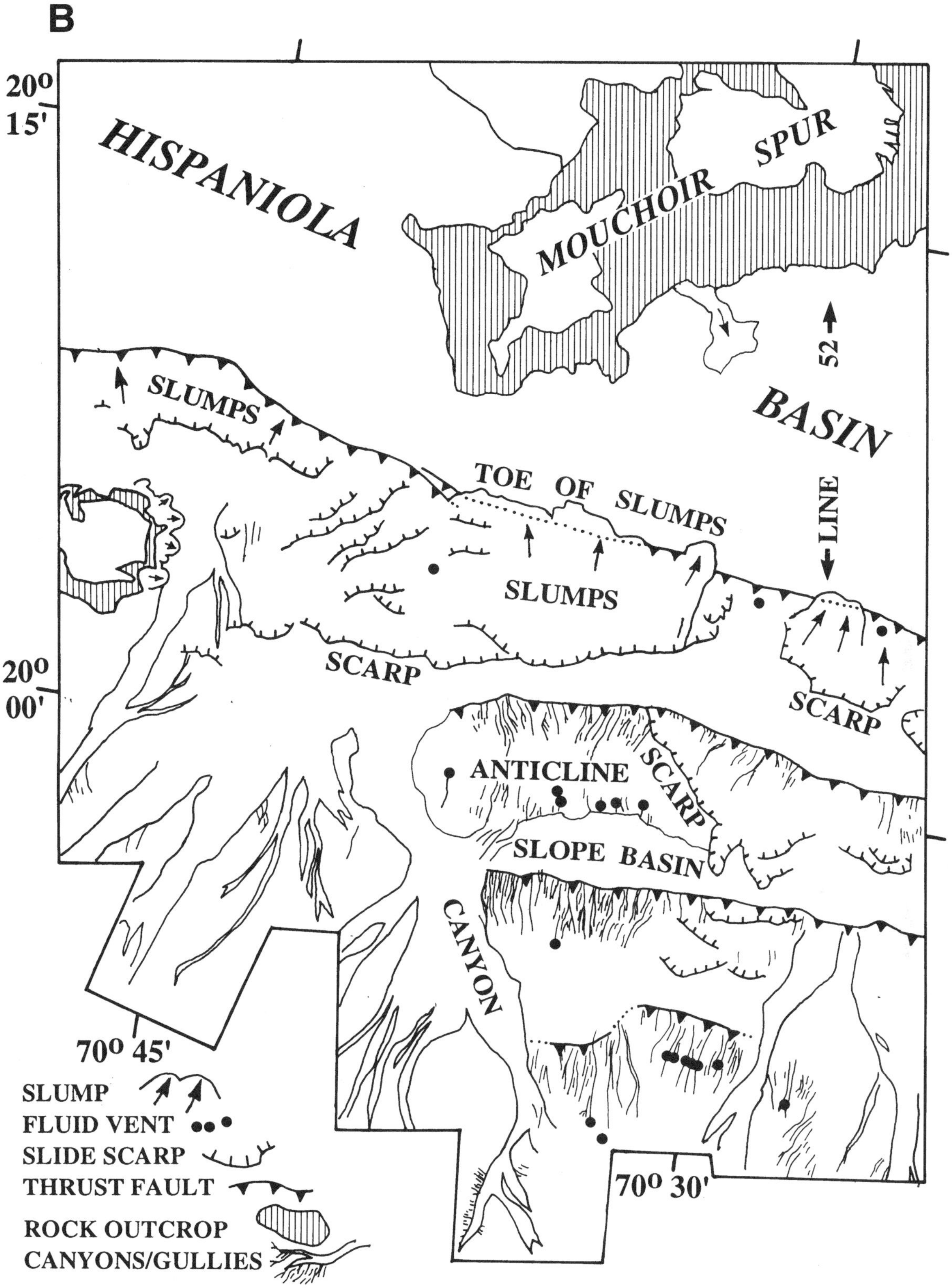

Figure 10B. Line drawing of major structural and sedimentary features observed on the side-scan sonar mosaic. Note the major zone of slumping off the frontal anticline just east of Mouchoir Spur. The basement-cored ridge exposed at the western edge of the image just north of 20° 00′N may represent a southwestern extension of Mouchoir Spur that has been thrust obliquely beneath the NHAP, causing collapse via slumping as it passes obliquely beneath the accretionary prism. Also note the occurrence of numerous, probable fluid vents associated with the development of gullies and canyons (shown by black dots).

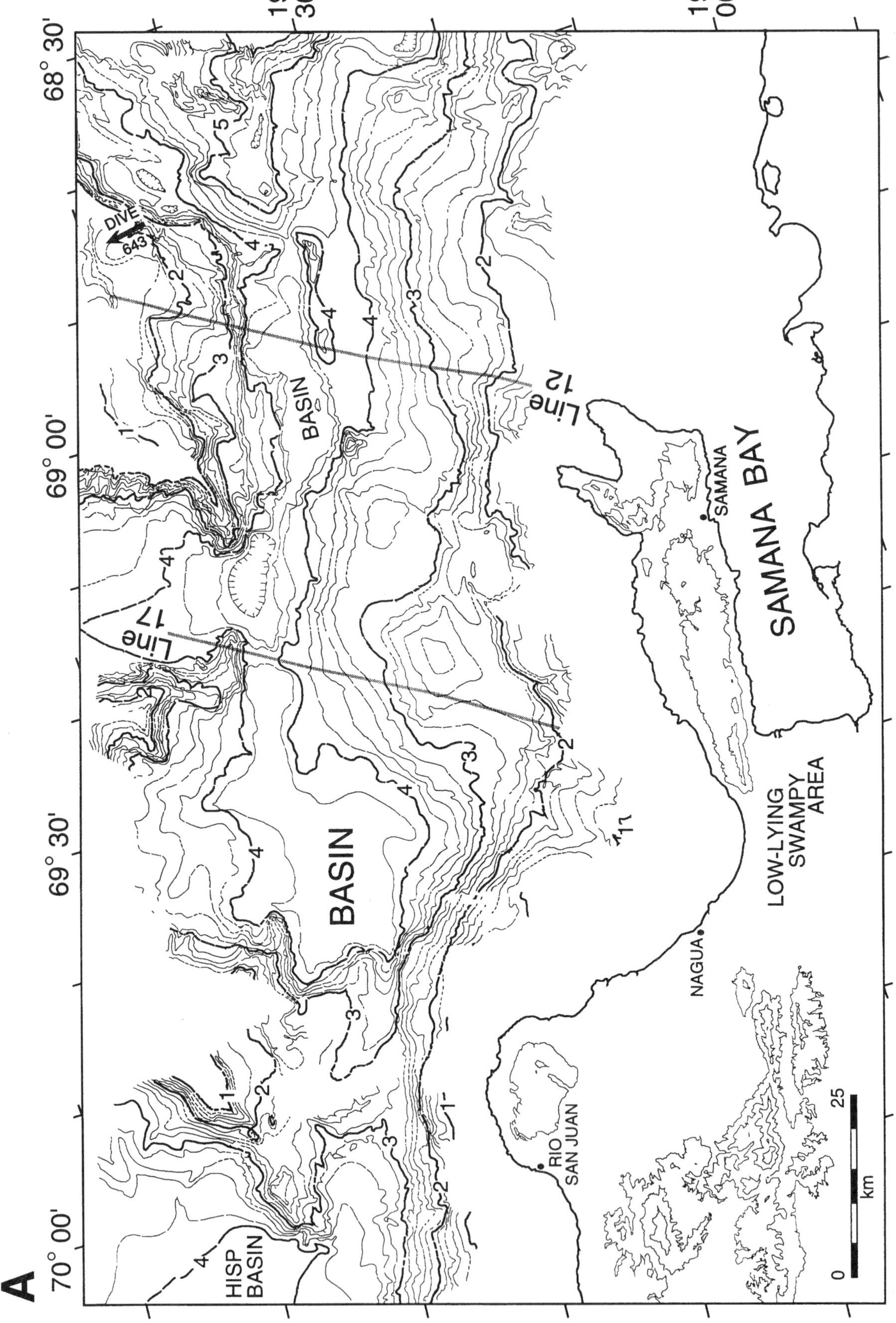

Figure 11A. Data and interpretation from the central survey area. SeaMARC II bathymetry of the central Hispaniola slope, including the Silver Spur collision zone. Contour interval is 200 m. Onland topography from compilation by de Zoeten et al. (1991). Gray lines show locations of seismic-reflection profiles shown in Figure 13. Black arrow shows submersible dive 643 discussed in text (from Heezen et al., 1985).

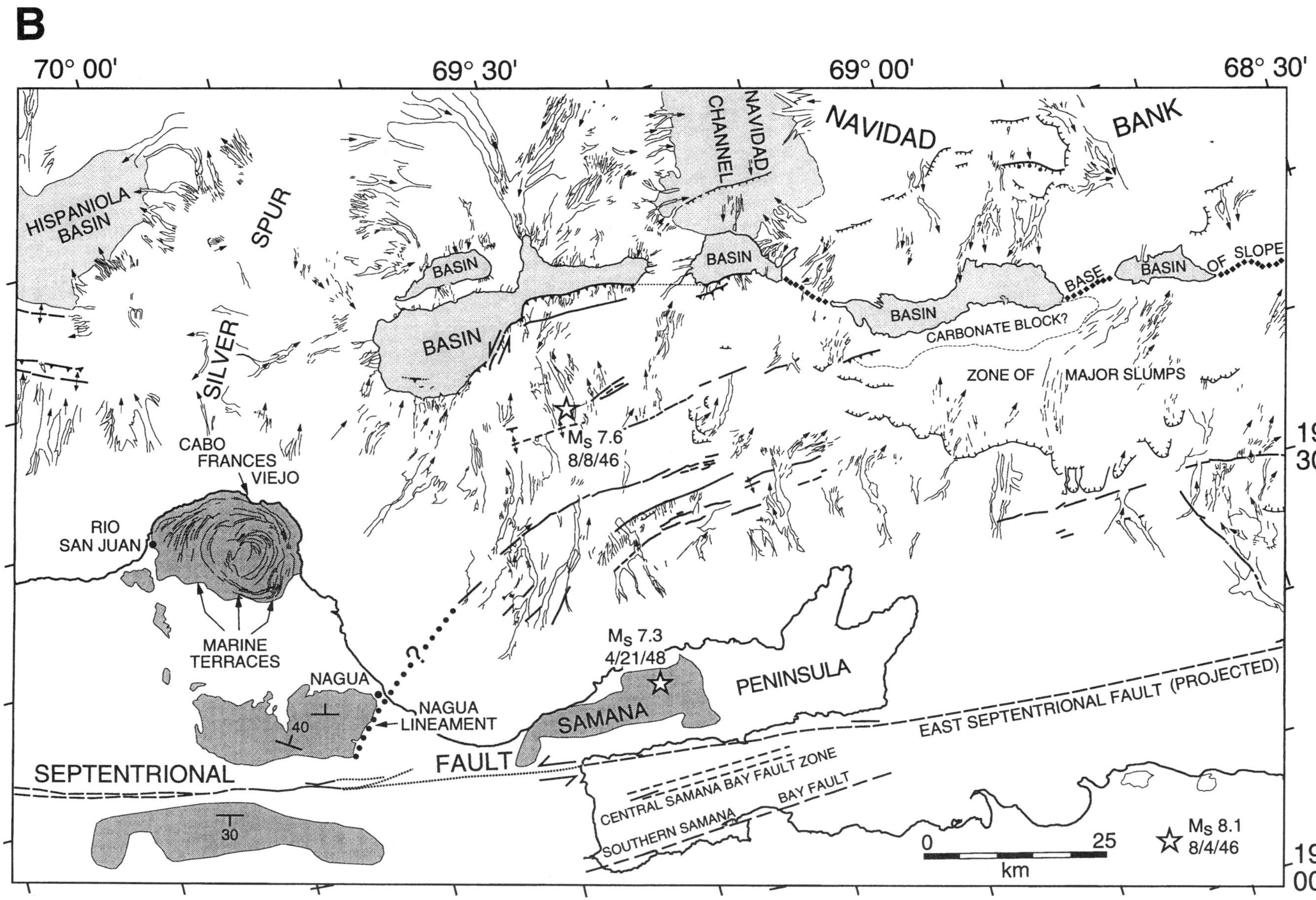

Figure 11B. Interpretation of major structures in the central part of survey area. Dark gray shading shows outcrop of Miocene-Pliocene Villa Trina Formation limestones (known as the La Canita Formation on the Samana Peninsula). The carbonate outcrop east of Rio San Juan encompasses younger (Pleistocene?) limestones that are lithologically similar to those of the Villa Trina Formation (see text for discussion). Onshore data from de Zoeten and Mann (1991) and de Zoeten et al. (1991), except for outcrop of the La Canita Formation, which is from Joyce (1991), and the pattern of Rio San Juan Peninsula terraces, which are from this study. Pale gray shading denotes flat-floored, turbidite-filled basins. Epicenters (stars) of 1946 Northeastern Hispaniola earthquake, its August 8, 1946, aftershock, and the April 21, 1948, Samana earthquake are from Kelleher et al. (1973). Symbols as in Figure 6.

Figure 12A. Offshore and onshore geology of Silver Spur collision zone. SeaMARC II side-scan sonar image of central part of the survey area showing the well-developed Silver Spur collision. Onshore data are an SAR image processed by Tim Dixon (used with kind permission).

B

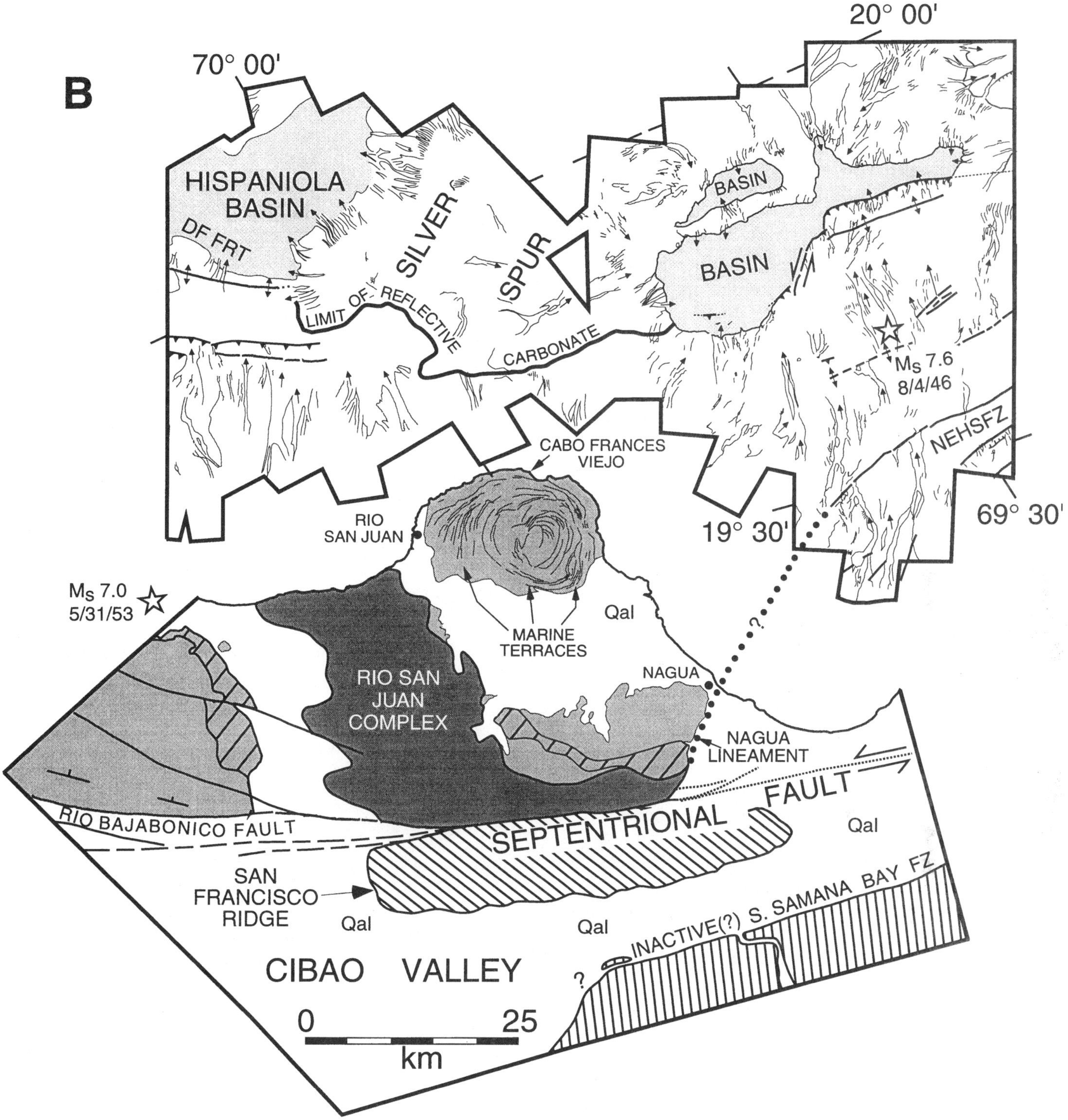

Figure 12B. Line drawing of major structural and sedimentary features on the side-scan sonar mosaic. DF FRT is deformation front of easternmost part of northern Hispaniola accretionary prism, where no frontal thrust fault is present. Deformation front within Silver Spur collision zone is poorly defined due to poor penetration of seismic reflection data but is taken to be at southern limit of reflective carbonate outcrop. Note curvature of western part of northeastern Hispaniola slope fault zone (NEHSFZ) into alignment with the north-northeast–trending Nagua topographic lineament. We postulate an oblique, left-lateral normal fault connecting these two features (dotted line). We speculate that this fault transfers slip from the Septentrional fault to the NEHSFZ. See text for discussion. Pale shading is outcrop of upper Miocene–lower Pliocene Villa Trina Formation (from de Zoeten and Mann, 1991; de Zoeten et al., 1991). Diagonal hatching denotes outcrop of resistant, reefal La Piedra member of Villa Trina Formation (modified from de Zoeten et al., 1991). Deposition of Villa Trina Formation limestones on Rio San Juan Peninsula probably continued into the Pleistocene. Vertical hatching denotes outcrop of Miocene–lower Pliocene Cevicos Formation limestone.

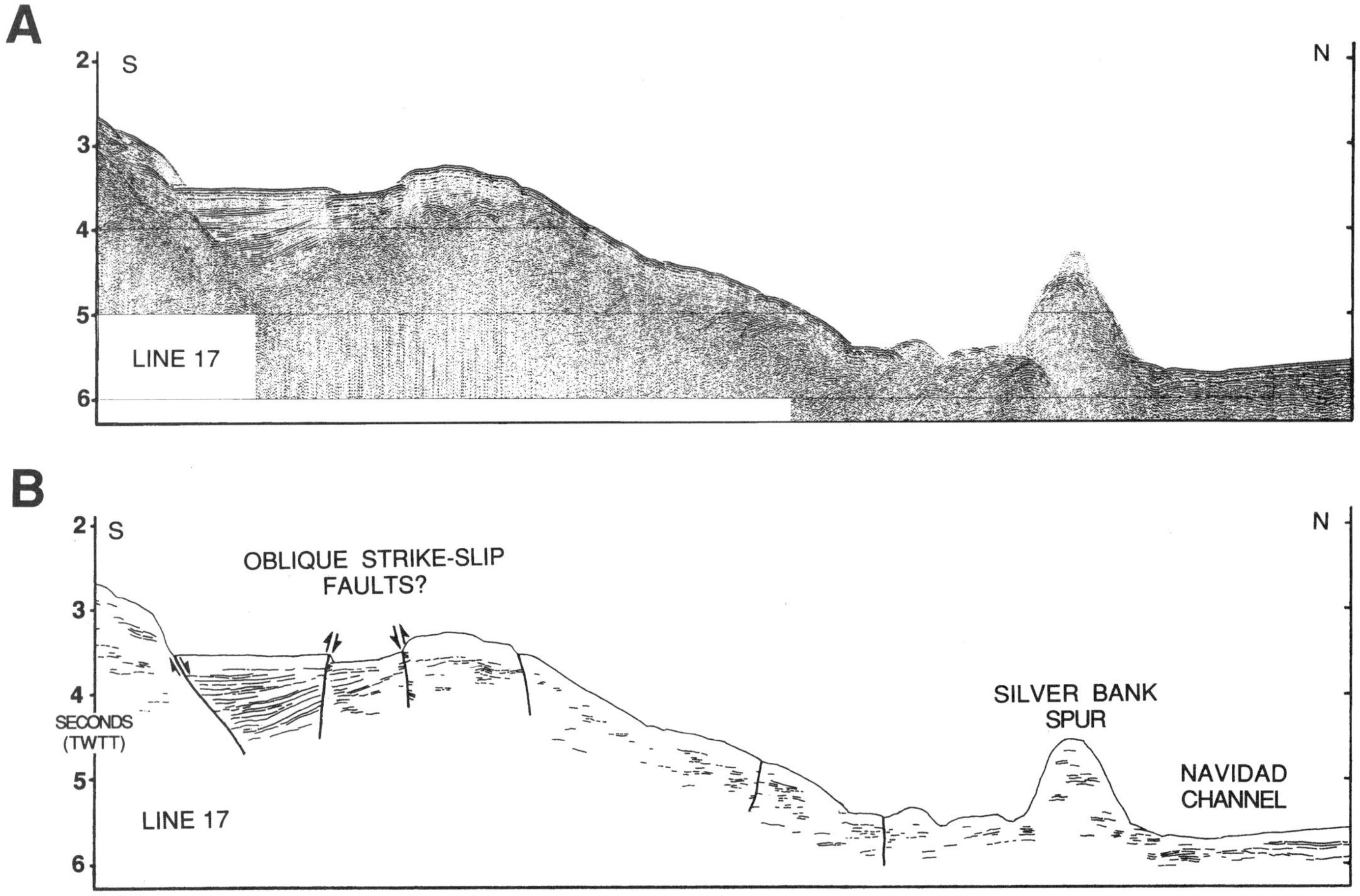

Figures 13A and B. Single-channel seismic reflection profiles from central part of survey area. Locations shown in Figure 4. All profiles have a vertical exaggeration of ~4:1. (A) Line 17 from the central part of the Silver-Navidad bank collision. Note the presence of what we interpret as oblique, left-lateral strike-slip faults cutting the upper slope, as well as the bathymetric irregularity of the colliding carbonate bank. (B) Line-drawing interpretation of Line 17.

The most prominent, southern two upper-slope faults curve westward into a southwest orientation where they appear to control the location of several linear submarine canyon segments near 69° 39′W (Fig. 11B). They project southwestward into a prominent, 220°-trending topographic lineament southwest of Nagua (Nagua lineament; Fig. 11B). We speculate that the southeast-facing lineament, which separates karst hills of Miocene-Pliocene Villa Trina Formation from low-lying fluvial and alluvial sediments, represents the onshore continuation of the left-lateral strike-slip faults observed along the slope. The lineament terminates just north of the Septentrional fault, the main left-lateral strike-slip onshore Hispaniola (discussed in following sections). The orientation of the lineament suggests that it may be a releasing bend connecting the offshore strike-slip faults with the Septentrional fault.

Most of the upper-slope strike-slip faults cannot be traced eastward beyond an irregular, generally west-northwest–trending lineament that runs parallel to the Hispaniola slope between 69° 12′W and 68° 40′W (Fig. 11B). In detail, this 65-km-long lineament consists of a series of cusp-shaped, down-to-the-north scarps. The lower Hispaniola slope north of the lineament appears as a mottled, generally low-reflectivity area on the side-scan mosaic, and single-channel and 3.5 kHz seismic reflection profiles show that much of this area exhibits hummocky surface topography. We interpret the lineament as the coalesced head scarps of a series of laterally extensive, but relatively shallow, gravitational slope failures that affected the entire lower Hispaniola slope from the Navidad collision zone (69° 05′W) to the eastern edge of the survey area (68° 00′W). The total relief of the head scarps is difficult to resolve exactly, but the slumps involved at least tens of meters, and possibly as much as 125 m of section. The lack of any discernible hemipelagic draping of the head-scarps and the hummocky nature of the sediment surface attest to the recent nature of this slope-failure event. This widespread slumping may have been caused by surface shaking related to the large earthquakes of the 1943–1953 north-central Caribbean sequence (Dolan and Wald, this volume). We interpret local slope-perpendicular lineaments within the slumped interval as

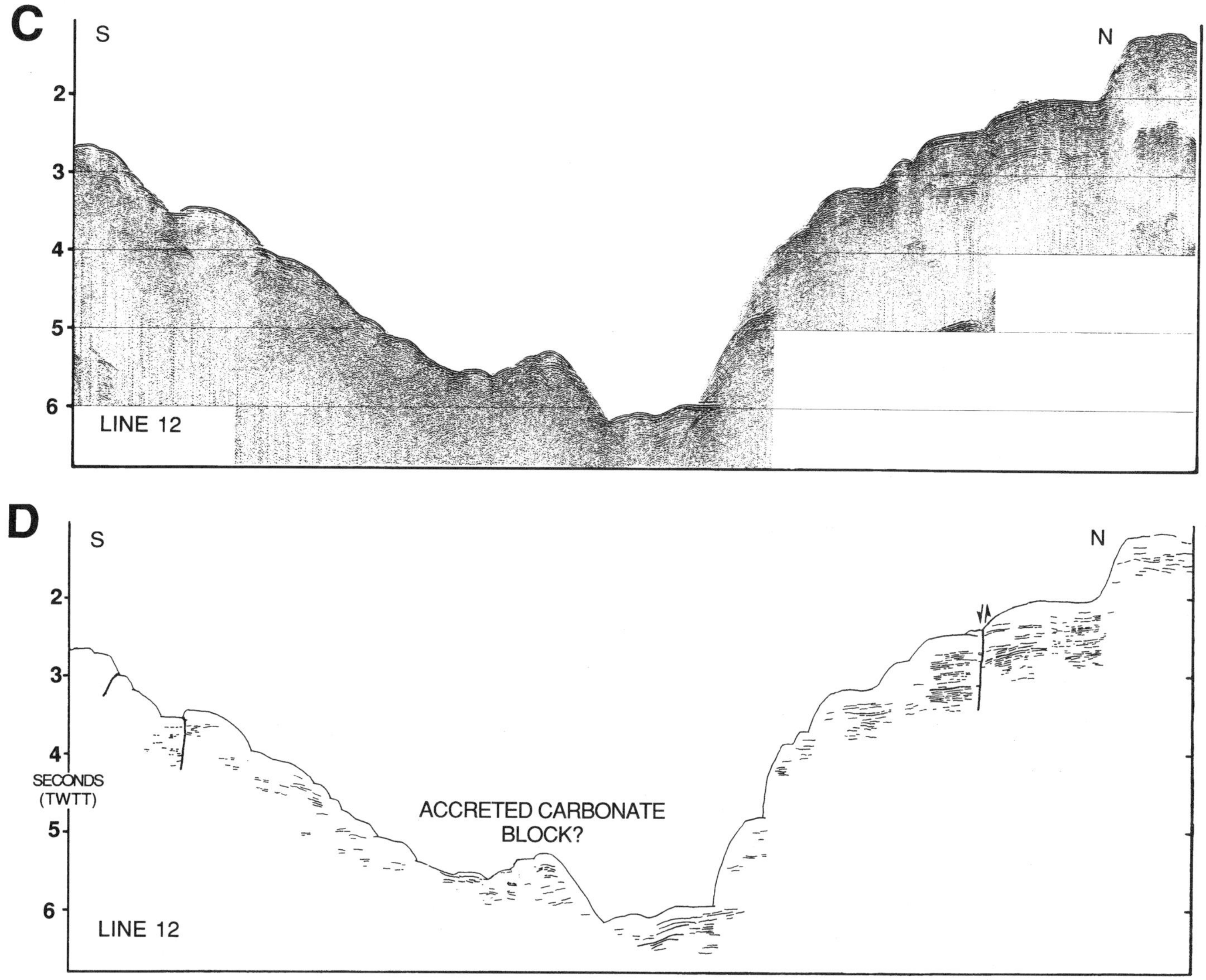

Figures 13C and D. Single-channel seismic reflection profiles from central part of survey area. Locations shown in Figure 4. All profiles have a vertical exaggeration of ~4:1. (C) Line 12 from the easternmost part of the Navidad bank collision. The "accreted carbonate block?" near the center of the figure may represent an effect of the extreme bathymetric irregularity of the southern margins of the carbonate banks. Thus, the apparent collisional deformation front along the north edge of the "accreted" bathymetric ridge may in fact represent only a bathymetric irregularity in the colliding bank. If so, then the true deformation front lies along the southern edge of the carbonate ridge. (D) Line-drawing interpretation of Line 12.

flow lines related to downslope sediment movement (Fig. 11B). The generally poor quality of seismic-reflection data in this area may result partially from scattering of the seismic energy source by the irregular, hummocky sediment surface.

Eastern Hispaniola slope subprovince (68° 30′W to 66° 30′W)

This zone includes data from both the southeastern part of the main SeaMARC II mosaic and the east-trending, 40 km × 190 km mosaic we collected across northern Mona Passage and along the northwestern Puerto Rico slope. The southeastern part of the main survey area contains several large, east-trending strike-slip faults (Fig. 14), which we discuss in detail in a following section.

As noted, east of 69° 15′W the lower part of the Hispaniola slope is characterized by generally low-reflectivity sea floor that exhibits extensive slumping. To the east of 68° 38′W the discrete, 65-km-long head scarp dies out. East of that point the entire lower slope down to the floor of the Puerto Rico trench appears to have slumped recently.

The deformation front along the southern edge of the Puerto

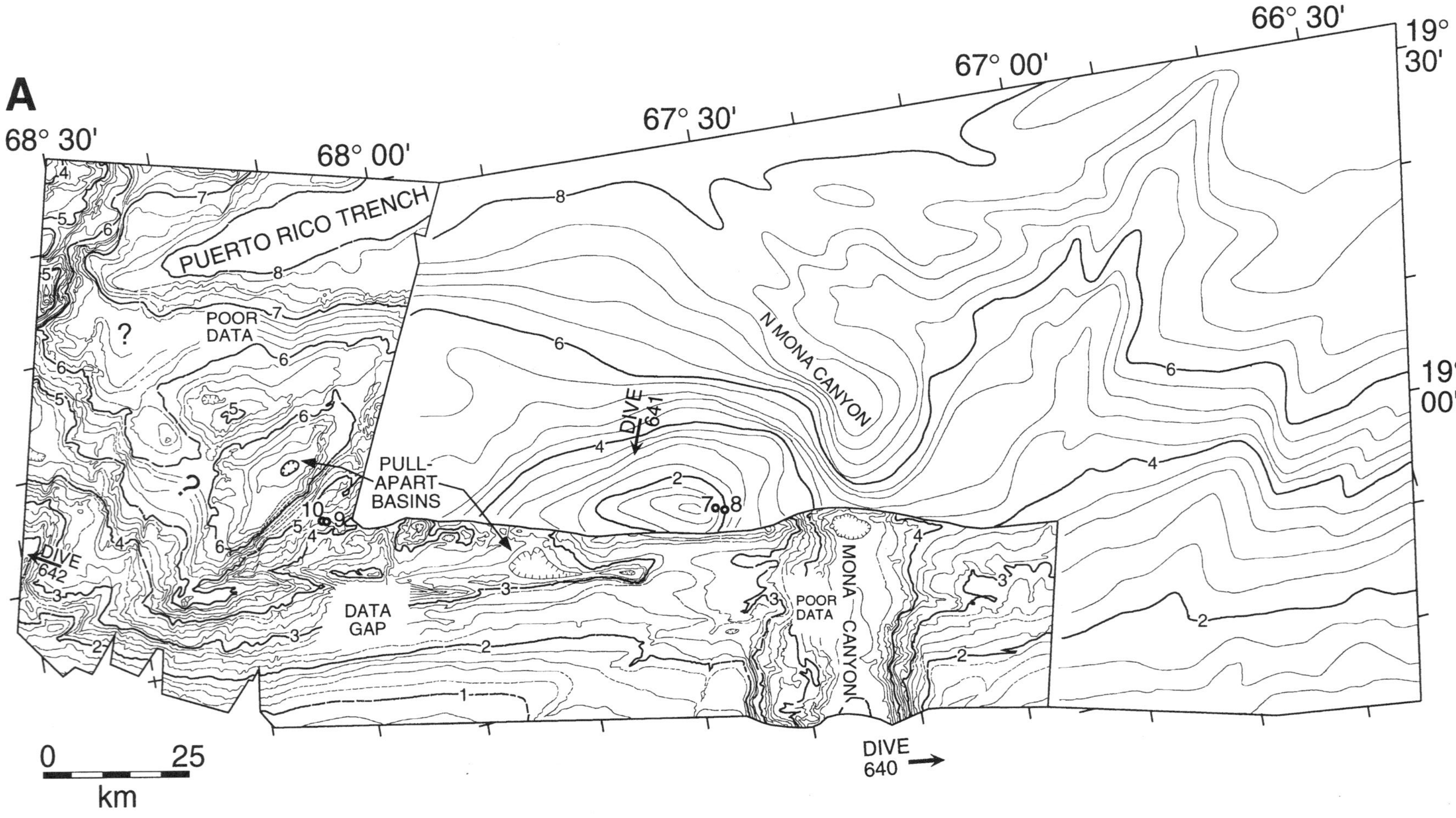

Figure 14A. Data and interpretation from the southeastern survey area. SeaMARC II bathymetry of the southeastern Hispaniola slope. Contour interval is 200 m. Black arrows show submersible dives 640 to 642 discussed in text (from Heezen et al., 1985). Black circles numbered 7 through 10 show locations of dredge hauls discussed in text (from Fox and Heezen, 1975).

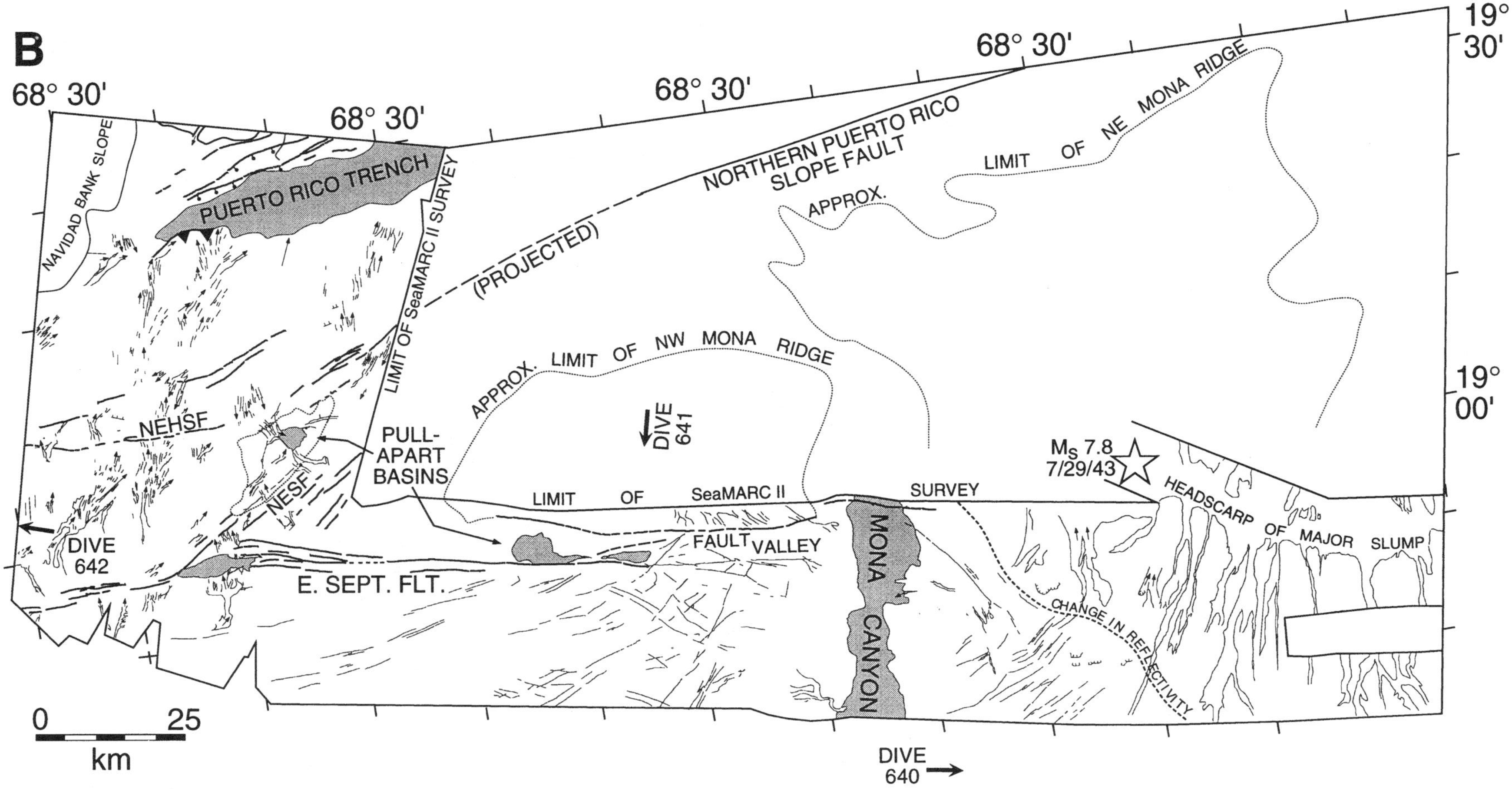

Figure 14B. Interpretation of major structures in the southeastern part of survey area. Gray shading shows limits of flat floors of Puerto Rico trench and Mona Canyon. Location of northern Puerto Rico slope fault (our terminology) from Masson and Scanlon (1991). Location of 1943 Northern Mona Passage earthquake (star) from Kelleher et al. (1973). Limits of Mona blocks modified from McCann and Sykes (1984). E. Sept. flt is eastern Septentrional fault zone. NEHSF is northern Hispaniola slope fault. NESF is northeastern Septentrional fault. Symbols as in Figure 6.

Rico trench is sharply defined on the side-scan mosaic along its entire length (Fig. 14). Mechanical problems limited the quality of the seismic reflection data in this region, but on several lines, Atlantic oceanic crust (discussed in following section) can be observed dipping beneath the Hispaniola slope.

A submersible dive (Heezen et al., 1985) along a scarp on the upper Hispaniola slope 55 km east of the Samana Peninsula encountered extremely fractured white, gray, and black marble with two vertical fracture sets, oriented east-west and north-south (Fig. 14; their dive 642). The metamorphic rocks, which are overlain by Tertiary (Eocene-Miocene) biopelmicrite, are part of a 500-km-long, marble-rich, high-pressure metamorphic terrane, herein referred to as the northern Caribbean metamorphic terrane (NCMT) (Fox and Heezen, 1975; Perfit et al., 1980). The NCMT encompasses several related but slightly different metamorphic assemblages thought to have formed in a Cretaceous subduction zone. The metamorphic terrane extends from north-central Hispaniola (Draper and Nagle, 1991; Pindell and Draper, 1991) eastward through the Samana Peninsula (Joyce, 1991), offshore along the northern part of Mona Passage between Hispaniola and Puerto Rico, and along the northern insular slope of Puerto Rico at least as far east as 65° 25′W (Fox and Heezen, 1975; Perfit et al., 1980).

Northwestern Puerto Rico slope–northern Mona Passage province (68° 10′W to 66° 30′W)

With the exception of Mona Canyon and the Septentrional fault-related features (described in following sections), the east-trending mosaic we collected across northern Mona Passage and along the northwestern Puerto Rico slope is characterized by a relatively reflective sea floor highlighted by two pervasive sets of well-developed faults (Fig. 14B). The faults are oriented east-northeast and west-northwest. In general the east-northeast–trending set appears to be predominant over most of the slope. Bathymetric data indicate that at least some of the lineaments have sea-floor expression, including scarps and deep, linear-to-slightly sinuous canyons. A north-south multichannel seismic-reflection profile across these features (Larue and Ryan, 1990; line VB-1N) shows that some of them are faults that exhibit a normal component of displacement, predominantly down-to-the-northwest, although down-to-the southeast displacements have also occurred locally. Although no direct evidence for strike-slip was observed along these faults, it is possible that they are predominantly transcurrent features and that the seismic-reflection data only resolve the vertical component of displacement. However, the degree of lateral displacement along these features remains unknown; although some features exhibit what appear to be lateral offsets on the side-scan data, there is no consistent offset direction and no discernible consistent sense of lateral offset of one set of lineaments by the other. The orientations and linearity of the two fault sets suggest that they may be conjugate strike-slip faults.

The reflective sea-floor and conjugate faults disappear ~25 km east of Mona Canyon across a relatively sharp, but irregular, northwest-trending lineament (labeled "change in reflectivity" near southeast edge of Fig. 14B). Northeast of this lineament the slope off northwestern Puerto Rico slope is characterized by a relatively smooth, low-reflectivity sea floor highlighted by numerous north-trending submarine canyons cut into high-reflectivity rocks (Fig. 14B). Previously collected GLORIA side-scan sonar data show that this same general sea-floor character extends eastward for several hundred kilometers across the Puerto Rico slope (EEZ SCAN 85 Staff, 1987; Masson and Scanlon, 1991). Seismic reflection and drill hole data indicate that the rocks exposed at the sea floor along the Puerto Rico slope are part of a north-dipping sequence of Neogene shallow-marine limestones that have been tilted northward down to depths as great as 4,000 m since ~4 Ma (Birch, 1986; Moussa et al., 1987).

Mona Canyon. The Mona Canyon is a deep, narrow trough in the northeastern part of Mona Passage off the northwest coast of Puerto Rico (Fig. 14). The generally north-trending canyon is divisible into three major elements: (1) a central segment with linear, north-trending walls and a flat basin floor at 4,800–5,000 m depth; (2) a deep northern segment that curves northward to a 330° orientation (EEZ SCAN 85 Staff, 1987); and (3) a southern segment that curves southeastward toward the Puerto Rico coast (EEZ SCAN 85 Staff, 1987). The northern segment is separated from the central segment by a steep, 2,500-m-high, north-facing scarp cutting perpendicular to the canyon (Fig. 14). Four crossings of the central segment show a gently north-sloping, flat canyon floor with a probable sediment pond at its northern end near 19° 00′N, 67° 22′W, just south of the apparent crossing of the Septentrional fault. The linear walls of the central segment of the canyon appear to be structurally controlled, and probably represent normal fault scarps associated with opening of the canyon. The east wall, in particular, appears to be a single large normal-fault scarp. It is not clear whether the northern and southern segments of the canyon are structurally controlled, and the steep slope in the northern segment of the canyon may represent a debris chute emanating from the Septentrional fault crossing at the northern end of the structurally controlled central canyon segment. A submersible dive up the extremely steep eastern wall of the canyon just south of our survey area revealed horizontally bedded volcanic or metavolcanic rocks from 3,300 m depth to 2,200 m depth (Heezen et al., 1985; their dive 640) (Fig. 14B). From 2,200 m to the top of the steep slope at 1,100 m, horizontally bedded limestones are exposed. These limestones exhibit abundant evidence for karst weathering down to ~1,500 m depth, with less well-defined evidence of karst development possibly observed as deep as 2,100 m (Heezen et al., 1985). The top of the slope east of the canyon is a broad plateau on which limestones with karst weathering features crop out (Heezen et al., 1985).

The northeast Mona block and northwest Mona ridge: Anomalously shallow areas of the northern Mona Passage. The northern part of Mona Passage is highlighted by two large, anomalously shallow blocks that flank the northern part of Mona Canyon (Fig. 14A). McCann and Sykes (1984) collectively called these two zones of shallow bathymetry their "anomalous block." The anomalous block is better expressed west of the canyon than to the east, and it is not clear that the bathymetry of the block east of the canyon is truly anomalously shallow with respect to the rest of the slope. We speculate that this area might simply represent part of the carbonate shelf sequence exposed along the north slope of Puerto Rico (Moussa et al., 1987). We refer to this part of McCann and Sykes' (1984) "anomalous block" as the northeast Mona block.

West of Mona Canyon, the anomalous block encompasses a 60-km-wide, oblong ridge that extends northward from the Septentrional fault for ~30 km to 19° 20′N (Fig. 14). We refer to this western shallow area as the northwest Mona ridge. The crest of this broad promontory is an east-trending, oblong hill just north of the Septentrional fault that reaches water depths shallower than 800 m. Dredge hauls and a submersible dive on the flanks of this hill indicate that it consists of marble, calcareous schist, mica schist, and greenschist of the NCMT overlain by Miocene to Pleistocene shallow-water limestones that exhibit karst weathering features (Fig. 14) (Fox and Heezen, 1975; Weaver et al., 1975; Heezen et al., 1976; Perfit et al., 1980).

Major faults in the southeastern survey area and northern Mona Passage

Data from the southeastern part of the main survey area and the northwestern Puerto Rico slope mosaic reveal several prominent, straight-to-slightly sinuous lineaments that are traceable for 10–100 km on the side-scan mosaic (Figs. 14B and 15). In particular, two major east-trending zones, one along ~19° 05′N and the other along ~19° 20′N, are traceable over much of the image (Fig. 14). The linearity of these features, coupled with the absence of any consistent sense of vertical separation along their lengths, strongly suggests that they are major strike-slip faults. Although we observed no geomorphic features indicative of strike-slip offset (e.g., offset canyons), the trend of the faults and the overall left-lateral nature of the plate boundary imply that they are sinistral strike-slip faults. The southern lineament is along strike of the easternmost part of the Septentrional fault system in the Samana Bay area 50 km west of our survey (Edgar, 1991) and probably represents the eastward continuation of the fault system (Fig. 14B); we therefore refer to this as the eastern Septentrional fault. We name the northern lineament, which corresponds to steep, linear slopes over much of its length, the northeastern Hispaniola slope fault zone (Fig. 14).

The more southerly, eastern Septentrional fault zone lineament bifurcates at 68° 25′W, with one strand continuing 105° toward Mona Canyon (eastern Septentrional fault) and the other strand (northeastern Septentrional fault) trending 050° toward the Puerto Rico trench, where it appears to connect with a major strike-slip fault along the lower Puerto Rico slope (Fig. 14B; EEZ SCAN 85 Staff, 1987; Masson and Scanlon, 1991). We name the latter fault the northern Puerto Rico slope fault. Together, the northeastern Septentrional–northern Puerto Rico slope fault system defines a major left step highlighted by a 7-km-deep, rhomb-shaped basin (discussed in following section).

Eastern Septentrional fault zone: Southeastern strand (eastern Septentrional fault). The Septentrional fault zone lineament corresponds to a deep canyon that extends eastward for >60 km from the southwest corner edge of the mosaic at 68° 42′W (Fig. 14). The canyon widens between 68° 27′W and 68° 18′W to form a narrow, 13 km × 3 km, elongate sedimentary basin. At the eastern end of the basin the fault-controlled canyon exhibits a marked change in strike, with an overall 085° trend west of the basin and a 102° trend to the east. These data suggest that the basin has developed in response to transtension along a minor releasing bend. West of the basin, the canyon is slightly sinuous, broad, and relatively shallow, with total relief ranging from 400 to 500 m relative to the surrounding sea floor. In contrast, the canyon east of 68° 18′W is more linear, and very narrow, and it exhibits total relief of 800–950 m. The canyon can be traced eastward on the side-scan mosaic to ~67° 55′W, where it ends. At this point the side-scan lineament can be traced eastward as a steep north-facing slope that extends east to 67° 38′W.

The bathymetric data indicate that the deep canyon is not the only slope-parallel, linear feature suggestive of strike-slip faulting in the southeastern survey area. Along the easternmost stretch of the eastern Septentrional fault canyon, a narrow, 300-m-high ridge parallels the northern edge of the canyon for at least 18 km. This ridge, which crops out 1.5–2 km north of the canyon, dies out eastward at about 67° 50′W (Fig. 14). Other east-west elongate features in the area include a subsidiary slope-parallel canyon, located 4 km south of the main canyon. These features suggest that the Septentrional fault system in this area consists of a zone of parallel faults at least 6 km wide.

The main Septentrional fault canyon and the steep north-facing slope along strike to the east form the southern boundary of a narrow, east-west elongate basin (Fig. 14). This 23-km-long basin, which extends from 67° 56′W to 67° 40′W, exhibits distinct western and eastern 3,500-m-deep subbasins. Basin width ranges from ~6 km to <3 km between the subbasins. The northern margin of the basin also forms a highly reflective lineament on the side-scan mosaic. This steep, south-facing slope appears to terminate near the northwestern end of the basin, at about 67° 51′W. Termination of this lineament at the northwest corner of the basin, coupled with the possible eastward termination of the southern lineament near the southeast corner of the basin, suggests that the basin may have formed as a pull apart along a 4–8-km-wide left step in the Septentrional fault system. The anomalously shallow (<800 m deep) hill that marks the surface expression of underthrust Mona bank forms the northern edge of the basin (Fig. 14).

The northern basin lineament can be traced eastward to Mona Canyon and appears to cross the canyon at ~19°N. A multichannel seismic reflection profile crossing a major fault just west of Mona Canyon at approximately the location of the southeastern Septentrional fault shows that the fault dips moderately southward and has a component of down-to-the-south normal displacement (Larue and Ryan, 1990; Larue et al., 1990; their Escarpment-19° fault). Although Larue and Ryan (1990) interpret this as a normal fault, the side-scan and bathymetric data indicate that it is a pre-

Figure 15 (this and opposite page). (A) SeaMARC II side-scan sonar image of the southeastern part of the main survey area. (B) Line drawing of major structural and sedimentary features observed on the side-scan sonar mosaic. Note zone of extensive slumping at northern corner of figure and 7-km-deep basin that we interpret as a pull apart along the major left-step between the eastern Septentrional fault and the northern Puerto Rico fault zone. NESFZ is northeastern Septentrional fault zone.

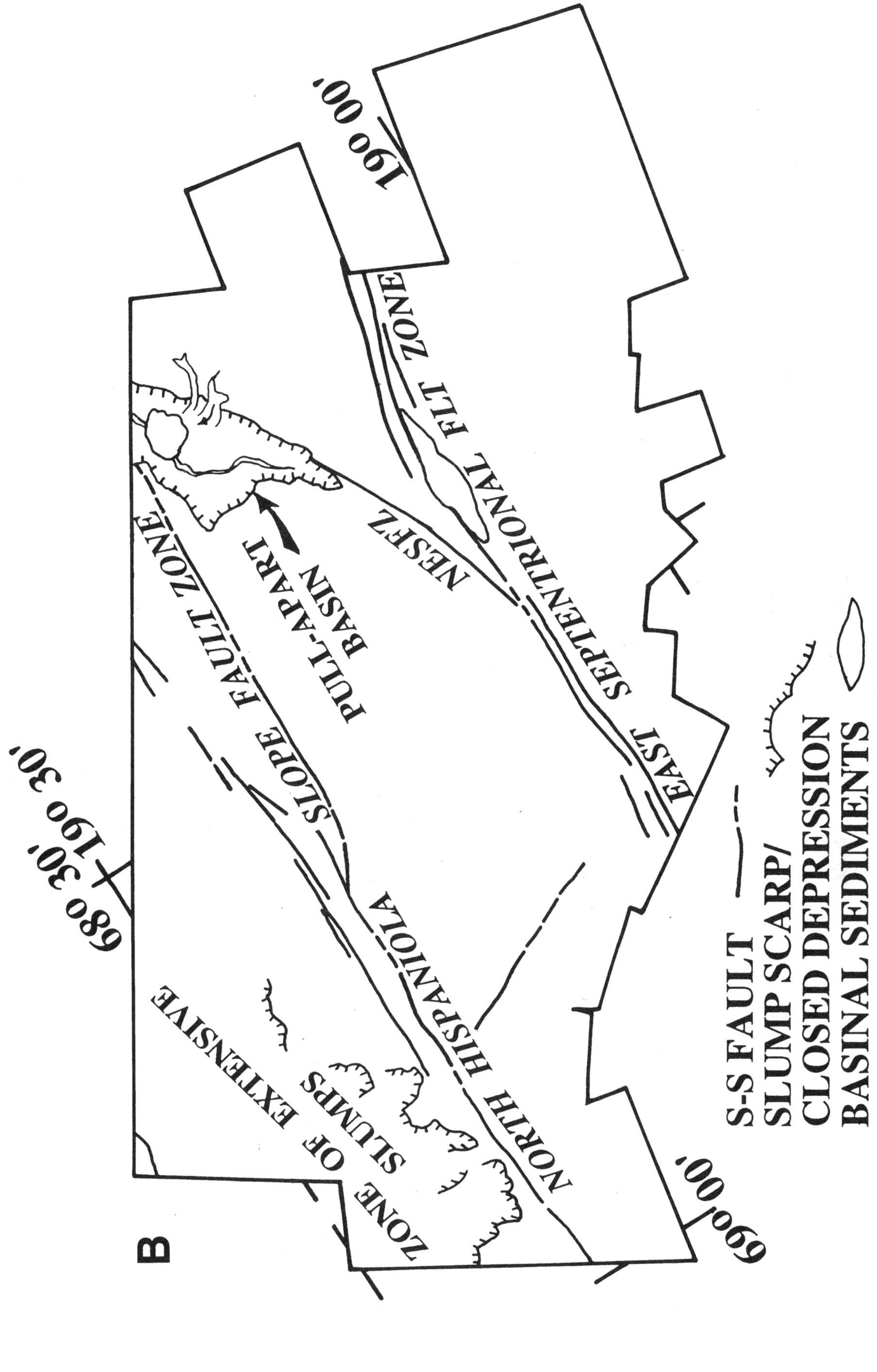
B
19° 00'
19° 30'
68° 30'
69° 00'
NORTH HISPANIOLA SLOPE FAULT ZONE
ZONE OF EXTENSIVE SLUMPS
PULL-APART BASIN
NESFZ
EAST SEPTENTRIONAL FLT ZONE
S-S FAULT
SLUMP SCARP/
CLOSED DEPRESSION
BASINAL SEDIMENTS

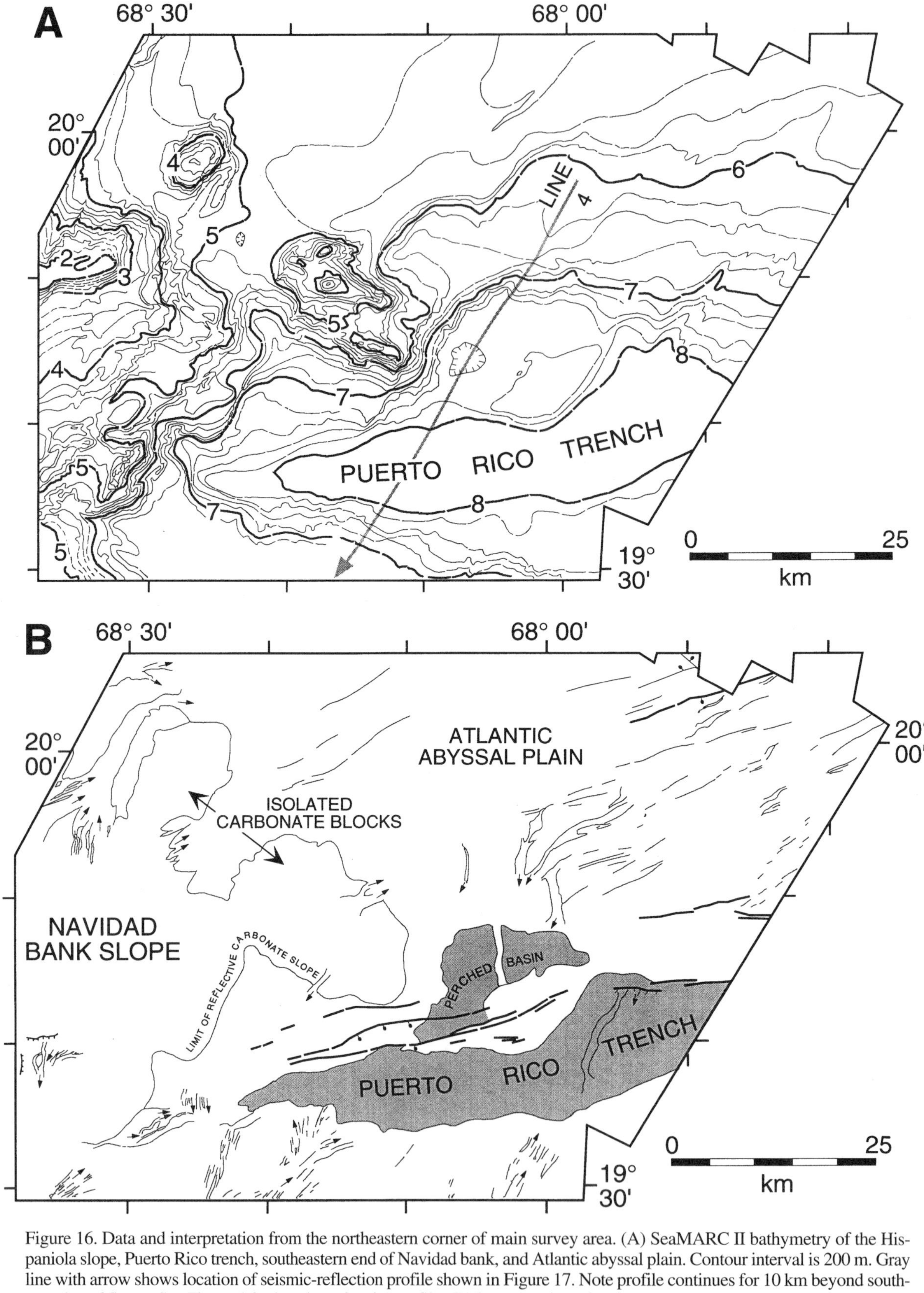

Figure 16. Data and interpretation from the northeastern corner of main survey area. (A) SeaMARC II bathymetry of the Hispaniola slope, Puerto Rico trench, southeastern end of Navidad bank, and Atlantic abyssal plain. Contour interval is 200 m. Gray line with arrow shows location of seismic-reflection profile shown in Figure 17. Note profile continues for 10 km beyond southern edge of figure. See Figure 4 for location of entire profile. (B) Interpretation of major structures in the northeastern corner of main survey area. Gray shading denotes flat floors of the Puerto Rico trench and the "perched" sedimentary basin to the north.

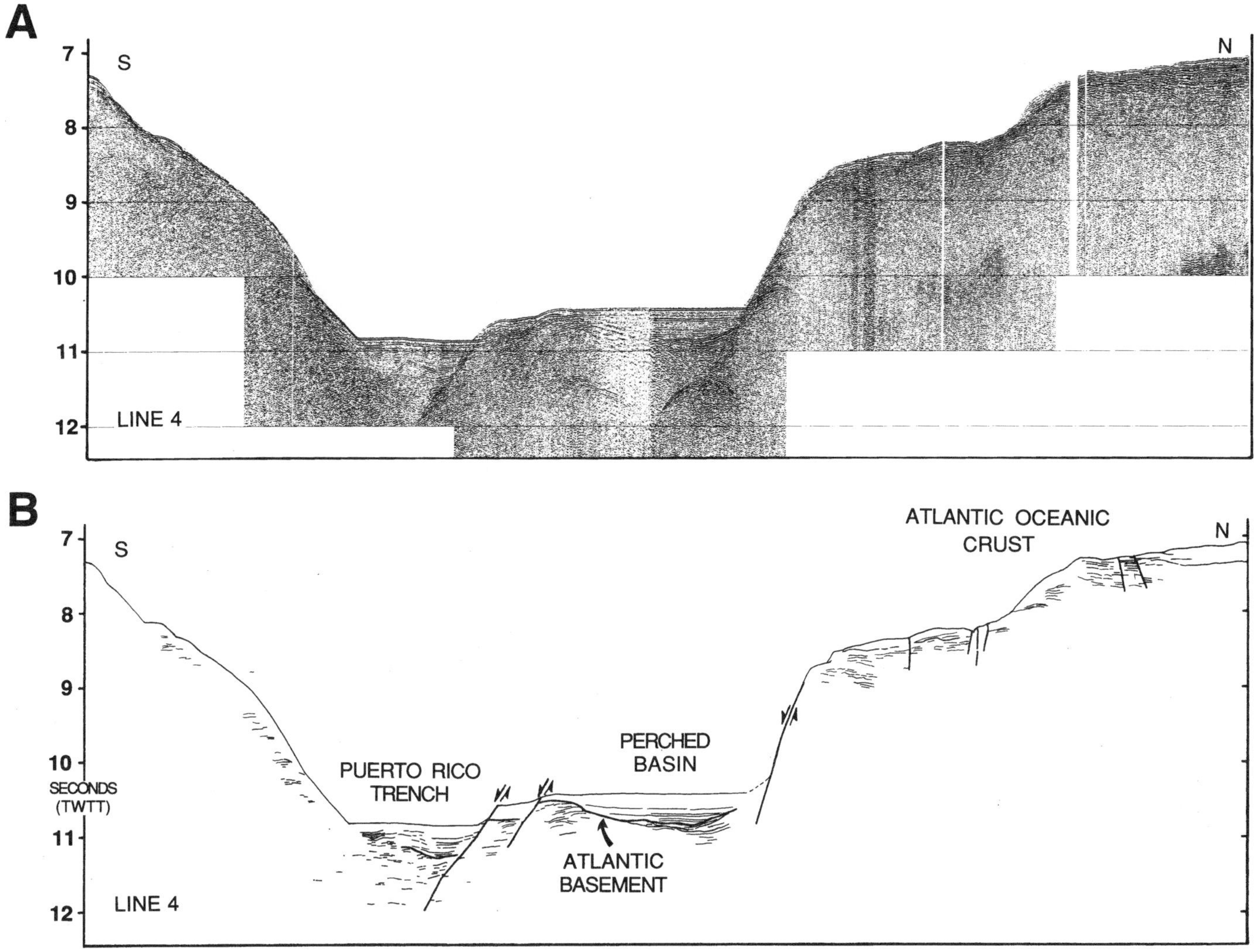

Figure 17. Single-channel seismic reflection profile Line 4 from northeastern part of main survey area. (A) Line 4 seismic-reflection profile. Location of profile shown in Figure 4. Profile has a vertical exaggeration of ~4:1. The northern edge of the 8-km-deep Puerto Rico trench in this region is characterized by laterally continuous, large-throw, oblique(?) normal faults. The deformation front along the southern edge of the trench is not well imaged on this line, although some disturbance of the sedimentary fill of the trench is apparent. (B) Line-drawing interpretation of Line 4.

dominantly strike-slip fault that exhibits a minor, local component of normal displacement where their profile is located. Alternatively, the fault they image may be one of the conjugate faults shown on the SeaMARC II mosaic, rather than the Septentrional fault. These data contradict one of the basic conclusions used by Speed and Larue (1991) to suggest that Puerto Rico is an "extensional allochthon"; that is, that Puerto Rico is decoupled from the underlying mantle above a deep, listric, south-dipping detachment fault that exhibits down-to-the-south normal slip.

The southeastern strand of the Septentrional fault continues for at least 15 km east of Mona Canyon, to about 67° 15′W, where the fault lineament trends off the side-scan mosaic. Unfortunately, the fault zone lineament is not readily discernible on lower resolution GLORIA side-scan sonar data collected from east of our SeaMARC II survey (EEZ SCAN 85 Staff, 1987; Masson and Scanlon, 1991). The presence of a possible eastward extension of the Septentrional fault system along the Puerto Rico slope within 125 km of the densely populated north coast of Puerto Rico suggests that this fault may represent an important, poorly understood seismic hazard to the population of Puerto Rico. In northern Hispaniola, the Septentrional fault has produced large prehistoric earthquakes that, if they were to recur today, would cause enormous damage (Prentice et al., 1993; Mann et al., this volume). However, the bifurcation of the fault system west of Mona Canyon indicates that motion accommodated along the single Septentrional fault strand in Hispaniola is split between at least two major strands north of Puerto Rico.

Eastern Septentrional fault: Northeastern strand (northeastern Septentrional fault). The northeastern Septentrional fault, which bifurcates from the eastern Septentrional fault at the minor releasing-bend basin at 68° 27′W, trends 050° and can be traced on the side-scan mosaic as a wide zone of highly reflective, parallel lineaments (Fig. 14). The main strand extends north to at least ~19° 11′N, where it forms the southeastern edge of a very deep (7 km), rhomb-shaped basin (centered at 19° 16.5′N, 68° 13.5′W). The basin is 32 km long and ranges in width from ~5 km in the northeast to 15 km near its southwestern edge. It is separated from the 8-km-deep Puerto Rico trench 30 km to the north by a west-trending, relatively shallow ridge whose crest ranges in depth from 5 to 6 km (Fig. 14A). Because of this ridge, drainage within the rhomb-shaped basin is completely internal, with canyons draining in from the southwest, northeast, and southeast. Furthermore, the basin is isolated from Hispaniola sediment sources by the deep canyon along the Septentrional fault. Consequently, the basin is sediment starved and exhibits only a very thin (~100 m thick) sedimentary fill ponded in a small (3 km × 5 km) area in the deepest part of the basin. Two dredge hauls (labeled 9 and 10 on Fig. 14A) from the steep eastern wall of the basin recovered marble with siliceous and micaceous stringers and mica schist with quartz and calcite bands, similar to other metamorphic rocks recovered from the NCMT (Fox and Heezen, 1975; Perfit et al., 1980).

The depth, morphology, and position of the 7-km-deep basin at a major 15-km-wide left-step in the Septentrional-north Puerto Rico slope fault system leads us to interpret it as a pull-apart basin that has developed during ongoing left-lateral strike-slip. The lack of pelagic mantling and significant sediment in the basin indicates that it is a young feature.

Northeastern Hispaniola slope fault zone (19° 20′W Lineament). The northeastern Hispaniola slope fault zone is a much less reflective feature than the eastern Septentrional fault lineaments described in previous section. This difference in reflectivity probably records mantling by pelagic and hemipelagic sediments, and may indicate that the Septentrional fault zone has been more recently active, or is moving faster, than the northeastern Hispaniola fault.

The fault zone changes strike near the northern end of the 7-km-deep, rhomb-shaped basin. East of the western edge of the basin at 68° 20′W, there are at least three well-defined lineaments, as well as several less pronounced lineaments, in a zone ~10 km wide. These lineaments trend 070°, parallel to the northern edge of the rhomb-shaped basin, which appears to be fault controlled. The two northern-most lineaments of the northeastern Hispaniola slope fault system trend along the axis of the east-trending ridge that separates the rhomb-shaped basin from the Puerto Rico trench (Fig. 14B). The zone of lineaments, as well as the northern edge of the rhomb-shaped basin and the high-standing ridge, parallels the southern edge of the Puerto Rico trench 20–30 km to the north. The presence of the most prominent fault lineaments along the axis of the ridge, rather than along the base of the ridge, supports our interpretation that the lineaments represent a zone of strike-slip faults.

West of the rhomb-shaped basin the northeast Hispaniola slope fault zone is linear and trends 275° for 80 km. The zone appears to coalesce westward at ~68° 40′W into a single main strand, although recent, widespread slumping of the Hispaniola slope north of 19° 22′N may obscure any more northerly strands in this area. The main strand is traceable westward toward the northeastern tip of the Samana Peninsula. The northern coast of the Samana Peninsula trends approximately east-west, roughly parallel to, and along trend with, the fault. Although we collected no data from the shallow portions of the northern Samana shelf, the similar orientations of the coastline and the fault suggest that the northern edge of the peninsula may be fault controlled.

ATLANTIC OCEANIC CRUSTAL PROVINCE

Atlantic oceanic crust crops out north of the Puerto Rico trench in the northeastern part of our survey area (Figs. 6 and 16). The most pronounced structural features imaged on the side-scan and seismic-reflection data are several large normal faults just north of the trench (Fig. 16B). These east-trending structures are traceable for ~60 km along strike, from the eastern edge of the mosaic almost to the southeastern corner of the Bahamas collision zone. They exhibit large, down-to-the-south normal separations of 200–1,500 m (Fig. 17). These faults form part of a near-continuous belt of normal faults that can be traced eastward for ~250 km on GLORIA side-scan sonar data along the northern margin of the Puerto Rico trench (Fig. 6; Masson and Scanlon, 1991). Dredge hauls from 6,500 to 8,100 m depth on the steep fault escarpments along the northern wall of the trench east of our survey area yielded an oceanic assemblage consisting of sepentinite, serpentinized peridotite, fresh- to greenschist-grade basalt (including pillow basalts), and Eocene(?) to Oligocene radiolarian chert mixed with abyssal clay and calcareous and siliceous ooze (Bowin et al., 1966; Nalwalk, 1969; Fox and Heezen, 1975; Perfit et al., 1980).

Between 67° 58′W and 68° 10′W, the normal faults form the southern edge of a large sedimentary basin perched more than 200 m above the Puerto Rico trench to the south (Fig. 16). In several instances the facing directions of the scarps reverse along strike, suggesting that these faults may also exhibit a component of strike-slip motion.

North of the trench-parallel normal faults, the most noteworthy structures are a set of 060°- to 070°-trending lineaments that crop out in the northeastern corner of the SeaMARC II mosaic (Fig. 16). These lineaments are oriented parallel to subparallel to the normal faults bounding the north wall of the trench and many of them are utilized as active drainages. The lineaments are subparallel to Cretaceous spreading lineations in the western Atlantic (Klitgord and Schouten, 1986) and may represent reactivated normal faults. Seismic-reflection profiles across most of these features are unrevealing, although several of the lineaments appear to be normal faults, with no consistent downthrown side. Similar lineaments, albeit with a more north-north-

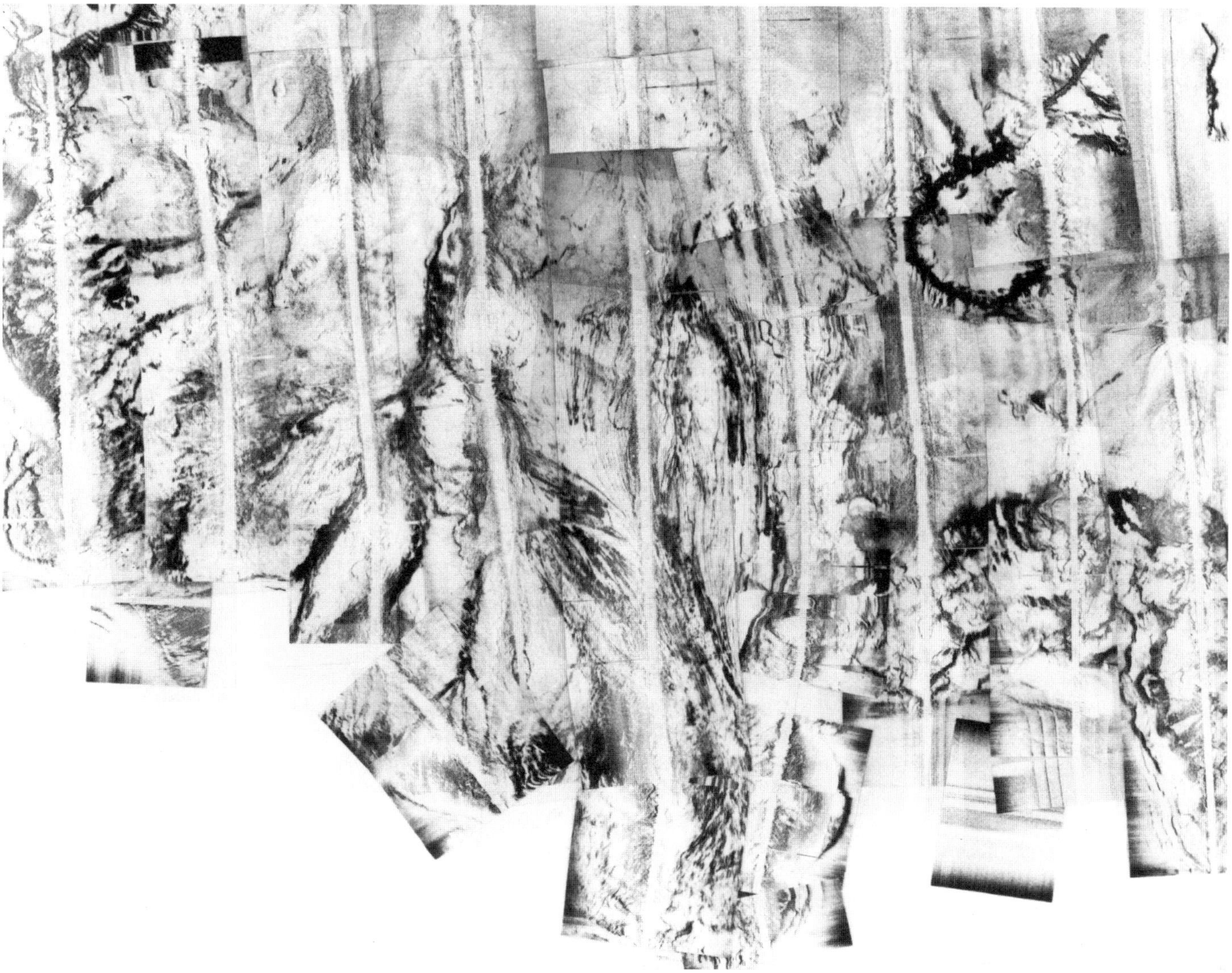

Figure 18. SeaMARC II side-scan sonar image of 50-km-long, 20-km-wide submarine avalanche chute on south side of Mouchoir bank. Compare with Figure 9 for location. We interpret the prominent lineaments visible within the chute as flow lines related to downslope transport of Mouchoir bank sediment.

east trend closer to the orientation of the Cretaceous spreading lineations, are visible on GLORIA data north of Puerto Rico (EEZ SCAN 85 staff, 1987; Masson and Scanlon, 1991).

SOUTHEASTERN BAHAMAS PROVINCE

The active, oblique collision between the southeastern Bahamas and Hispaniola provides an ideal opportunity to study the response of shallow-water carbonate platforms to collision. Our geophysical data indicate that Mouchoir, Silver, and Navidad banks exhibit little primary tectonic deformation, except along their southernmost edges. However, the collision has caused dramatic sedimentary effects on the banks, with large-scale step-back and collapse of the platform margins (Fig. 6). These sedimentary features are discussed in detail in two companion papers (Mullins et al., 1991; 1992).

Both Silver and Mouchoir banks have experienced large-scale step back of their bank margins, albeit at different times and in response to different causes. The bathymetric and seismic-reflection data reveal a large (5,000 km^2) deep-water (3,000–4,000 m) plateau north of Silver bank (Fig. 5). This plateau formed in response to 50–60 km of retreat (step back) of the northern edge of the bank during mid-Cretaceous time (Mullins et al., 1991; 1992). The timing of this step-back event correlates with a mid-Cretaceous increase in the rate of ocean-crust formation (Larson, 1991) that may have been responsible for a eustatic sea-level high-stand that peaked at ~80–100 Ma (Haq et al., 1987). Thus, step back and retreat of the northern margin of Silver bank may be the result of a tectonically driven rise in relative sea level that exceeded the capacity of the platform to maintain itself in the photic zone.

Mouchoir bank is a small, irregular carbonate platform that has been built on top of a much larger, deep-water (1,000 m) plateau (Fig. 3; Mullins et al., 1991; 1992). Bathymetric data, coupled with biostratigraphic ages from several dredge hauls collected during our cruise (reported in Mullins et al., 1991; 1992), indicate late Tertiary, pre–mid-Pliocene drowning and step back of Mouchoir bank. Along Caicos bank, west of our survey area, the amount of this step-back ranges from 10 to >25 km (Mullins and Hine, 1989). This late Tertiary step back event contrasts markedly with extensive carbonate bank progradation and aggradation during the same period in the northwestern Bahamas (Eberli and Ginsburg, 1987; 1989), indicating that the southeastern Bahamas step-back event was local in origin. We suggest that the different sedimentary response of the southeastern Bahamas banks reflects their proximity to the active collision zone. Northward onlap of the flat-lying Hispaniola basin turbidite fill suggests southward tilting of the drowned southern margin of Mouchoir bank prior to most deposition in the basin. Similarly, multichannel seismic reflection data from just west of our survey area document that the Caicos-Turks banks are tilted to the south and may possibly be downfaulted (Wanless et al., 1989).

Although sedimentary collapse features are common on the slopes of all three banks, these features are best expressed along the southern slope of Mouchoir bank. The southern margin of Mouchoir bank is highlighted by several large (10–25 km wide), amphitheater-shaped scarps that expose the highly reflective carbonates of the eroded bank margin (Fig. 9). Collapse of the platform margin within the largest of these reentrants, which we interpret as very large slump scars, has embayed the platform margin by more than 20 km. A very wide (25 km), shallow chute, similar to terrestrial avalanche chutes, emanates from this reentrant. Numerous well-developed lineations within the chute, which we interpret as flow lines from a very large debris flow, are spectacularly imaged on the SeaMARC II mosaic (Fig. 18). These flow lines extend more than 50 km downslope, from the base of the head scarp of the reentrant at about 1,000 m, to the edge of the Hispaniola basin at 4,000 m. In addition to the amphitheater-shaped slump scars, numerous submarine canyons and smaller gullies channel bank-top carbonate sediment to the adjacent basins.

Two of the most prominent bathymetric features of the southeastern Bahamas are the deep (4,000 m), linear troughs separating Mouchoir and Silver banks and Silver and Navidad banks (Fig. 5). We investigated these narrow troughs, known as Silver bank channel and Navidad channel, respectively (Uchupi et al., 1971), to determine if any active differential motion was occurring between the banks during collision. Oblique seismic-reflection crossings of the margins of the channels reveal no evidence of active deformation within the channels. Similarly, the side-scan data show no disruption of surficial channel sediments suggestive of active deformation, save in two instances (discussed in following section). Furthermore, no tectonic disruption was observed along the northern margins of these banks. These results suggest that the three southeasternmost Bahama banks are behaving as a single mechanical unit during the ongoing collision. This conclusion conflicts with that of Pierson (1982), who used differences in the thickness of upper Cenozoic, bank-top shallow-water carbonates in the southeastern Bahamas to infer that the banks are behaving as individual tectonic elements. The absence of faulting between banks suggests that differential subsidence and uplift of the banks must be accommodated by some diffuse process, such as large-scale warping or lateral variations in the amount of southward tilting of the banks associated with collisional underthrusting. Thus, if the interbank channels are of tectonic origin, they must predate the collision. We suggest that they may be long-lived features related to Triassic-Jurassic rifting (Mullins and Lynts, 1977).

In addition to the sedimentary collapse and step-back features observed along the banks, a 20-km-long, down-to-the-south normal fault occurs along the southern margin of Navidad bank (Figs. 11B and 13D). This slightly arcuate, east-trending fault, which represents the only potentially tectonic feature that we observed on Navidad bank, may be related to similarly oriented normal faults that we observed along the northern margin of the Puerto Rico trench to the east. Alternatively, it may be a relatively coherent, gravitationally driven block glide associated with bank-margin collapse.

Using a submersible, Heezen et al. (1985; their dive 643) traversed a steep east-trending escarpment along the southeastern edge of Navidad bank (Fig. 11A). They encountered numerous young-looking fractures and at least two 20–70-m-tall, sheer normal fault surfaces with vertical, 1-cm-deep grooves near the middle of the escarpment (at 3,470–3,450 and 3,170–3,100 m depth). The normal faults crosscut exposed Cretaceous dolomites (lagoonal deposits?) and appeared to be active, as evinced by fresh exposures of white dolomite and limestone not yet coated by the ubiquitous black manganese crusts that cover all other exposures.

We observed two other examples of probable recent faulting displacing sea-floor sediments. At the far western end of our survey, the southern edge of Mouchoir bank carbonate apron is defined for at least 25 km by a 290°-trending, down-to-the-south normal fault (Fig. 7B). Farther east, a 25-km-long, east-trending, down-to-the-south escarpment cuts across the southern end of Navidad channel (Fig. 11B). The geometry of this linear fault at depth is not known, although we suggest that it may represent a small south-side-down normal fault similar to those observed farther east along the southern edge of Navidad bank and the northern edge of the Puerto Rico trench.

GEOLOGY OF ONSHORE NORTHERN HISPANIOLA

Septentrional fault zone

The Septentrional fault zone (SFZ), which trends west-northwest across northern Hispaniola for 250 km, represents the primary left-lateral strike-slip fault in the Hispaniola part of the NCPBZ (Mann et al., 1984, this volume; Pindell and Barrett, 1990; Prentice et al., 1993). At its northwestern end, the SFZ trends offshore along the northern margin of Haiti and extends into the Oriente fault zone south of Cuba (Fig. 2; Calais et al., 1989, 1992, this volume; Edgar, 1991; Calais and Mercier de Lépinay, 1995). The Oriente fault zone, which represents the northern boundary of the Cayman spreading center, presently accommodates ~10–15 mm/yr left-lateral motion (Rosencrantz et al., 1988). Mann et al. (this volume) present evidence for maximum possible slip rates on the Septentrional fault for Holocene time of ~15–17 mm/yr. Paleoseismologic investigations along the central segment of the fault confirm late Holocene activity and show that the Septentrional fault represents a significant seismic hazard to the region (Prentice et al., 1993; Mann et al., this volume).

Where it extends offshore in eastern Hispaniola, the SFZ runs along the southern margin of Samana Peninsula and trends eastward toward the Mona Passage (Fig. 11B; Edgar, 1991). As mentioned, the fault zone here forms a prominent bathymetric lineament on the SeaMARC II mosaic. Pindell and Barrett (1990) proposed a regional plate model in which the SFZ has accommodated an unknown fraction of 400 km of left-lateral motion between Hispaniola and Cuba since early Miocene time; earlier (middle Eocene–early Miocene) strike-slip motion along the plate boundary was probably accommodated by fault systems located along the southern (Peralta Belt) and northern (Tavera Belt) margins of the extinct Greater Antilles island arc (Dolan et al., 1991; Pindell and Draper, 1991). The total amount of strike-slip that has occurred along the Septentrional fault system is only weakly constrained. Erikson et al. (1998) used reconstructions of amber-bearing sedimentary basins to estimate ~55–125 km of total post–early Miocene offset along the SFZ, with a preferred estimate of 85 km. Eberle et al. (1982) proposed an estimate of ~100 km based on cross-fault correlation of alluvial gold deposits. Both of these estimates suffer from the absence of well-defined piercing lines—well-defined sedimentary basin edges in the case of the Erikson et al. (1998) estimate and a narrow gold-bearing zone in the case of the Eberle et al. (1982) estimate—and should therefore be viewed as very general estimates of the total slip on the Septentrional fault system.

Lateral changes in strike of the Septentrional fault define a 200-km-long restraining bend in northern Hispaniola, and the more northwesterly orientation of the fault in central Hispaniola results in restraining bend uplift of the coastal Cordillera Septentrional (Fig. 2; Mann et al., 1984). In central Hispaniola the active trace of the Septentrional fault crops out along the northern edge of the 10–30-km-wide, low-lying Cibao Valley, with the peaks of the Cordillera Septentrional located parallel to, and just to the north of, the fault (Fig. 7; Mann et al., this volume). The Cordillera Septentrional are highest in north-central Hispaniola (between ~70°W and 71°W), where peaks reach as high as 1,200 m in elevation. In general, elevation diminishes both to the west and west of the zone of high peaks, and much of the range exhibits maximum elevations of a few hundred meters (de Zoeten et al., 1991). To the east the mountains give way to a flat, marshy lowland only a few meters above sea level just west of Samana Bay (Figs. 7A and 11A). The diminishing elevation of the range correlates with lateral changes in fault strike into more east-west orientations at the east and west ends of the Hispaniola restraining bend (Mann et al., 1984, this volume).

In the zone of high peaks adjacent to the SFZ, the main trace of the SFZ cuts across the northern Cibao Valley, 1–2 km south of the mountain front (Mann et al., 1984; Prentice et al., 1993; Mann et al., this volume). Miocene to lower- (mid?) Pliocene shallow-marine limestones of the Villa Trina Formation cap 400–1,200 m peaks in the Cordillera Septentrional, indicating significant post–mid-Pliocene uplift of the range (de Zoeten and Mann, 1991; Dolan et al., 1991; de Zoeten et al., 1991). In addition, the Villa Trina and older formations in the Cordillera Septentrional have all been either tilted down to the north-northeast, or folded into west-northwest–trending anticlines (Fig. 7B) (de Zoeten and Mann, 1991). A north-south seismic-reflection profile crossing the Cibao Valley south of the highest part of the Cordillera Septentrional, coupled with industry well data, indicates that the Cibao Valley overlies a deep, asymmetric sedimentary basin (Edgar, 1991). The basin fill is thicker to the north, where total sedimentary thickness is >5 km; Plio-Pleistocene basin fill reaches 3.6 km in thickness near the northern edge of the basin (Edgar, 1991). Erikson et al.(1998) use outcrop data to model the subsidence history of the Cibao basin. Their age estimates, coupled with consistent downsection steepening of bedding dips to the north, suggest that synsedimentary subsidence of the northern portion of the Cibao basin may have accompanied early mid-Pliocene(?) initiation of Cordillera Septentrional uplift. The northward tilting of Neogene sedimentary rocks of the Cibao basin records thrust loading, supporting the hypothesis of north-dipping contractional faults beneath the Cordillera Septentrional. The seismic reflection data have been interpreted to show several north-dipping faults beneath the northern end of the Cibao basin that may merge northward at depth with the Septentrional fault (Edgar, 1991). These data suggest that older strands of the Septentrional fault zone may have had a northward dip. Erikson et al. (1998) use gravity data to infer a moderate northward dip (possibly as shallow as 50°) for the Septentrional fault along the northern edge of the deepest part of the Cibao basin between ~70° 30′ W and 71° W. They estimate at least 5 km of north-south shortening and 5–7 km of vertical motion along the central part of the Septentrional fault. This implies a long-term dip-slip-only rate of ~2 mm/yr averaged since Miocene time. Calais et al. (1989) have suggested oblique-contractional slip along the western part of the Septentrional fault in northwestern Hispaniola. Taken together, these data suggest that a north-dipping Septentrional fault may have experienced a component of north-side-up reverse motion that resulted in uplift of the Cordillera Septentrional.

However, major, recent reverse displacements along the north-dipping Septentrional fault system appear to be at odds with young geomorphic features along the mountain front and the results of recent excavations across the fault in the northern Cibao Valley south of the zone of high Cordillera Septentrional peaks. These geomorphic and paleoseismologic data indicate that at the surface in this area the Septentrional fault is a near-pure left-lateral strike-slip fault that locally exhibits an oblique-normal component of displacement (Prentice et al., 1993; Mann et al., this volume). This result indicates that, if uplift of the Cordillera Septentrional is active, then the shortening responsible for the uplift must occur on faults other than the Septentrional fault, at least in the near-surface realm.

It is possible that oblique-reverse slip at depth on the Septentrional fault is partitioned into the near-pure strike-slip fault observed at the surface and a parallel set of contractional structures. However, the southern mountain front of the Cordillera Septentrional is highly irregular, deeply incised, and does not appear to exhibit any obvious features associated with recent surficial thrust faulting (Mann et al., this volume). This suggests that either: (1) uplift of the Cordillera Septentrional did occur on the oblique-reverse, north-dipping Septentrional fault following deposition of Villa Trina Formation limestones (i.e., since post–mid-Pliocene time), but this uplift has recently ceased; or (2) ongoing restraining bend uplift may be accommodated along north-dipping blind thrust faults that diverge from an oblique-slip Septentrional fault at depth beneath the range. However, the absence of any signs of uplift of the active trace of the Septentrional fault argues against the idea that the recent trace of the strike-slip fault is underlain by an active, north-dipping blind thrust fault. Alternatively, the absence of active north-dipping thrust faults beneath the southern edge of the Cordillera Septentrional may indicate that uplift is currently accommodated along south-dipping backthrusts beneath the range. Although active, onshore contractional structures could represent a potentially significant seismic hazard to this densely populated region, this possibility has not been investigated in detail and remains poorly understood.

In summary, in the near-surface realm, the active strand of the Septentrional fault in the region of highest Cordillera Septentrional peaks (and therefore presumably of maximum north-south shortening) is a pure strike-slip fault that cannot be used to explain uplift of the Cordillera Septentrional. Uplift of these mountains, if ongoing, must occur on faults other than the shallow part of the Septentrional fault.

Camu fault zone

North of the Cordillera Septentrional, the poorly understood Camu fault zone (CFZ) is traceable for ~120 km along strike, generally parallel to the SFZ (Fig. 7B). It is not known if the CFZ is seismically active or if it has experienced Quaternary movement (Pindell and Draper, 1991). However, Draper and Nagle (1991) suggest that the fault has accommodated at least 60 km of left-lateral strike-slip motion since Eocene time (at least partially since mid-Miocene time). Interpretation of the Camu fault as a major sinistral fault is supported by its linear surface trace and steep dip in outcrop (Pindell and Draper, 1991). The western, offshore extension of the CFZ is not traceable onto our SeaMARC II mosaic or GLORIA side-scan sonar data to the west (Dillon et al., 1992), suggesting the possibility that the CFZ may extend westward parallel to the coast before merging with the SFZ offshore northwestern Hispaniola (Fig. 2). The eastward extension of the CFZ remains enigmatic. The fault is not traceable onto our side-scan data and it may merge eastward with any of several northwest-trending oblique(?)-reverse faults southwest of the Rio San Juan Peninsula (Draper and Nagle, 1991). Alternatively, we speculate that the fault may have originally been a major left-lateral fault extending through what is now the northern part of the Rio San Juan Peninsula. The upper-slope strike-slip faults north of the Samana Peninsula are along trend of the Camu fault (Figs. 6, 7B, and 11B), and we suggest that these two fault systems may have once been contiguous before the Silver Spur collision shut off motion along the section of the ancestral Camu–Northeast Hispaniola slope fault zone that originally extended through the Rio San Juan Peninsula.

Uplifted marine terraces of northern Hispaniola

Coral-reef terraces occur along much of the northern coast of the Dominican Republic. However, by far the most prominent marine terraces occur on the Rio San Juan Peninsula (SJP), just south of Silver Spur collision (Fig. 12). There a 420-m-high, 17 km × 13 km dome-shaped hill is highlighted by a spectacular flight of at least 15 uplifted marine terraces. The terraces tilt down to the southwest, probably reflecting the westward propagation of the Silver Spur collision. At Cabo Frances Viejo, along the northern SJP, the inferred oxygen isotope stage 5e (~125,000 B.P.) coral terrace occurs at an elevation at ~18–19 m, indicating uplift of ~12–13 m. This yields an uplift rate of ~0.1 mm/yr, which appears to be a maximum, or near-maximum, uplift rate for the collision zone; the 125 ka terrace decreases in height, both to the east and west of the northern San Juan Peninsula, to only 8–10 m, yielding an uplift rate of only 0.012–0.033 mm/yr. Similarly, 8–10 m 125 ka terrace elevations outside the collision zone (e.g., at Sosua) indicate extremely low late-Quaternary uplift rates along the north Hispaniola coast (F. Taylor, personal communication, 1991; Dolan, unpublished data).

GEOLOGY OF NORTHERN PUERTO RICO AND THE NORTHERN PUERTO RICO SLOPE

Like Hispaniola, Puerto Rico is a remnant of the Cretaceous-Eocene Greater Antilles island arc (Pindell and Barrett, 1990).

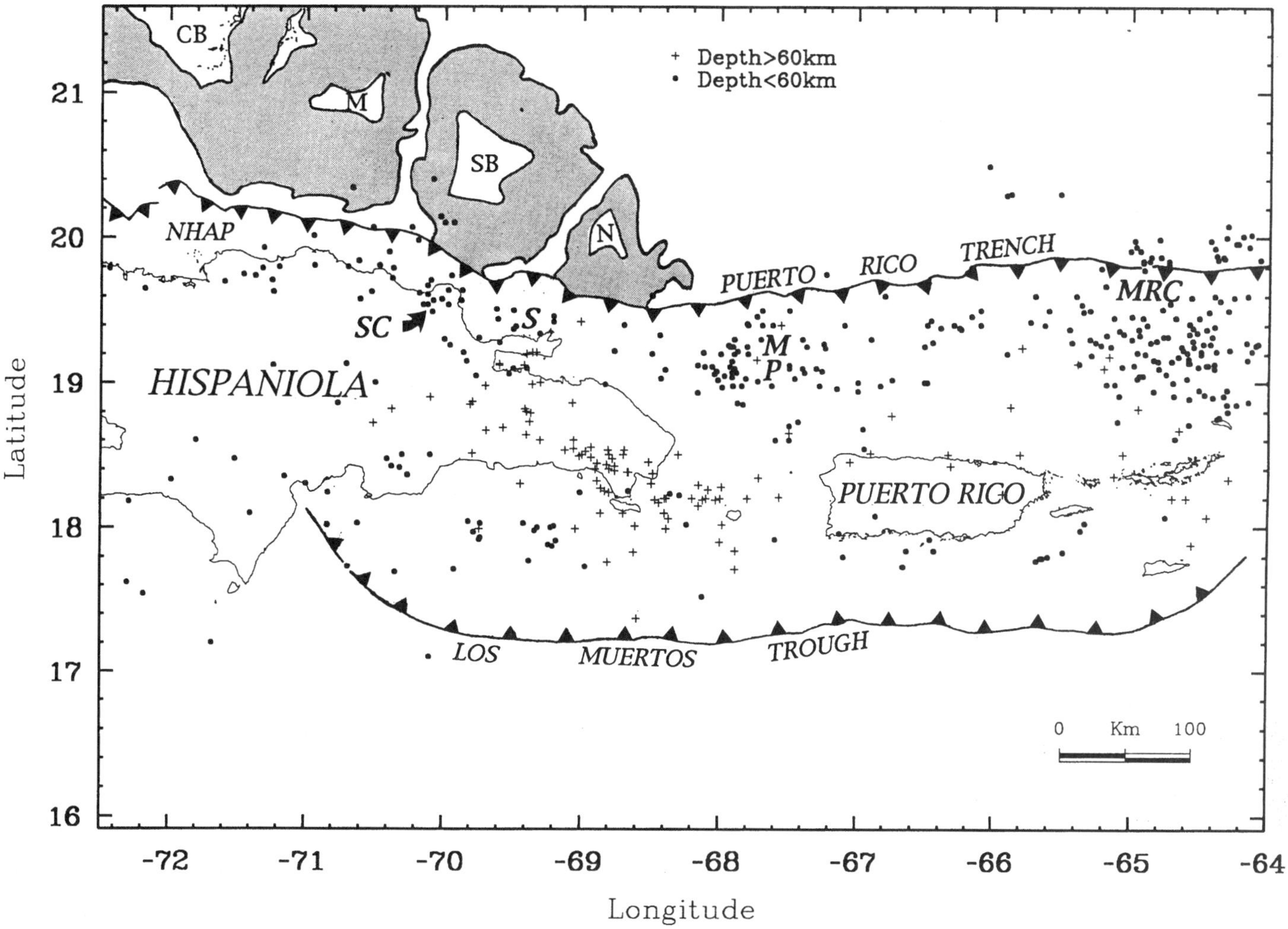

Figure 19. Map of recent seismicity in the north-central Caribbean (all seismicity data from National Earthquake Information Center catalog for 1962–1992). Lines with teeth denote deformation fronts of the Puerto Rico trench–northern Hispaniola deformed belt and Los Muertos deformed belt fault systems. Location of deformation fronts based on Case and Holcombe (1980); Masson and Scanlon (1991); Dillon et al. (1992); and this study. Note clusters of seismicity associated with south-dipping underthrusting of Atlantic oceanic lithosphere and general absence of seismicity to north and south of northern Caribbean plate boundary zone (NCPBZ) deformation fronts. The only significant seismicity outside the NCPBZ occurs north of Main Ridge near 64°W, which is a relatively poorly surveyed region, and at the southwest corner of Silver bank, at the leading (western) edge of the Bahamas-Hispaniola collision zone. Also note west-northwest–trending zone of deep seismicity beneath southeastern Hispaniola and southern Mona Passage. *MP* is northern Mona Passage seismicity cluster; *S* is Samana Peninsula cluster; *SC* is Sosua cluster; *MRC* is Main Ridge cluster; CB is Caicos bank; M is Mouchoir bank; SB is Silver bank; N is Navidad bank.

Deformed arc rocks in Puerto Rico are overlain with angular unconformity by a planar sequence of upper Oligocene siliciclastic rocks (see summary in Dolan et al., 1991) and Miocene-Pliocene limestones (Moussa et al., 1987). The limestones are similar to the Villa Trina Formation of northern Hispaniola. Since deposition of the Neogene carbonate cap, Puerto Rico has been arched into a broad anticlinorium, with the core of Mesozoic-Tertiary arc rocks flanked on the north and south by tilted limestone units (see Lewis et al., 1990, and Dolan et al., 1991, for summary). Paleomagnetic studies of the Neogene limestones of northern Puerto Rico indicate that the island rotated 24.5° counterclockwise between ~10 and 4.5 Ma (Reid et al., 1991). Since ~4.5 Ma the northern carbonate sequence has been tilted northward at an angle of ~4° to depths as great as 4 km (Birch, 1986; Moussa et al., 1987). Birch (1986) attributes this subsidence event to tectonic erosion of the northern Puerto Rico margin during subduction of Atlantic lithosphere.

GLORIA side-scan and seismic-reflection data from off-

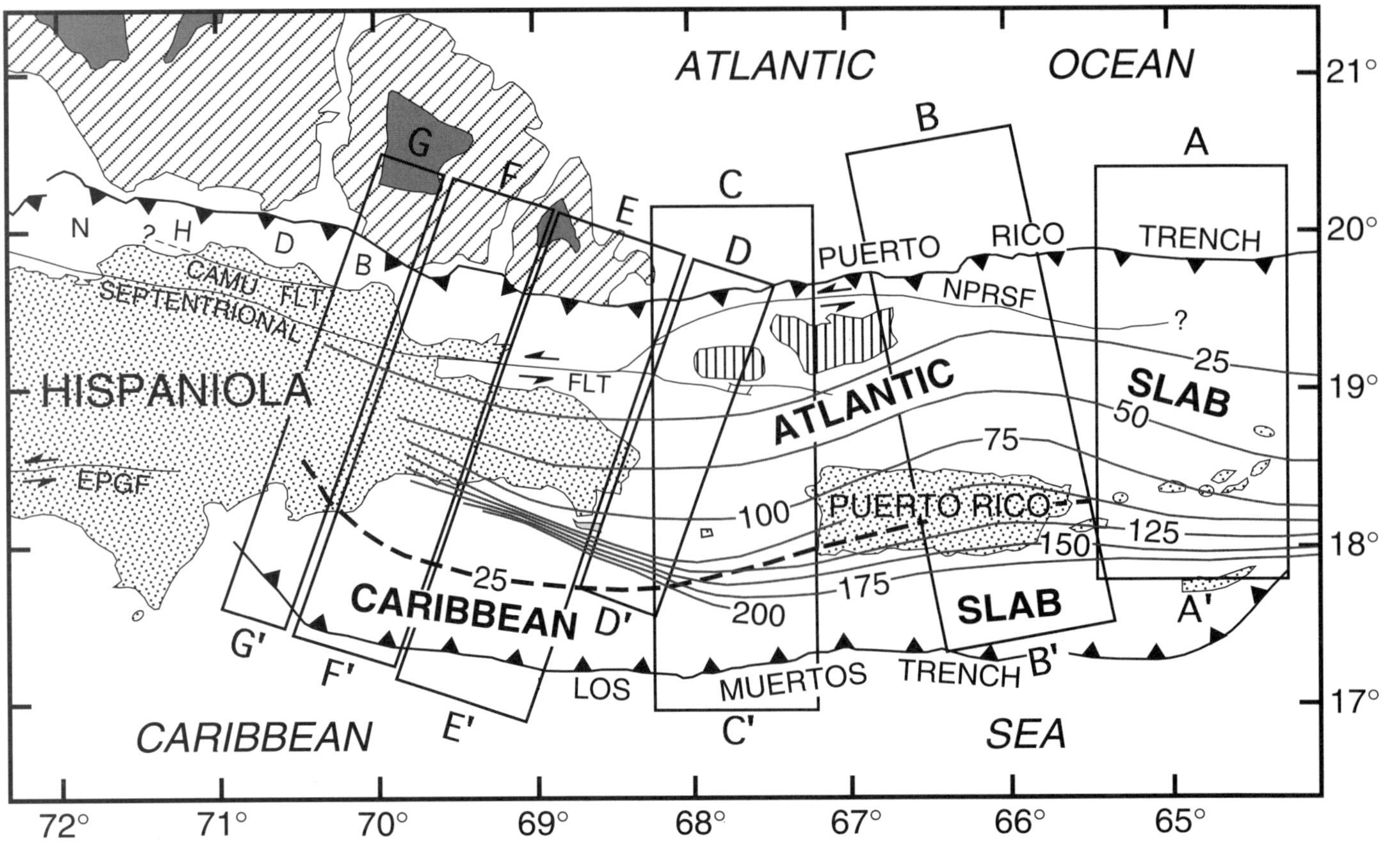

Figure 20. Structure contour map and location map showing limits of seismicity used in constructing cross sections in Figure 21. Contours are for top of underthrust Atlantic (south-dipping) and Caribbean (north-dipping) slabs. Note near-vertical zone of seismicity beneath eastern Hispaniola and southern Mona Passage, as well as changes in strike of both the shallow and deep parts of the underthrust Atlantic slab at the active Hispaniola-Bahamas collision zone. NHDB is northern Hispaniola deformed belt; NPRSF is north Puerto Rico slope fault; EPGF is Enriquillo-Plantain Garden fault; Flt is fault.

shore northern Puerto Rico show several prominent, east-trending strike-slip faults on the lower slope south of the Puerto Rico trench, including the north Puerto Rico slope fault, which we interpret as an eastward extension of the northeastern strand of the Septentrional fault (Fig. 14B; Masson and Scanlon, 1991). In addition, Larue et al. (1990) have proposed the existence of a strike-slip fault along the northern slope of Puerto Rico (their Escarpment-19° fault) on the basis of seismic reflection data and the steepness of the insular slope.

Multichannel seismic-reflection profiles across the lower Puerto Rico slope reveal transpressional flower structures, overthrusting of trench turbidites by slope rocks, and areas of possible accreted sediment (Larue and Ryan, 1998). Similarly, our reinterpretation of seismic-reflection data collected during GLORIA acquisition (EEZ SCAN 85 Staff, 1987) suggests evidence of contractional deformation. For example, a seismic-reflection profile from this area presented by Masson and Scanlon (1991; their Fig. 5D) shows southward-dipping strata in the Puerto Rico trench with increasing dip downsection. This fanning of bedding dips is consistent with thrust loading of trench strata from the south, and strongly suggests a long-lived component of convergence across the trench. Furthermore, our reinterpretation of the EEZ SCAN Staff (1987) seismic-reflection data from the area south of the trench indicates the presence of probable anticlines and possible backthrusts, consistent with a small amount of convergence across this zone.

Evidence for contractional structures extends eastward from 67°W to at least 64° 30′W, indicating that the entire 085°-trending plate margin north of Puerto Rico is transpressional. Masson and Scanlon (1991) proposed a model in which Puerto Rico was actively rotating counterclockwise. This model predicts an east-to-west increase in extension offshore northern Puerto Rico. In contrast, both the seismic reflection data and the focal mechanism of the 1943 earthquake (Dolan and Wald, this volume) reveal active contractional deformation along the plate interface northwest of Puerto Rico. These data imply that Puerto Rico is not actively rotating as a single, rigid block and suggest that the late Miocene–early Pliocene counterclockwise rotation of Puerto Rico was probably a relatively short-lived event. Evidence of consistent transpressional deformation also rules out models of active transtension along the Puerto Rico margin, as proposed by Larue and Ryan (1990) and Speed and Larue (1991).

SEISMICITY

Almost all earthquakes in the north-central Caribbean occur south of the deformation front defined by the SeaMARC II data, with essentially no seismicity beneath the southeastern Bahamas, except for a few earthquakes beneath the western edge of the Silver bank collision zone (Fig. 19). Shallow- to deep-focus earthquakes occur beneath eastern Hispaniola and Puerto Rico, but only shallow-focus events occur west of about 71°W (McCann and Pennington, 1990). This change is roughly coincident with the western (leading) edge of the Silver-Navidad bank collision zone.

Three-dimensional structure of the plate boundary zone

Three major zones of seismicity have been defined in the north-central Caribbean. Data from earthquake networks in Puerto Rico, the Dominican Republic, and the Virgin Islands define a south-dipping slab beneath Puerto Rico and the Virgin Islands (Sykes and Ewing, 1965; McCann and Sykes, 1984; and McCann and Pennington, 1990). This slab appears to be continuous with the slab of Atlantic oceanic crust being subducted beneath the Lesser Antilles island arc along the eastern margin of the Caribbean plate (McCann and Pennington, 1990). In addition to the east-trending, south-dipping seismic zone, McCann and Sykes (1984) defined a northwest-striking, nearly vertical zone of earthquakes beneath Mona Passage and eastern Hispaniola. The earthquakes in this zone, which extend from 75 to 210 km depth, are characterized by near-vertical T-axes (McCann and Sykes, 1984; Calais et al., 1992). Byrne et al. (1985) recognized a third major zone of earthquakes that define a gently north-dipping plane associated with northward underthrusting of the Caribbean slab beneath Hispaniola and Puerto Rico along the Los Muertos trough (Fig. 2). Focal mechanisms of earthquakes near the top of the Caribbean slab show shallowly north-dipping thrust faulting (Byrne et al., 1985; Harvard centroid moment tensor (CMT) catalog).

Seismicity in the western part of our study area is relatively poorly defined. Poor spatial resolution of earthquakes in northwestern Hispaniola precludes assigning any of these events to individual faults, but these predominantly shallow events probably occur along the strike-slip Septentrional fault system and associated contractional structures onshore and offshore northern Hispaniola (Fig. 2).

To better define the geometries of the three zones of seismicity in the Hispaniola–Puerto Rico region, we have plotted a map and cross sections of small-to-moderate earthquakes (M<6.1) that have occurred in the north-central Caribbean during 1962 to 1992 (data from National Earthquake Information Center [NEIC] catalog). The seismicity map reveals that recent seismicity associated with the underthrust Atlantic slab is not evenly distributed over the area (Fig. 19). Rather, recent earthquakes beneath northern Hispaniola and Puerto Rico are concentrated in five distinct clusters, herein named (from east to west) the Main Ridge cluster, the northern Mona Passage cluster (subdivided into northwest and northeast subclusters), the Samana cluster, the Sosua cluster, and the Hispaniola basin cluster.

We have also plotted a series of cross sections of seismicity in the region, using only those recent earthquakes with well-located focal depths (i.e., excluding earthquakes assigned default depths of 10 and 33 km; Figs. 20 and 21). Although this leaves out a significant portion of recent earthquakes, it provides a clearer picture of the different tectonic elements of the region. Based on these cross sections, we have constructed a structure contour map of the top the underthrust Atlantic and Caribbean slabs, as well as the southern (presumably upper) edge of the near-vertical zone of seismicity (Fig. 20). This contour map shows pronounced lateral changes in both the strike and dip of all three features.

The well-defined, south-dipping Benioff zone revealed by local network data in the eastern Puerto Rico–Virgin Islands area dips about 45° S to a depth of ~150 km (Fig. 21A; McCann and Sykes, 1984). The underthrust Atlantic slab shallows slightly to the west beneath central Puerto Rico, where it dips ~30° S to a depth of ~100 km (Fig. 21B). The slab may steepen at 100 km, but the scarcity of well-located deep earthquakes beneath central Puerto Rico makes this uncertain. The Main Ridge cluster is centered beneath the Main Ridge, a northwest-trending bathymetric high interpreted by McCann and Sykes (1984) as an aseismic ridge developed along a fracture zone.

Within the northern Mona Passage cluster the south-dipping seismic zone is traceable to a depth of only ~40 km (Fig. 21C and D). The dip of the underthrust Atlantic slab in this area is 20°–25°, and the zone of seismicity projects to the surface at the deformation front along the southern edge of the Puerto Rico trench. This observation is consistent with our interpretation that the top of the south-dipping zone of seismicity represents the top of the underthrust Atlantic slab. In this area, the base of the deepest part of the seismically defined underthrust slab is separated from the top of the deeper, near-vertical zone of seismicity by an 80-km-wide zone of seismic quiescence (Fig. 21C, D, and E; McCann and Sykes, 1984). Slightly farther west, beneath eastern Hispaniola, the underthrust slab is poorly defined by recent seismicity, although aftershocks of the 1946 northeast Hispaniola earthquake, relocated by Russo and Villaseñor (1995), crudely define a shallowly south-dipping (~20°–25°) slab extending from the sea floor at the deformation front to the mainshock of the earthquake at ~20 km depth (Fig. 21E). The deep slab, which is well defined by recent seismicity below 90 km depth, dips nearly vertically down to a depth of at least 210 km. Almost no earthquakes occur between ~25 and 90 km depth beneath easternmost Hispaniola.

In contrast to the eastern Hispaniola and Mona Passage clusters, earthquakes within the Samana cluster do not exhibit a zone of upper mantle seismic quiescence (Fig. 21F). Rather, they define a somewhat diffuse, but continuous, south-dipping slab extending from ~20 to 150 km depth. Numerous aftershocks of the 1946 earthquake (relocated by Russo and Villaseñor, 1995) define a shallowly south dipping rupture plane beneath the Samana Peninsula that extends to a depth of ~15 km (Fig. 21E).

West of the Samana cluster, the Sousa cluster also exhibits a

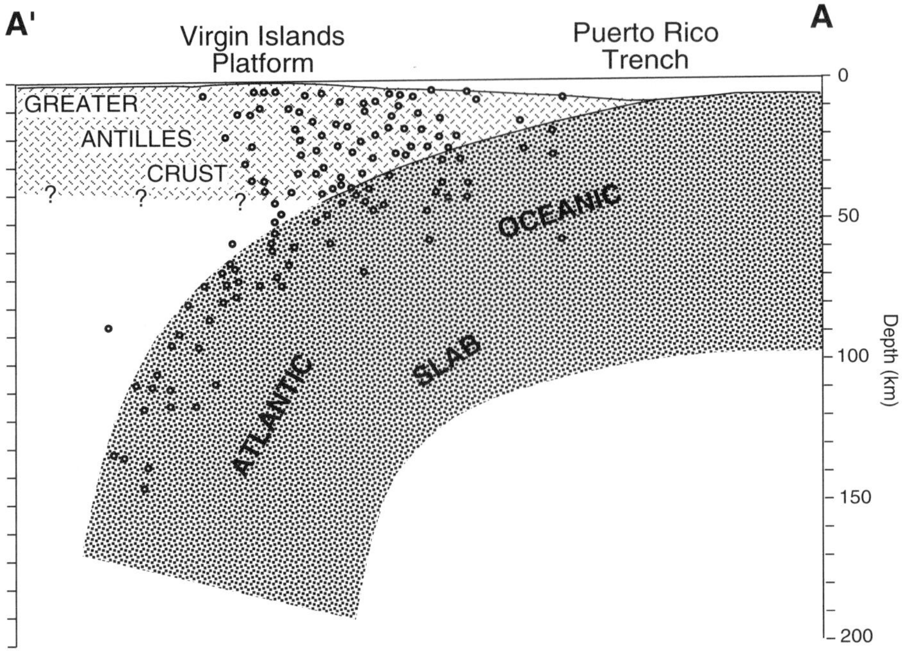

Figure 21 (this and next three pages). Cross sections of recent seismicity in the north-central Caribbean region. Heavy line at top of figure represents bathymetry-topography along center line of each cross section. No vertical exaggeration. Note that although profiles are oriented approximately perpendicular to regional strike, the profiles are oblique to some tectonic and topographic features. Shading denotes approximate thickness of different lithospheric plates based on assumptions that the North America (Atlantic) plate consists of normal oceanic lithosphere ~90–100 km thick, and that the thicker crust of the Caribbean plate (Edgar et al., 1971; Case et al., 1990) results in a thicker lithosphere. All seismicity data from National Earthquake Information Center (NEIC) catalog for 1962–1992 except as noted. Only earthquakes with well-defined depths have been used; earthquakes assigned default depths of 10 and 33 km were not used in constructing these sections. (A-A') Well-defined, south-dipping Wadati-Benioff zone beneath the Virgin Islands east of Puerto Rico. Data from Fischer and McCann (1984), who used local network data. Cross section is north-south, centered on 65°W. (B-B') 350° profile showing Wadati-Benioff zone beneath eastern Puerto Rico extending to ~150 km depth. Note cluster of shallow events at Los Muertos trench associated with northward underthrusting of Caribbean crust beneath Hispaniola. (C-C') North-south section through the northern Mona Passage seismicity cluster. Note weakly defined, shallowly north-dipping zone of seismicity associated with Los Muertos trough underthrusting and diffuse, steeply dipping zone of seismicity beneath southeastern Hispaniola. (D-D') 017° section through the northern Mona Passage seismicity cluster. Note diffuse, near-vertical zone of seismicity beneath southeastern Hispaniola. Cross section oriented normal to strike of the zone of deep seismicity. (E-E') 017° section through easternmost Hispaniola showing well-developed, vertical zone of deep earthquakes extending down to 210 km depth. Triangles denote aftershocks of the 1946 northeastern Hispaniola earthquake that were relocated by Russo and Villaseñor (1995). Note the near-total absence in the NEIC data of well-located earthquakes associated with the south-dipping plate interface between underthrust Atlantic oceanic crust and Hispaniola that ruptured during the M_s 8.1 mainshock on August 4, 1946 (Dolan and Wald, this volume). In contrast, the relocated 1946 aftershocks define a shallowly southwest-dipping plane approximately along our proposed 1946 mainshock rupture surface. Also note that the deeper relocated 1946 aftershocks occur within the zone of deep seismicity defined by the past 30 years of NEIC data. Focal mechanisms are shown for the 1946 mainshock (Dolan and Wald, this volume) and for the M_s 6.7 earthquake of June 24, 1984, which occurred along the north-dipping Los Muertos trough zone of underthrusting (from Byrne et al., 1985). (F-F') 017° cross section through Samana seismicity cluster. Triangles denote 1946 aftershocks relocated by Russo and Villaseñor (1995). Note the very well defined, shallowly southwest-dipping zone of 1946 aftershocks. These shallow aftershocks extend almost to the deformation front defined by the SeaMARC II data, suggesting that the 1946 mainshock may have locally ruptured all the way to the sea floor in this collisional region of enhanced mechanical coupling. Also note the diffuse, southward-dipping zone of seismicity associated with underthrusting of Atlantic oceanic crust and Bahamas platform carbonates. The gentle dip of the 1946 aftershock zone, coupled with its location north of the Septentrional fault, confirms that the 1946 mainshock occurred along the shallowly southwest-dipping plate interface, rather than along a strand of the Septentrional fault system, as has been suggested by Russo and Villaseñor (1995). (G-G') 017° section through central Hispaniola and Sosua seismicity cluster. Triangles denote 1946 aftershocks relocated by Russo and Villaseñor (1995). The paucity of well-located earthquakes precludes identification of any well-defined planes in this region. This diffuse pattern of shallow- to intermediate-depth seismicity is characteristic of the northern part of the plate boundary zone in western Hispaniola west of about 71°W (McCann and Pennington, 1990). Northward dip of septentrional fault from Erikson et al. (1998).

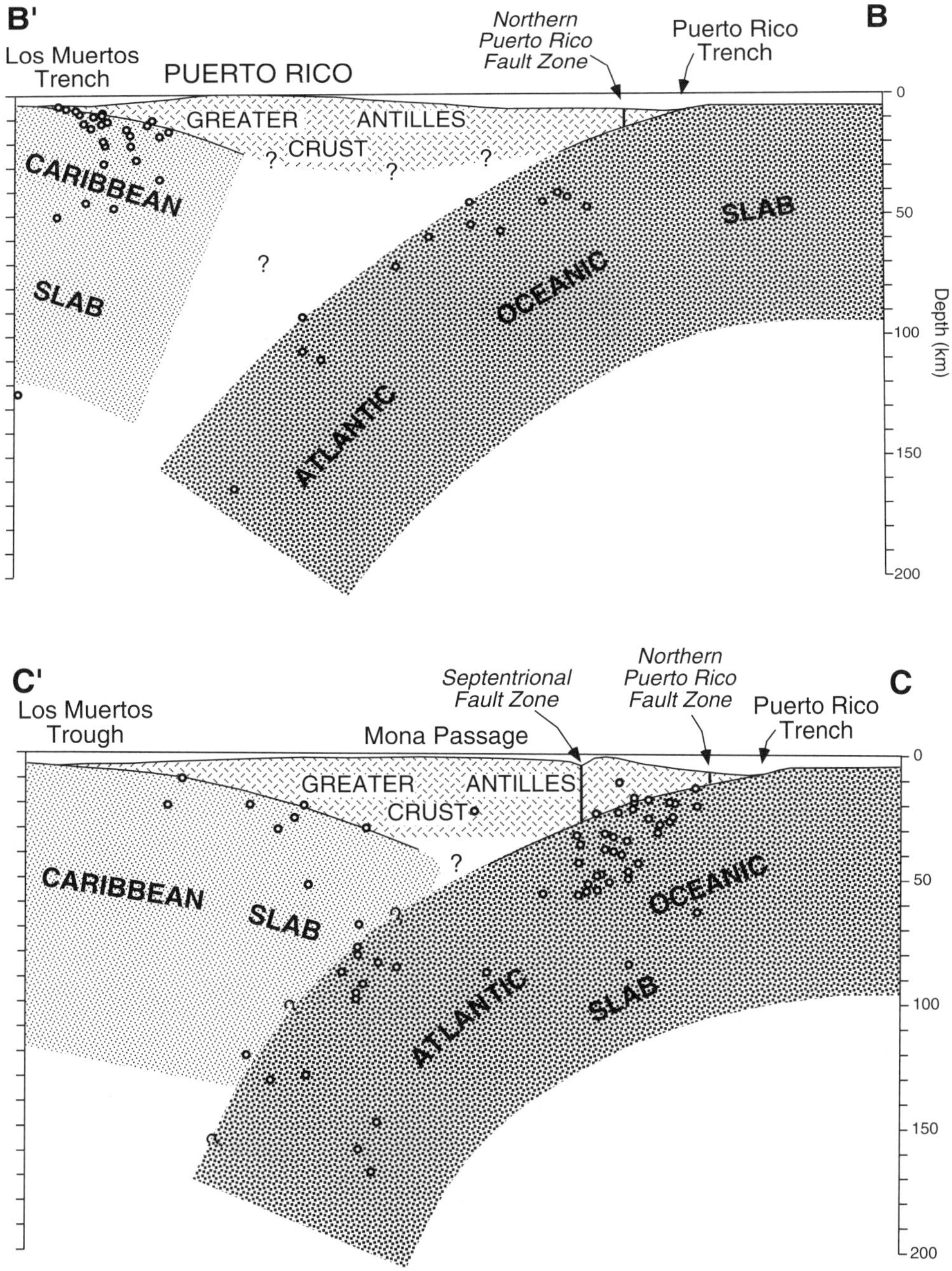

crudely defined upper limit of seismicity that dips gently southward (Fig. 21G). The common, deep earthquakes discernible on all cross sections to the east are absent from the Sousa cluster; almost all events in the region are shallower than 60 km. A few earthquakes occur beneath Silver Spur and the southwestern corner of Silver bank just north of the Sousa cluster. These events represent the only significant seismicity north of the deformation front, and we speculate that they are probably related either to bending of the slab as it obliquely underthrusts the Hispaniola margin or to incipient thrust faulting within the colliding bank as it begins to deform internally in response to the collision (Fig. 19). Earthquakes in the Hispaniola basin cluster (~71° 20′W to 71° 30′W) are generally shallow, but there are too few well-located events to construct a meaningful cross section.

In contrast to the laterally discontinuous intermediate-depth seismicity, the near-vertical zone of deep seismicity is continuous along strike from 70° W to ~67° 30′W, where it appears to die out (Fig. 19). Isolated deep earthquakes along strike to the east, however, suggest that the deep slab is continuous beneath Puerto Rico and connects with the deep part of the Wadati-Benioff zone under eastern Puerto Rico and the Virgin Islands. If so, the strike of the deep slab changes from 300° to nearly east-west at ~67° 30′W (Fig. 20). The near-vertical zone of earthquakes is best defined beneath Mona Passage and easternmost Hispaniola, where the dip

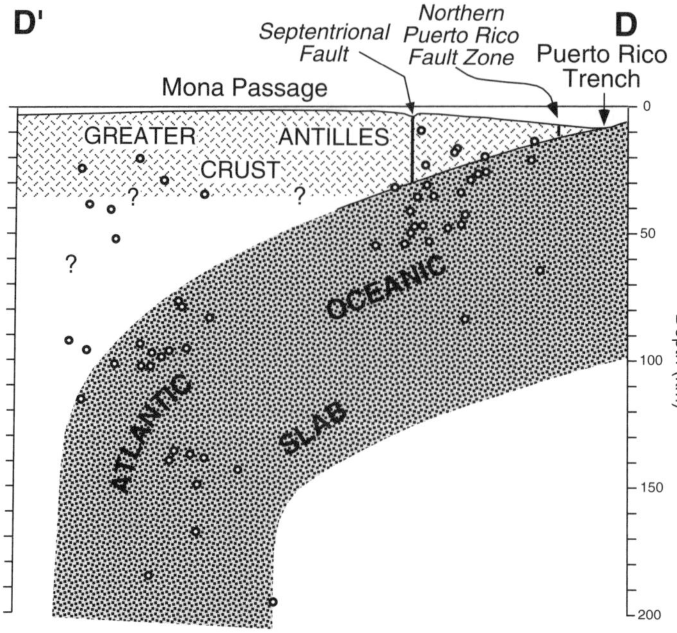

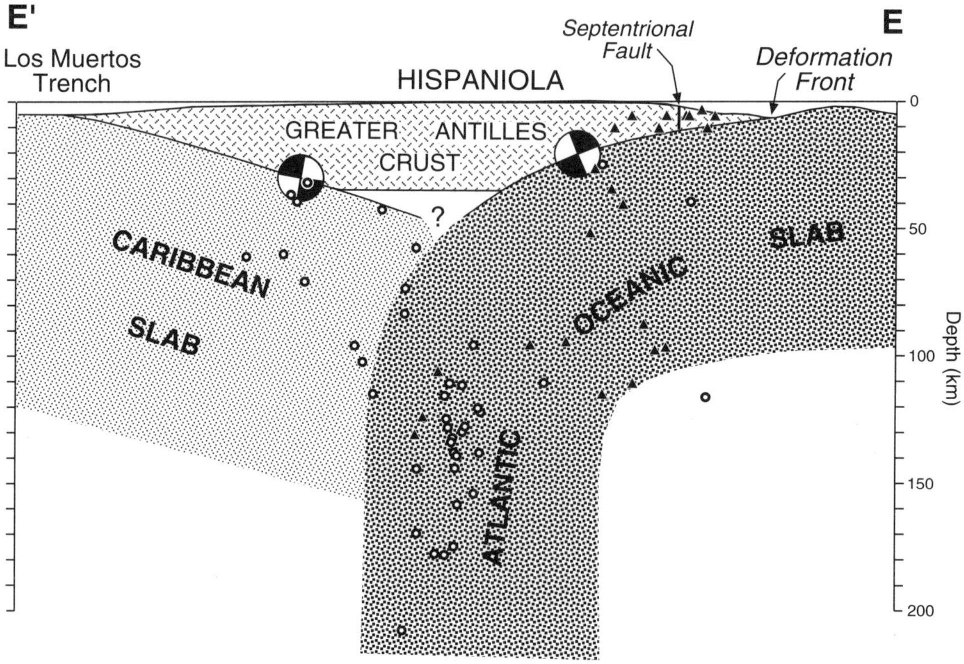

of the southern edge of the zone is essentially vertical, and may even locally be steeply north-dipping (Fig. 21E). Although the 300° strike of the zone does not change significantly to the northwest, the zone is much more diffuse beneath north-central Hispaniola, where it appears to dip steeply southward (Fig. 21E). The maximum depth of earthquakes shallows northwestward along the western part of the zone of deep seismicity, from 210 km at 69°W, to 140 km at 69° 45′W, to 115 km at 70°W (Fig. 20).

The dipping zone of seismicity associated with northward underthrusting of Caribbean lithosphere along Los Muertos trough (Byrne et al., 1985) dips ~15° northward at 67° 30′W, directly above the vertical slab recognized by McCann and Sykes (1984; Figs. 20 and 21C). The June 24, 1984, M_s 6.7 San Pedro basin event, the largest earthquake to occur along the Los Muertos trough zone during the instrumental period, exhibited near-pure thrust motion along a plane dipping 9° north at 32 km depth (Fig. 21E; Byrne et al., 1985).

The seismicity-based structure contour map of the top of the underthrust Atlantic and Caribbean slabs reveals pronounced lateral changes in both the strike and dip of these fea-

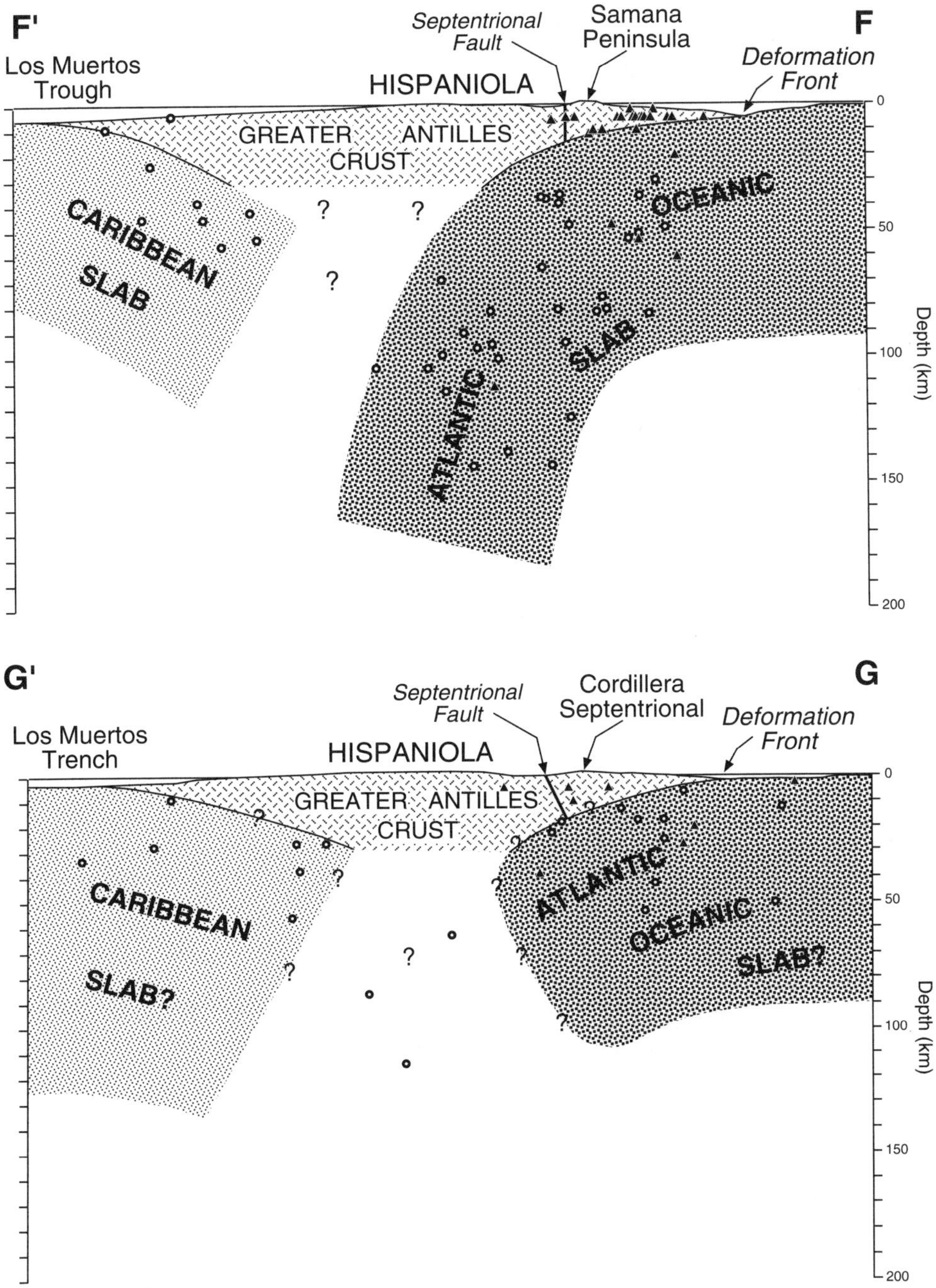

tures (Fig. 20). The most pronounced change in strike of the shallow part of the underthrust Atlantic slab is centered near the Mona Passage. To the east the slab dips 45° south and strikes east-northeast. To the west the slab shallows in dip and the strike changes to northwest. This abrupt change coincides with the location of the Silver-Navidad bank and postulated Mona bank collision zones. As described, the near-vertical slab also exhibits a similar change in strike at the same location. The dip of the deep part of the slab appears to steepen westward from eastern Puerto Rico, where it dips ~45° south (Fig. 21A) to vertical at about 68°W (Fig. 21D and E). As discussed, farther west the deep-slab dip appears to shallow somewhat to ~75° south (Fig. 21F). The zone of steepest slab dip lies just west of the site of the closest approach of Caribbean lithosphere to the underthrust Atlantic slab at 67° 30′W (Fig. 21D and E).

Large earthquakes and the Bahamas-Hispaniola collision

A series of large earthquakes occurred in the north-central Caribbean during the decade 1943–1953 (Fig. 22). Dolan and Wald (this volume) have studied this sequence in detail, and the following discussion is taken from their study. The north-central Caribbean region was first settled during Columbus' second voyage in 1493, and consequently has the longest historical earth-

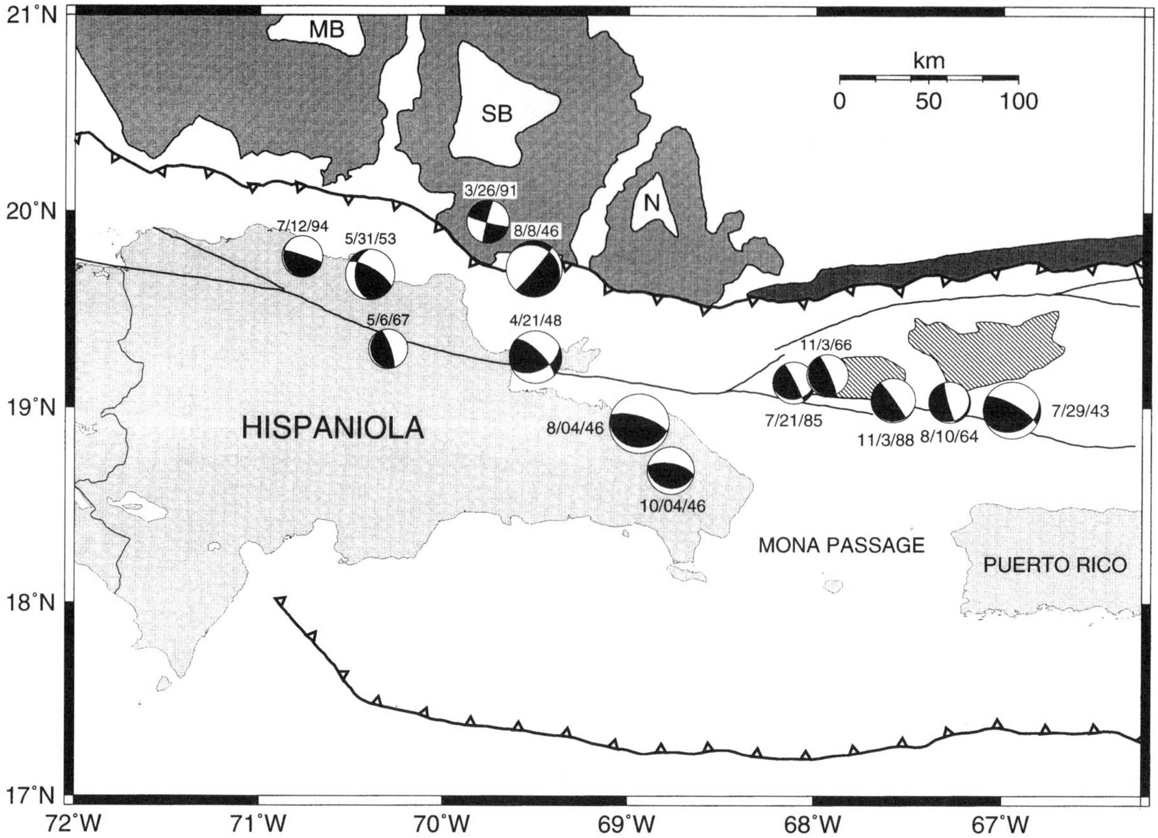

Figure 22. Focal mechanisms of moderate and large earthquakes in the north-central Caribbean for period 1943–1994 (from Dolan and Wald, this volume). Large focal mechanisms are from Dolan and Wald (this volume); small focal mechanisms are compiled from Molnar and Sykes (1969), Kafka and Weidner (1979), and Harvard data base. All focal mechanisms are lower hemisphere projections; black areas show compression, white areas show dilation. Earthquake locations of 1943–1953 sequence are from Kelleher et al. (1973). Diagonal hatching denotes zones of anomalously shallow bathymetry that we interpret as the sea-floor manifestation of ongoing collisional underthrusting of Mona bank. Pale gray shading denotes land. Medium gray shading denotes carbonate basement of southeastern Bahamas banks. Dark gray shading shows extent of flat, turbidite-filled floor of the Puerto Rico trench.

quake record in the New World. The historical record indicates that the 1943 to 1953 sequence was anomalous for the region; the M_s 8.1 August 4, 1946, northeast Hispaniola earthquake was probably the largest event to strike the area since at least 1564 (Chalas-Jimenez, 1989; W. R. McCann and L. Feldman, written communication, 1995). In the following discussion, we use the surface-wave magnitude (M_s) estimates of Kelleher et al. (1973), despite the fact that subsequent studies suggest that these estimates are systematically too large. For example, Kelleher et al. (1973) estimate an M_s of 8.1 for the 1946 mainshock, whereas Abe (1981) estimates M_s 8.0, and both Pacheco and Sykes (1992) and Russo and Villaseñor (1995) estimate M_s 7.8. Nevertheless, for the sake of consistency we use the Kelleher et al. (1973) estimates because only their study estimated magnitudes for all of the events we discuss. Where available, we include more recent magnitude estimates in parentheses after the Kelleher et al. (1973) estimate. For example, the 1946 northeast Hispaniola earthquake would be referred to as M_s 8.1 (Kelleher et al., 1973; M_s 8.0 in Abe, 1980; M_s 7.8 in both Pacheco and Sykes, 1992 and Russo and Villaseñor, 1995).

The two largest events in the 1943–1953 sequence were an M_s 7.8 earthquake (Kelleher et al., 1973; M_s 7.7 in Abe, 1981; M_s 7.5 in Pacheco and Sykes, 1992) that occurred offshore northwest Puerto Rico on July 29, 1943, herein referred to as the northern Mona Passage earthquake, and the aforementioned M_s 8.1 (Kelleher et al., 1973; M_s 8.0 in Abe, 1981; M_s 7.8 in both Pacheco and Sykes, 1992 and Russo and Villaseñor, 1995) northeastern Hispaniola earthquake, which occurred near the coast of the northeastern Dominican Republic (Fig. 22). The aftershock zone of the 1946 earthquake, based on 53 events that were relocated by Kelleher et al. (1973), exactly matches the limits of the Silver-Navidad collision zone as defined by the SeaMARC II data (Fig. 23). Similarly, the 1943 earthquake was located at the site of our postulated Mona bank collision.

When compared with the structure contour map of the underthrust Atlantic slab beneath Hispaniola (Figs. 20 and 22),

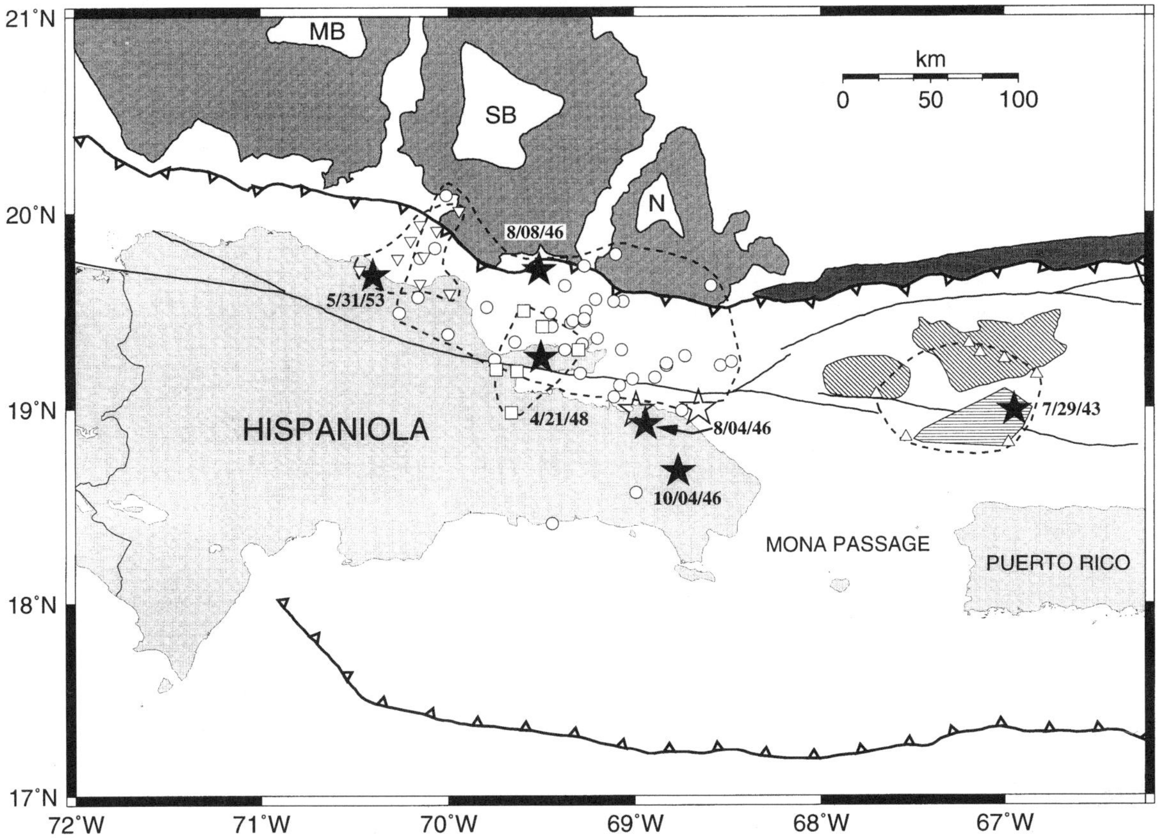

Figure 23. Relocated mainshocks and aftershocks of the 1943 northern Mona Passage earthquake and the 1946 northeastern Hispaniola earthquake from Kelleher et al. (1973). Mainshocks and large aftershocks, including the four largest aftershocks of the 1946 event (August 8, 1946, October 4, 1946, April 21, 1948, and May 31, 1953), are denoted by black stars. Smaller aftershocks are denoted by various open symbols. Open stars show relocations of the 1946 mainshock and October 4, 1946, aftershocks by Russo and Villaseñor (1995). Russo and Villaseñor's (1995) relocation of the 1946 mainshock epicenter does not change significantly from that of Kelleher et al. (1973). Russo and Villaseñor's (1995) relocation of the October 4, 1946, aftershock does change significantly, however, shifting 20 km to the north-northeast of the Kelleher et al. (1973) location. Note coincidence of 1946 aftershock zone with the 200-km-long Silver-Navidad bank collision zone. Solid lines with teeth are deformation fronts of the Puerto Rico trench–northern Hispaniola accretionary prism system (north of Hispaniola and Puerto Rico) and the Los Muertos trench accretionary prism (south of the islands). Dark shading denotes limits of southeast Bahamas carbonate banks (Dolan and Wald, this volume). The two irregular zones of diagonal hatching north-northwest of Puerto Rico represent areas of anomalously shallow bathymetry that we interpret as the surficial manifestation of collisional underthrusting of Mona bank. The triangular zone of horizontal hatching north of Puerto Rico denotes the rupture plane of the 1943 northern Mona Passage earthquake (Dolan and Wald, this volume). Note the proximity of the 1943 rupture zone to the proposed Mona bank collision. Solid lines without teeth are major, left-lateral strike-slip faults of the Septentrional fault system. Darkest shading represents basinal sediments in the Puerto Rico trench. MB is Mouchoir bank; SB is Silver bank; N is Navidad bank.

the focal plane, focal depth, the shallow southward dip of aftershocks (relocated by Russo and Villaseñor, 1995), and the west-northwest strike of the 1946 mainshock faulting plane are all consistent with the earthquake having occurred on the plate interface along the top of the underthrust Atlantic oceanic slab. This observation, coupled with the spatial coincidence of the aftershock zone with the Silver-Navidad bank collision zone, suggests that the earthquake ruptured the strongly coupled interface between Hispaniola and the underthrust portions of the colliding Bahama banks. The focal mechanism, focal depth, and location of the 1943 earthquake suggest that this event also probably occurred at a collisional asperity along the plate interface, in this case our postulated Mona bank collision.

The 1946 mainshock was the first event in a sequence of five large earthquakes that struck northern Hispaniola between 1946 and 1953 (Figs. 22 and 23). This sequence included two very large aftershocks, one offshore northern Hispaniola (August 8, 1946; M_s 7.6 in Kelleher et al., 1973; M_s 7.4 in Pacheco and Sykes,

1992; M_s 7.3 in Russo and Villaseñor, 1995) and one southeast of the mainshock beneath eastern Hispaniola (October 4, 1946; M_s 7.0 in Kelleher et al., 1973; M_s 6.1 in Russo and Villaseñor, 1995) (Figs. 22 and 23). In addition to the two early aftershocks, two other large aftershocks occurred west of the mainshock, an M_s 7.3 earthquake (April 21, 1948; Kelleher et al., 1973) that occurred beneath the Samana Peninsula, in the center of the 1946 rupture, and a 1953 M_s 7.0 (Kelleher et al., 1973) event located along the northern Hispaniola coast near the city of Sosua (Figs. 22 and 23). The coincidence of the western limit of the 1953 rupture and the leading edge of the Silver Spur collision suggests that the 1953 earthquake ruptured the area of strong coupling between Hispaniola and the underthrust portion of Silver Spur. The 1953 rupture plane extended ~10–15 km west of the western limit of the 1946 aftershocks (Fig. 23) (Sykes and Ewing, 1965; Kelleher et al., 1973), indicating that the 1953 earthquake may have extended the 1946 rupture to the extreme western (leading) edge of underthrust carbonate basement (Fig. 12B).

Earthquake focal mechanisms. Dolan and Wald (this volume) calculate mechanisms for the six large earthquakes that occurred during the 1943–1953 sequence (Fig. 22). In addition to the focal mechanisms calculated by Dolan and Wald for the six largest events of the 1943–1953 sequence, focal mechanisms of seven moderate earthquakes that have occurred in the region since 1962 have been published (Molnar and Sykes, 1969; Kafka and Weidner, 1979; Harvard centroid moment tensor database) (Fig. 22). Most of the 13 mechanisms show predominantly thrust motion along shallowly south- and southwest-dipping planes (Fig. 22). Although these earthquakes exhibit thrust faulting focal mechanisms, rake angles indicate that motion during most of them was highly oblique to the strike of the plate margin. Obliquity of slip during these earthquakes to the plate margin ranges from ~40° to near 90°, indicating a range of motion from mixed thrust/left-lateral displacement to left-lateral strike-slip offset nearly parallel to the strike of the underthrust Atlantic slab (Figs. 20 and 22). In general, the focal mechanisms for the smaller events show more highly oblique motion (Fig. 22).

Four focal mechanisms have been determined for the northern Mona Passage cluster (Molnar and Sykes, 1969; Kafka and Weidner, 1979; Harvard CMT catalog). Three of these are from the northeastern subcluster. All three occurred at 16–23 km depth along the shallowly south-dipping plane that defines the upper limit of seismicity in the area, which we interpret as the interface between the underthrust Atlantic slab and the overriding Hispaniola–Puerto Rico block (Fig. 20). Although all three of these earthquakes exhibit shallowly south- or southwest-dipping thrust focal mechanisms, rakes show that displacement was predominantly left-lateral, nearly parallel to the strike of the underthrust Atlantic slab at this location (Figs. 20 and 22). The one focal mechanism determined from the northeastern Mona subcluster shows a shallowly southwest-dipping thrust focal mechanism with a rake indicating highly oblique left-lateral motion parallel to the plate margin (Molnar and Sykes, 1969). This event, however, is much deeper (48 km) than the northeastern subcluster earthquakes; the hypocenter is located well below the inferred slab interface as defined by the planar, upper limit of seismicity. This earthquake therefore indicates that at least some plate-boundary motion is taken up by intra-slab, oblique-thrust faulting. We suggest that as subduction of the slab slows in response to the increased mechanical coupling within the collision zone, that the slab begins to break up internally, causing the 20–60-km-deep zone of intra-slab seismicity in the northern Mona Passage cluster (Fig. 21C).

Molnar and Sykes (1969; their event 105) calculated a focal mechanism for a 1967 M_s 5.3 event that occurred at a depth of 39 km beneath the northeastern Cibao Valley ~45 km southwest of the Rio San Juan Peninsula (Fig. 22). Their mechanism shows near-pure thrust motion on either a shallowly west-southwest–dipping plane or a steeply northeast-dipping plane. Based on: (1) the angular discordance between the strikes of the Septentrional fault (285°) and the northeast-dipping focal plane (340°), (2) the south-southwest dip of the underthrust Atlantic slab to the southeast, and (3) the south-southwest dip of offshore thrust faults in the accretionary prism to the north, we interpret the 1967 event as having occurred on the shallowly west-southwest–dipping plane. The rake derived from the focal mechanism indicates that motion during the earthquake was predominantly left-lateral, subparallel to the plate boundary in this area.

A moderate earthquake occurred beneath the city of Puerto Plata on July 12, 1994. The focal mechanism for this event shows nearly pure thrust motion on either a near-vertical or a shallowly south-dipping plane (Fig. 22) (Harvard CMT catalog). We interpret this event to have occurred on the downdip projection of the south-dipping thrust fault beneath the northern Hispaniola accretionary prism.

Molnar and Sykes (1969; their event 104) also calculated a focal mechanism for a moderate earthquake that occurred on April 20, 1962, within the Dillon et al. (1992) GLORIA survey area. Their mechanism indicates that the earthquake occurred at a shallow depth just north of the deformation front of the NHAP as defined on the GLORIA side-scan sonar mosaic (Dillon et al., 1992). The P-axis indicates that the earthquake was an oblique left-lateral/reverse event that occurred on a moderately northwest-dipping plane. The epicenter is near the center of the Hispaniola (Caicos) basin, near a prominent left-step in the deformation front (Dillon et al., 1992). The location of the event north of the accretionary prism indicates that the earthquake occurred within carbonate basement of the Caicos bank. The left-lateral motion along a northeast-striking plane revealed by the mechanism and the prominent left-step in the deformation front at this location suggest a prominent structural discontinuity in the underthrust Caicos bank carbonate basement.

DISCUSSION

The data reveal an extremely complex plate boundary that exhibits active collision of several bathymetrically high-standing blocks, two opposing underthrust slabs (the northern one characterized by highly oblique underthrusting), and considerable structural variation along strike. One of the most important controls on these lateral variations in structural style is the direction of plate motion relative to the local orientation of the plate boundary structures.

Relative plate motion in the north-central Caribbean region

Most Caribbean researchers agree that motion along the northern Caribbean plate boundary is predominantly left-lateral, with much of the left-lateral component of motion in the Hispaniola region accommodated along the Septentrional fault (e.g., Mann et al., 1984, 1990, this volume; Pindell and Barrett, 1990; Calais et al., 1992; Prentice et al., 1993). However, the exact direction of relative plate motion in the Hispaniola–Puerto Rico area, and in particular the degree of obliquity of motion, have been the focus of a long-standing debate. A range of plate-motion models for the northern Caribbean margin has been proposed, from pure east-west strike-slip (Mann et al., 1984), to slightly oblique convergence (Bracey and Vogt, 1970; Schell and Tarr, 1978) to northeast-southwest transpression (McCann and Sykes, 1984; Deng and Sykes, 1995) to transtension (Larue and Ryan, 1990; Speed and Larue, 1991), to microplate rotations (Masson and Scanlon, 1991). The geophysical, geological, and seismological data described earlier place important constraints on the orientation of the plate motion vector.

In the north-central Caribbean region the plate boundary exhibits pronounced changes in strike, from 085° off Puerto Rico and the Virgin Islands, to 110° within the Hispaniola restraining bend, to 085° off southern Cuba near the eastern end of the Cayman trough. Evidence for active accretionary prism development parallel to the left-lateral Septentrional-Oriente fault system over the entire length of the margin from southern Cuba eastward to the Silver-Navidad collision zone indicates that relative plate motion in this region is transpressional (Austin, 1983; Calais et al., 1992, 1995; Dillon et al., 1992; Dolan and Wald, this volume; this study). Dolan and Wald (this volume) used earthquake focal mechanisms for the 1943 and 1946 mainshocks—two of the largest earthquakes to strike the plate margin during the last several hundred years—to show that the south-dipping plate-interface thrust fault off northern Hispaniola and northwestern Puerto Rico accommodates oblique motion, with subequal amounts of left-lateral and thrust motion.

Evidence for a contractional component of deformation is less pronounced north of Puerto Rico, and the slope there is dominated by several large strike-slip faults (Masson and Scanlon, 1991). Seismic reflection data, however, do reveal evidence for a small component of north-south shortening across the margin at least as far east as 65°W (Larue and Ryan, 1998). These data indicate that in the Hispaniola–Puerto Rico area relative plate motion must be oriented at least slightly more northerly than the 085° orientation of the plate margin north of Puerto Rico. But the absence of evidence for large-scale contractional deformation along the northern Puerto Rico slope (e.g., accretionary prism development), suggests that the relative plate motion vector is oriented only slightly more northerly than the 085° trend of the plate boundary. The combined geological and seismological data thus indicate that relative motion between the North America plate and Hispaniola and Puerto Rico is east-northeast, supporting the early transpressional model of Bracey and Vogt (1970), as well as later, similar models such as those of Schell and Tarr (1978), McCann and Sykes (1984), and Deng and Sykes (1995). The data also rule out proposals of transtensional deformation off Puerto Rico (Larue and Ryan, 1990; Speed and Larue, 1991) and pure east-west left-lateral motion in northeastern Hispaniola (Russo and Villaseñor, 1995; see Dolan and Wald, 1997; this volume for a complete discussion). Based on the geological and seismological constraints described in previous sections, our best estimate of the relative plate motion direction along the stretch of the plate boundary between north-central Hispaniola and the Virgin Islands is ~075° ± 5°.

Hispaniola and Puerto Rico, however, constitute part of a series of separate microplates caught up within the 250-km-wide northern Caribbean plate boundary zone (NCPBZ), which exhibits active deformation zones along both its northern and southern boundaries (Fig. 2; Ladd and Watkins, 1978; Mann et al., 1984, 1995; Bryne et al., 1985; Ladd et al., 1990; Masson and Scanlon, 1991; Dillon et al., 1992). Therefore, the relative motion vector between the North America plate and the Hispaniola and Puerto Rico blocks of the NCPBZ does not define the overall motion direction between the North America plate and the true Caribbean plate, which lies south of the Los Muertos deformed belt and the Enriquillo-Plantain Garden fault zone (Fig. 2). Because the Los Muertos deformed belt along the southern edge of the NCPBZ exhibits active shortening (Ladd and Watkins, 1978; Byrne et al., 1985; EEZ Scan 85 Staff, 1987; Ladd et al., 1990; Masson and Scanlon, 1991), overall relative motion between the Caribbean plate (sensu stricto; i.e., the plate south of the NCPBZ) and the North America plate must be more convergent than the ~075° direction we estimate for the northern deformation belt of the NCPBZ. The degree to which the North America–Caribbean (sensu stricto) relative plate motion vector is oriented more northeasterly than ~075° depends on the ratio between slip rates on the left-lateral fault systems (Septentrional and Enriquillo-Plantain Garden fault systems) and slip rates on the contractional/oblique-contractional fault systems (northern Hispaniola deformed belt structures, including the northern Hispaniola accretionary prism thrust faults, and Los Muertos deformed belt accretionary prism thrust faults). The higher the contractional/oblique–contractional fault slip rates, the more northerly the relative plate motion vector will be oriented.

As we will discuss, the obliquity of motion along the northern margin of the NCPBZ has important consequences for the recent tectonic evolution of the region, particularly because our model predicts diachronous, westward-younging collisional underthrusting of the Bahamas ridge.

Lateral variation in structure along the plate boundary

The lateral change in strike of the northern plate boundary between Puerto Rico and central Hispaniola is reflected in all major structural features in the area, including the strike of the underthrust Atlantic slab, the transpressional deformation front, and the surface trace of the left-lateral Septentrional fault system. The change in plate boundary strike is also recorded by the pronounced lateral change in structural style revealed by the SeaMARC II data across Hispaniola–Puerto Rico insular slope.

The northwesterly orientation of the boundary in the western part of our survey area results in a local increase in convergence and the consequent development of the abundant contractional structures, including both the small, northern Hispaniola accretionary prism (NHAP) and the contractional fault system responsible for uplift of the onshore Cordillera Septentrional. As the plate boundary curves into a more east to east-northeast orientation eastward, the compressional component of plate motion is reduced, resulting in the predominantly strike-slip deformation that we observed in the eastern survey area.

In addition to the changing strike of the plate boundary, lateral changes in the thickness of basinal sediments at the deformation front represent a second reason for the differing structural style of the Hispaniola slope to the west and east of the Silver-Navidad collision. Offscraping of thick Hispaniola basin turbidites facilitates the development of the large, concentric folds observed in the NHAP. In contrast, the discontinuous nature and thin sediment infilling of the small basins within the Bahamas collision zone to the east preclude the development of large-scale folds, despite seismological evidence for a significant component of north-south shortening along this part of the plate boundary during the 1946–53 northeastern Hispaniola earthquake sequence (Dolan and Wald, this volume).

We originally assumed that the small size of the NHAP was a result of near-margin–parallel underthrusting and little consequent offscraping. But the evidence for active north-south shortening provided by Dolan and Wald's (this volume) earthquake focal mechanisms implies that the small size of NHAP is due to a rather short-lived period of offscraping, indicating that the NHAP is a relatively young feature. As discussed in previous sections, the upper Hispaniola slope south of the active NHAP is deeply incised and does not exhibit well-developed thrust-related anticlines. The marked difference in structural style between the upper and lower Hispaniola slopes west of the collision zone suggests that they may have formed during different periods. We speculate that the upper slope represents the remnants of an older phase of accretion that was interrupted by either a period of nonaccretion or of subduction erosion of the margin, possibly related to oblique underthrusting of a now-subducted, southward-projecting, high-standing spur of Mouchoir bank.

Strain partitioning in northern Hispaniola and northern Mona Passage

The parallelism of the offshore, south-dipping northern Hispaniola slope thrust system and the left-lateral Septentrional fault system indicates effective partitioning of contractional and translational deformation in northern Hispaniola and the northern Mona Passage. At the plate-margin scale, oblique plate motion is rarely perfectly partitioned onto parallel strike-slip and contractional fault systems. Commonly, both the strike-slip and contractional structures exhibit some degree of oblique slip. This is true of the NCPBZ in northern Hispaniola, where the focal mechanisms of the 1946 mainshock and its 1948 and 1953 aftershocks indicate oblique left-lateral/thrust motion on the southwest-dipping plate interface (Figs. 21 and 22) (Dolan and Wald, this volume). Similarly, the Septentrional fault locally exhibits a component of contractional deformation (Calais et al., 1992). Nevertheless, in Hispaniola, the Septentrional appears to accommodate much of the left-lateral component of motion along the NCPBZ (Mann et al., 1984; Mann et al., this volume). Furthermore, the geomorphology of the active fault zone, coupled with several paleoseismological trench exposures, indicates that motion along the most recently active fault trace is predominantly strike-slip (Prentice et al., 1993; Mann et al., this volume).

Thus, much of the left-lateral component of motion along the plate boundary is accommodated along the Septentrional fault, with some left-lateral motion also accommodated as oblique slip along the south-dipping plate interface (northern Hispaniola slope fault). In contrast, the Septentrional fault appears to accommodate very little contractional deformation, with only local evidence of oblique convergence along the active trace of the fault (Calais et al. 1992). In fact, in at least some localities, the most recently active trace exhibits oblique-normal displacements (Prentice et al., 1993; Mann et al., this volume). Much of the contractional component of motion along the margin is accommodated along the offshore plate interface. These data indicate effective, although not complete, strain partitioning along the Hispaniola part of the plate boundary.

We propose that the Septentrional fault merges with the shallowly dipping thrust interface at depths of ~15–30 km (Fig. 21 E and F). In this model, the northern part of Hispaniola, north of the Septentrional fault, is a thin crustal sliver underlain by the south-dipping thrust interface. This sliver, which we refer to as the Septentrional terrane, is moving eastward at a much slower rate than the bulk of Hispaniola south of the Septentrional fault. The difference in these rates is equal to the slip rate on the Septentrional fault.

The seismicity data indicate that this situation holds at least as far west as the 1953 earthquake at the western edge of the active Bahamas collision. West of the collision zone, however, the deep seismicity that characterizes the plate boundary from central Hispaniola eastward is absent. The lack of deep seismicity shows that Atlantic oceanic lithosphere is apparently not being actively underthrust to great depths beneath northwestern Hispaniola. We propose in a later section of this chapter that the location and geometry of the underthrust Atlantic slab was controlled by the distribution and relative buoyancy of the major tectonic elements of the northern Caribbean region.

East of ~70°W, we have proposed that the geometry of this

fault interaction is relatively straightforward, with the Septentrional fault merging downward into the plate interface along the top of the underthrust Atlantic slab (Fig. 21). We suggest that this same fault geometry extends all the way from the western edge of the intermediate-depth seismicity we have used to define the underthrust Atlantic slab (~70° W) eastward at least as far as central Puerto Rico (the geometry of the strike-slip system offshore northeastern Puerto Rico and the Virgin Islands has not yet been well-enough identified to characterize the state of strain partitioning east of the area discussed in this chapter).

But what is the geometry of the deep plate boundary faults in northwestern Hispaniola, west of the well-defined, deeply underthrust slab? The rarity of deep- and intermediate-depth earthquakes in the region prevents direct imaging of the geometry of the deep plate-boundary structures. Nevertheless, the lateral continuity of the south-dipping northern Hispaniola slope fault zone eastward into areas where the slab is well defined suggests that the Septentrional fault zone merges with a south-dipping thrust fault at a few tens of kilometers depth beneath northwestern Hispaniola. However, without the basic mechanical control exerted by the interface along the top of the underthrust slab, the geometry of the deep plate-boundary structure does not necessarily conform to the pattern we have suggested farther east. Whatever the geometry of the deep plate interface beneath northwestern Hispaniola, the deep boundary fault must accommodate the total oblique thrust–left-lateral motion of the combined Septentrional and offshore thrust fault systems.

West of the Windward Passage, the plate boundary in southeastern Cuba also exhibits partitioning of strain onto parallel strike-slip and contractional structures. There, the Oriente left-lateral strike-slip fault system lies parallel to, and north of, the north-dipping thrust faults of the Santiago deformed belt (Calais et al., 1992, this volume; Calais and Mercier de Lépinay, 1995). As in northwestern Hispaniola, there is no seismic evidence for a deeply underthrust slab beneath southeastern Cuba. Thus the geometry of the deep plate boundary structures remains uncertain. But, in both cases, the deep plate boundary must accommodate the total oblique slip manifested at the surface by the parallel strike-slip and contractional fault systems.

Although the plate boundary in southeastern Cuba is structurally similar to that in northwestern Hispaniola, the dip of the contractional structures is opposite. The oppositely dipping contractional fault systems do not appear to overlap; the south-dipping faults of the northern Hispaniola deformed belt terminate westward at ~73° 30′W, whereas the north-dipping faults of the Santiago deformed belt terminate eastward at about the same longitude (Fig. 2). Thus, the switch in dip occurs beneath the Windward Passage. We suggest that this switch in dip is controlled by the distribution of Greater Antilles island-arc basement, which has acted as a mechanical backstop during post-Miocene transpression; the arc basement lies south of the plate boundary in Hispaniola, resulting in a south-dipping system of offshore contractional faults, including the plate-interface northern Hispaniola slope fault. The opposite geometry occurs in southeastern Cuba, resulting in the north-dipping contractional structures of the Santiago deformed belt.

In summary, the entire northern deformation zone of the NCPBZ, from the Windward Passage east to at least central Puerto Rico, appears to be characterized by partitioning of strain between a south-dipping thrust interface and the parallel left-lateral strike-slip faults of the Septentrional–northern Puerto Rico slope fault system. Furthermore, the data described in this chapter show that the underthrust Atlantic slab appears to be continuous, although of varying depth, from ~70°W eastward to at least north of the Virgin Islands. McCann and Sykes (1984) suggested that the slab continues even farther to the east, eventually merging with the subducted Atlantic slab beneath the Lesser Antilles island arc.

Origin of anomalously shallow bathymetry north of Mona Passage: Model for late Cenozoic underthrusting of a southeastern Bahama bank

The origin of the two anomalously shallow regions of the northern part of Mona Passage—the northeast Mona block and northwest Mona ridge—may be related to collisional underthrusting of the Bahamas ridge. McCann and Sykes (1984) proposed a model in which these shallow regions (collectively referred to as their "anomalous block") represented the remnants of a previously unrecognized carbonate bank of the southeastern-most Bahamas. They proposed that this bank was accreted to the northern edge of the Caribbean plate during a northward jump in the deformation front several million years ago. Their model was based on three lines of reasoning: (1) the two regions of shallow bathymetry are anomalous with respect to the surrounding insular slope; (2) the anomalous block lies southeast of, and along trend with, the banks of the southeastern Bahamas, which are themselves aligned parallel to fracture zones in the Atlantic ocean crust (Klitgord and Schouten, 1986); and (3) dredge hauls from the slopes of the anomalous block west of Mona Canyon consist of shallow-marine and pelagic limestones similar to those expected along the margins of the Bahama carbonate banks.

We agree with McCann and Sykes (1984) insofar as: (1) the bathymetry of their anomalous block is indeed anomalous, particularly the area west of Mona Canyon; and (2) the anomalous block probably does record the effects of a collision of a southeastern Bahama bank. However, we propose an alternative model for the origin of the high-standing block that involves ongoing underthrusting of the bank and consequent disruption of the overriding Caribbean margin. Unlike the McCann and Sykes (1984) model, our model does not involve either accretion of the carbonate bank to the Caribbean plate or a northward jump in the trace of the deformation front.

We base our arguments primarily on a reinterpretation of the tectonostratigraphic significance of the dredge results and on a comparison of the Mona Passage example with several active analogs of seamount collisions. The dredge results reveal two distinct rock suites. The first, cited by McCann and Sykes (1984), consists of shallow-marine limestones as well as pelagic limestones, similar to those that would be expected from a shallow-water carbonate bank (Fox and Heezen, 1975; Heezen et al., 1985). However, submersible dives show that these rocks were deposited atop metamorphic rocks of the NCMT (Heezen et al., 1985). The metamorphic rocks that make up the NCMT are generally thought to have formed in a Cretaceous subduction zone or collisional setting along the leading (northern) edge of the Caribbean plate (Joyce, 1991; Pindell and Draper, 1991; Perfit et al., 1980; Draper and Nagle, 1991). These results indicate that the NCMT, including the basement of Mona bank, has been mechanically part of the Caribbean plate throughout Tertiary time; it is not part of the North America (Atlantic) plate upon which the Bahama banks developed. Thus, rather than constituting a carbonate bank, as suggested by McCann and Sykes (1984), we interpret the dredged limestones as having been deposited on top of uplifted metamorphic rocks of the NCMT.

We propose that the northeast Mona block and northwest Mona ridge record uplift associated with oblique underthrusting of a now-subducted, southeastern Bahama carbonate bank, which we refer to as Mona bank. As discussed earlier, several studies have shown that underthrusting of high-standing ridges or seamounts commonly results first in uplift of the overriding margin, and then collapse of the uplift in response to oversteepening of the slope (Lallemand and Le Pinchon, 1987; Collot and Fisher, 1991; Hagen and Moberly, 1994). We hypothesize that as Mona bank impinged obliquely on the northern edge of the Caribbean plate, it caused diachronous uplift of the insular slope, from east to west. Continued collision resulted in the configuration we see today, with the limestone-capped, metamorphic rocks uplifted as much as 3 km above their expected depth. We suggest that the crest of Mona bank lies beneath the metamorphic terrane, where it provides basal support for the anomalously shallow bathymetry north of Mona Passage. As we will discuss, this interpretation has important implications for the seismic behavior of the plate margin.

A submersible dive (Fig. 14) (Heezen et al., 1985; their dive 641) shows that the northern edge of northwest Mona ridge is a steep normal fault with large blocks of Pleistocene reef-lagoon limestones slumped down northward to depths of 3,600 m. These data indicate that the north side of the anomalous block is collapsing toward the trench. We suggest that this collapse is a result of the removal of topographic support of the Mona block uplift as the underthrust Mona bank passes obliquely westward beneath the insular slope.

Effects and implications of collisional underthrusting of the southeastern Bahamas beneath northern Hispaniola

Perhaps the most striking feature revealed by the SeaMARC II data is the active Silver-Navidad bank collision zone. Collision of these bathymetrically high-standing banks has had several profound effects on the plate margin. In terms of hazards to life and property in the north-central Caribbean, the most important consequence of the collision is the localization of large thrust earthquakes along the plate interface. Dolan and Wald (this volume) show that the M_s 7.8 (Kelleher et al., 1973; M_s 7.7 in Abe, 1981; M_s 7.5 in Pacheco and Sykes, 1992) northern Mona Passage earthquake in 1943 and the M_s 8.1 (Kelleher et al., 1973; M_s 8.0 in Abe, 1981; M_s 7.8 in both Pacheco and Sykes, 1992 and Russo and Villaseñor, 1995) northeastern Hispaniola earthquake in 1946 occurred at the sites of active collisional underthrusting of bathymetrically high-standing Bahama banks. They propose that collision of the banks results in increased mechanical coupling between the underthrust Atlantic slab and the overriding Caribbean margin. This increased mechanical coupling results in large asperities capable of generating great thrust earthquakes along the plate interface. Based on the termination of the 1943–53 westward-propagating earthquake sequence at the west edge of the collision zone, we speculate that areas of the plate interface not undergoing collision exhibit a lower degree of interplate coupling, resulting in a consequently smaller percentage of seismically released strain (Dolan and Wald, this volume). Based on the locations of the 1943 and 1946 earthquakes at collisional asperities, we hypothesize that large earthquakes occurring on the plate interface generally only nucleate at sites of active collisional underthrusting. This does not necessarily mean, however, that large earthquakes that nucleate at the collisional asperities might not sometimes rupture through regions that are not undergoing collision.

Morphologically, one of the most obvious effects of the Silver-Navidad bank collision is the western termination of the Puerto Rico trench, which extends for more than 500 km to the east, where it changes strike and becomes the Lesser Antilles trench. Offshore northeast Hispaniola the trench abruptly terminates against the eastern edge of the Navidad bank collision. Seafloor depth along the trench axis shallows by almost 4,000 m, from 8,100 m north of the Mona Passage to 4,300 m within the Navidad collision zone, in less than 30 km along strike.

The collision is also responsible for many of the structural and geomorphic features of northern Hispaniola. The occurrence of shallow-marine carbonates of the Villa Trina Formation atop 400–1,200-m-high peaks in the Cordillera Septentrional indicates post-early Pliocene uplift of northern Hispaniola. We suggest that onset of this uplift event records initial collision of Hispaniola with the southeastern Bahamas in early to mid-Pliocene time. This age is supported by the pre–mid-Pliocene southward tilting and stepback of Mouchoir bank, which we interpret as being due to thrust loading of the southern carbonate apron of Mouchoir bank during initial collisional contact with the Hispaniola margin (Mullins et al., 1991). An early to mid-Pliocene initiation of collision is also supported by age estimates of Cibao basin strata that are interpreted as synorogenic deposits (Erikson et al., 1998).

One of the most spectacular effects of the collision is the concentric flight of uplifted marine terraces on the Rio San Juan Peninsula (Figs. 11B and 12). The location of the Rio San Juan Peninsula along the southward projection of Silver Spur indicates that the marine terraces have been uplifted during collisional underthrusting of Silver Spur. This collision marks the leading edge of the oblique Bahamas ridge collision zone, and the termination of the westward-propagating 1946–1953 northeastern Hispaniola earthquake sequence beneath the westernmost edge of the collision supports our interpretation that large thrust earthquakes along the plate interface are localized at sites of active collision.

Oblique, collisional underthrusting of Silver Spur also probably explains the anomalous north-south trend of the coastline along the eastern edge of the Rio San Juan Peninsula, which contrasts markedly with the overall west-northwest trend of the rest of the northern Hispaniola slope. We hypothesize that the north-south margin formed when west-southwestward oblique underthrusting of high-standing Silver Spur removed support from beneath that part of the slope. As Silver Spur passed westward, the slope collapsed, possibly along as-yet unidentified, north-south–trending normal faults parallel to the eastern, trailing edge of the Silver Spur collision.

The Rio San Juan Complex, a group of Mesozoic–early Tertiary metamorphic and sedimentary rocks, forms the basement for the marine terraces of the Rio San Juan Peninsula (Draper and Nagle, 1991). The western edge of the metamorphic complex appears to be faulted over Miocene shallow-water carbonates along a series of north-northwest–trending reverse faults (Draper and Nagle, 1991). We suggest that thrust emplacement of the Rio San Juan Complex over late Tertiary rocks may be caused by basal tractions associated with the Silver Spur collision.

Furthermore, we speculate that shortening related to oblique underthrusting of Silver Spur may have terminated strike-slip along the Camu fault in the Rio San Juan Peninsula area. The Camu fault lies along strike with the northeastern Hispaniola slope fault system to the east, and the two fault segments may originally have connected through the present site of the collision. We suggest that collisional underthrusting and consequent shortening in the Rio San Juan Peninsula area shut off strike-slip motion, which presumably transferred southward to the Septentrional fault. It is not clear whether the Camu fault to the west of the collision is still active, although it was clearly an important structure in Tertiary time (Pindell and Draper, 1991; Draper and Nagle, 1991). We hypothesize that at least a small component of Septentrional fault motion is transferred eastward onto the northeastern Hispaniola slope fault zone via the northeast-trending Nagua lineament, which we interpret as a transtensional transfer fault (Fig. 11B).

Deep structure of the north-central Caribbean plate boundary: Opposing underthrust slabs and upper mantle collision

The recent seismicity of the north-central Caribbean reveals three distinct subducted slab elements beneath eastern Hispaniola and Puerto Rico: (1) underthrust Atlantic oceanic lithosphere defined by a south-dipping Wadati-Benioff zone beneath eastern Hispaniola, Puerto Rico, and the Virgin Islands; (2) a diffuse, vertical zone of seismicity beneath eastern Hispaniola and Mona Passage; and (3) Caribbean lithosphere defined by a gently north-dipping zone of earthquakes extending down-dip from the Los Muertos trough south of the islands (Figs. 20 and 21; McCann and Sykes, 1984; Byrne et al., 1985; McCann and Pennington, 1990). Recognition of these three distinct slab elements raises several important questions about their origins and interactions with one another.

Although the south-dipping and north-dipping zones of seismicity appear to be clearly associated with underthrusting of Atlantic and Caribbean lithosphere, respectively (McCann and Sykes, 1984; Byrne et al., 1985; Dolan and Wald, 1997, this volume), the origin of the vertical slab is less clear-cut. McCann and Sykes (1984) proposed that the vertical zone of seismicity represents a slab of underthrust Atlantic lithosphere. They hypothesized that the vertical slab is detached from the shallow, underthrust Atlantic slab, and that it is sinking into the aesthenosphere, thus explaining the vertical T-axes of earthquakes in this zone. Furthermore, they suggest that the deep slab became detached during Pliocene(?) collision with a southeastern Bahama bank (their "anomalous block"), which clogged the subduction zone, resulting in a northward jump in the subduction trace to the present Puerto Rico trench. Although we agree with their interpretation that the vertical zone of earthquakes represents underthrust Atlantic lithosphere, we propose a different model for the origin of the deep slab beneath eastern Hispaniola and Mona Passage.

As supporting evidence for their model, McCann and Sykes (1984) use the presence of the 40-km-wide zone of seismic quiescence that separates the deep, vertical slab and the shallow, gently south-dipping slab beneath Mona Passage as evidence that the deep slab is detached from the shallow slab and is sinking into the aesthenosphere. They note that a sinking-slab model is compatible with the focal mechanisms of earthquakes in the deep zone of seismicity, most of which exhibit near-vertical T-axes suggestive of vertical extension in the slab.

However, the seismicity data indicate that the zone of seismic quiescence identified by McCann and Sykes (1984) is of limited lateral extent, with continuous, seismically defined, underthrust Atlantic lithosphere discernible both to the east (to 150 km depth; Fig. 21A and B) and to the west (to ~140 km depth; Fig. 21F). Thus, if the zone of seismic quiescence does record detachment of the slab, then detachment must have occurred only in a narrow region beneath Mona Passage. We suggest that the lack of earthquakes in the quiescent zone simply represents the highly discontinuous nature of recent seismicity in the region; most of the seismicity occurs in clusters that we speculate are related to areas of collision and consequent enhanced interslab mechanical coupling (Fig. 19).

Furthermore, in the year following the large 1946 northeast Hispaniola earthquake, 15 earthquakes occurred within the deep, vertical zone of seismicity; during the preceding and succeeding years an average of only one event per year occurred in this zone (Russo and Villaseñor, 1997). We interpret this increased activity along the deep slab after the 1946 earthquake as evidence of mechanical continuity between the shallow and deep slabs. We speculate that the increase in deep earthquakes reflects mechanical readjustment of the subducted slab in response to major coseismic underthrusting during the earthquake. Specifically, many of the deep 1946 aftershocks occur near the locus of maximum curvature of the slab (Fig. 21E and F). This part of the slab would be most directly affected by a coseismic pulse of slab bending, with extension near the top of the slab and compression near its base. We also note that although these deep events are technically aftershocks of the 1946 event, having occurred within one rupture length of the 1946 event during a postmainshock period of heightened seismicity, they are part of a long-lived pattern of intermediate and deep-focus earthquakes that is ongoing. In contrast, the planar cluster of shallow, gently south-dipping aftershocks was clearly a temporally and spatially anomalous event that was related directly to the 1946 mainshock. Thus, the short-lived cluster of shallow, gently south-dipping aftershocks outline the actual mainshock rupture plane, whereas the long-lived, diffuse zone of deep aftershocks did not occur directly on the mainshock rupture plane. Rather, they are part of a long-lived sequence of earthquakes occurring within the underthrust slab as it mechanically readjusts to its changing geometry in response to ongoing underthrusting during large oblique, thrust-interface earthquakes such as the 1946 earthquake.

Based on the continuity of the deep and shallow parts of the underthrust Atlantic slab both to the east and west of Mona Passage and the highly temporally anomalous increase in the number of deep earthquakes in the vertical slab after the 1946 northeast Hispaniola earthquake, we suggest that the vertical slab is Atlantic oceanic lithosphere that is still mechanically attached to the shallow underthrust slab, except possibly in a small region beneath Mona Passage. This interpretation raises an obvious question: What caused the abrupt, 65° bend in slab dip at the top of the vertical zone of seismicity?

The seismicity data indicate that south-dipping, Atlantic seismic zone and the north-dipping Caribbean seismic zone abut each other beneath eastern Hispaniola (~69°30′W) to westernmost Puerto Rico (67°00′W). This relationship is best illustrated on Figure 21C, D, and E, where earthquakes associated with the rather planar Caribbean slab abut, and may even overlie, the southernmost earthquakes of the shallow, south-dipping Atlantic slab. Beneath eastern Hispaniola the point of slab contact lies directly above the vertical zone of earthquakes (Fig. 21E). Based on these data, we propose that the underthrust Caribbean and Atlantic lithospheric slabs are actively colliding beneath the eastern Hispaniola and Mona Passage. Using these same data, Dillon et al. (1994) simultaneously and independently arrived at the same conclusion.

We suggest that collision of Caribbean lithosphere with Atlantic lithosphere has resulted in the sharp bend in the underthrust Atlantic slab. The Caribbean slab, being thicker (Edgar et al., 1971) and (presumably) more buoyant, may locally override the Atlantic slab. Once the Atlantic slab has been overthrust by the Caribbean slab, it no longer actively collides, and it sinks into the aesthenosphere. The vertical T-axes of earthquakes from the vertical zone of seismicity indicate downdip extension, but do not necessarily imply detachment of the deep Atlantic slab from the shallow part of the underthrust slab. These data are thus compatible with the increase in deep earthquakes after the shallow 1946 earthquake and the continuous nature of the deep and shallow zones of seismicity both west and east of Mona Passage, both of which indicate that the vertical slab has not detached from the shallow slab, except possibly locally beneath Mona Passage.

Dillon et al. (1994) suggested that overthrusting of the Atlantic slab by the Caribbean slab may have caused downbowing of the Atlantic slab. They speculate that this thrust loading by the Caribbean slab is responsible for the great depth of the Puerto Rico trench north of Puerto Rico, as well as the very large gravity low associated with the trench. This is an intriguing idea that needs to be tested with more and better constrained earthquake locations and focal mechanisms.

Model for tectonic evolution of the north-central Caribbean during the late Cenozoic

The Greater Antilles islands of Cuba, Hispaniola, Puerto Rico, and the Virgin Islands represent the remains of a Cretaceous–early Tertiary volcanic arc that has been dismembered during the past ~50 Ma by left-lateral motion along the northern Caribbean plate boundary zone (Pindell and Barrett, 1990). In the following section we propose a model in which the dismemberment and transportation of the arc fragments to their present positions was controlled by a combination of lateral changes in the shape of the Bahamas province and major lateral changes in the architecture and buoyancy of the Bahamas-Tiburon ridge.

Bahamas-Tiburon ridge. The Bahamas carbonate province is thought to have originated along the Bahamas fracture zone during Jurassic–Early Cretaceous opening of the Atlantic Ocean (Sheridan et al., 1988). This is reflected in the orientation of the Bahama Escarpment, which parallels the west-northwest–trending fracture-zone fabric of the western central Atlantic (Fig. 1; Klitgord and Schouten, 1986). The northwestern Bahamas (northwest of Exuma Sound) developed above rifted continental crust (Ladd and Sheridan, 1987), whereas the central and southeastern Bahamas may have developed above oceanic crust (Uchupi et al., 1971; Ladd and Sheridan, 1987; Sheridan et al., 1988). Alternatively, the southeastern Bahamas may have been built atop highly extended and intruded continental crust (Mullins and Lynts, 1977).

McCann and Sykes (1984) proposed that the 1,000-km-long Bahamas carbonate province and the Tiburon Rise, one of several west-northwest–trending aseismic ridges that extend beneath the Lesser Antilles accretionary prism east of the Lesser Antilles island of Guadeloupe, represent two ends of a continuous, 2,000-km-long basement ridge (Fig. 1). In their model, much of this ridge lies beneath the northeastern Caribbean plate, having been overridden by Eocene to Holocene eastward motion of Caribbean. Alternatively, the Bahamas ridge and the Tiburon Rise may be disconnected from one another as part of a discontinuous series of basement ridges that developed above the Bahamas fracture zone.

If the McCann and Sykes (1984) model is correct, then the Bahamas-Tiburon ridge must exhibit remarkable differences in crustal architecture, width, sedimentary style, and buoyancy along strike. In the northwest, the active Bahamas platforms are >400 km wide (measured perpendicular to the Bahama Escarpment) and are capped by >6 km of shallow-water carbonate deposits (Eberli and Ginsburg, 1987, 1989). Similarly, the southeastern Bahamas exhibit a 5.5-km-thick carbonate cap. Gravity data indicate a total crustal thickness in the southeastern Bahamas of ~16–21 km, relative to ~6 km for adjacent Atlantic oceanic crust to the north (Uchupi et al., 1971). Thus, the southeastern and northwestern Bahamas, although they exhibit major differences in width and sedimentary character, are both fundamentally buoyant features.

In contrast, the Tiburon Rise is a typical aseismic ridge. Such features are not isostatically compensated and their high ridge topography is sustained by the rigidity of the underlying oceanic plate (Abrams et al., 1988; Bouyesse and Westercamp, 1990; Muller and Smith, 1993; Pockalny et al., 1996). The absence of a crustal root indicates that aseismic ridges are not buoyant features. Furthermore, the Tiburon Rise is a much smaller, narrower feature than the Bahamas; the Tiburon Rise is only 40–50 km wide and ~1–1.5 tall, and does not exhibit a thick carbonate cap like the Bahamas. Because they are not buoyant features, aseismic ridges are typically subducted, although the topographic effects of underthrusting the several-kilometer-tall ridges can cause major disruption of the overlying block (Lallemand and Le Pinchon, 1987; Collot and Fisher, 1991; Hagen and Moberly, 1994). Thus, the major mechanical difference between the Bahamas province northwest of Puerto Rico and the Tiburon Rise to the east of the Lesser Antilles subduction zone is that the Bahamas are buoyant and will resist subduction, whereas the Tiburon Rise is not buoyant and will not significantly resist subduction.

In addition to the lateral difference in buoyancy between the Bahamas and the area along strike to the southeast, lateral changes in the width of the Bahamas also have played a major role in the tectonic development of the northern Caribbean. Although much of the southwestern margin of the Bahamas has been subducted beneath the northern Caribbean, geological and seismological observations allow us to reconstruct the probable original limits of buoyant Bahamas crust. In the northwest, the Bahamas were originally even wider than they are today. Detailed mapping and subsurface data indicate that the underthrust Bahamas block extends at least 20 km southwest of the present coastline of Cuba (Draper and Barros, 1994). Some authors have even proposed that metamorphosed carbonates exposed in tectonic windows in southwestern Cuba (e.g., Escambray Complex) represent remnants of the Bahamas platform rocks that have been thrust beneath a cover of island arc rocks (Mann et al., 1995). Although the exact extent of underthrusting of the Bahamas beneath Cuba remains the subject of debate, the Bahamas are at least 400 km wide in the central Cuba area.

Northwest of Hispaniola a large, concave-to-the-south reentrant occurs along the southern margin of the Bahamas, narrowing the exposed part of the carbonate province from ~350 km off the eastern tip of Cuba to as little as 150 km near Caicos bank, 250 km to the east (Fig. 2). We refer to the area south of this narrow region of the Bahamas as the Caicos reentrant, because it encompasses the present Caicos basin, as well as the Hispaniola basin to the east (Fig. 24). Although much of the present size and morphology of the reentrant is due to large-scale sedimentary collapse (Mullins and Hine, 1989), the basic shape of the reentrant may reflect the original shape of the southwestern Bahamas margin. In marked contrast to the very wide northwestern Bahamas carbonate platform tops, east of Cuba the shallow bank tops and islands of the southeastern Bahamas are located parallel to, and within 50 km of the Bahama Escarpment (Fig. 2). Side-scan sonar data show that in the southeastern part of the Caicos reentrant, off northwestern Hispaniola, the southwestern edge of the Bahamas is not actively colliding with the island slope (Dillon et al., 1992).

The SeaMARC II data indicate that at Silver Spur the southwestern edge of the Bahamas bends sharply southward, extending beneath the Hispaniola coast. East of Silver Spur, the exact location of the original southwestern margin of the Bahamas is poorly known, having been obliquely subducted beneath Hispaniola as part of the ongoing collision. The 1943–53 earthquakes, however, appear to have been localized at active collisional asperities, suggesting that the location of the earthquake rupture planes can be used as a proxy for the extent of underthrust buoyant Bahamas basement (Dolan and Wald, this volume). Dolan and Wald (this volume) use these data to suggest that the Bahamas widen abruptly to the east of Silver Spur, from ~110 km just west of Silver bank to at least 200 km just east of the Silver Spur collision.

The lack of sufficient data from the northeastern Caribbean region precludes determining whether the Bahamas ridge extends as a continuous feature all the way to the Tiburon Rise. The Tiburon Rise is a discontinuous feature, however, terminating eastward ~160 km east of the Lesser Antilles deformation front. This observation suggests that the Bahamas-Tiburon ridge may not be a continuous feature beneath the northeastern Caribbean, in conflict with McCann and Sykes' (1984) original model. Pockalny et al. (1996) have even suggested that the Tiburon Rise is maintained by active faulting related to proximity of the fracture zone to the Lesser Antilles subduction zone. In either event, the buoyant Bahamas province ends somewhere between Mona Passage and the nonbuoyant Tiburon Rise. The proposed Mona bank collision is the southeasternmost part of the buoyant Bahamas province that can be inferred from seismological data (Dolan and Wald, this volume), and in the following model we suggest that Mona bank marks the southeasternmost extent of buoyant Bahamas crust.

Dismemberment of the Greater Antilles island arc in response to repeated, diachronous attempts to subduct the buoyant Bahamas Ridge. The data discussed in this chapter allow us to propose a model that explains many fundamental questions

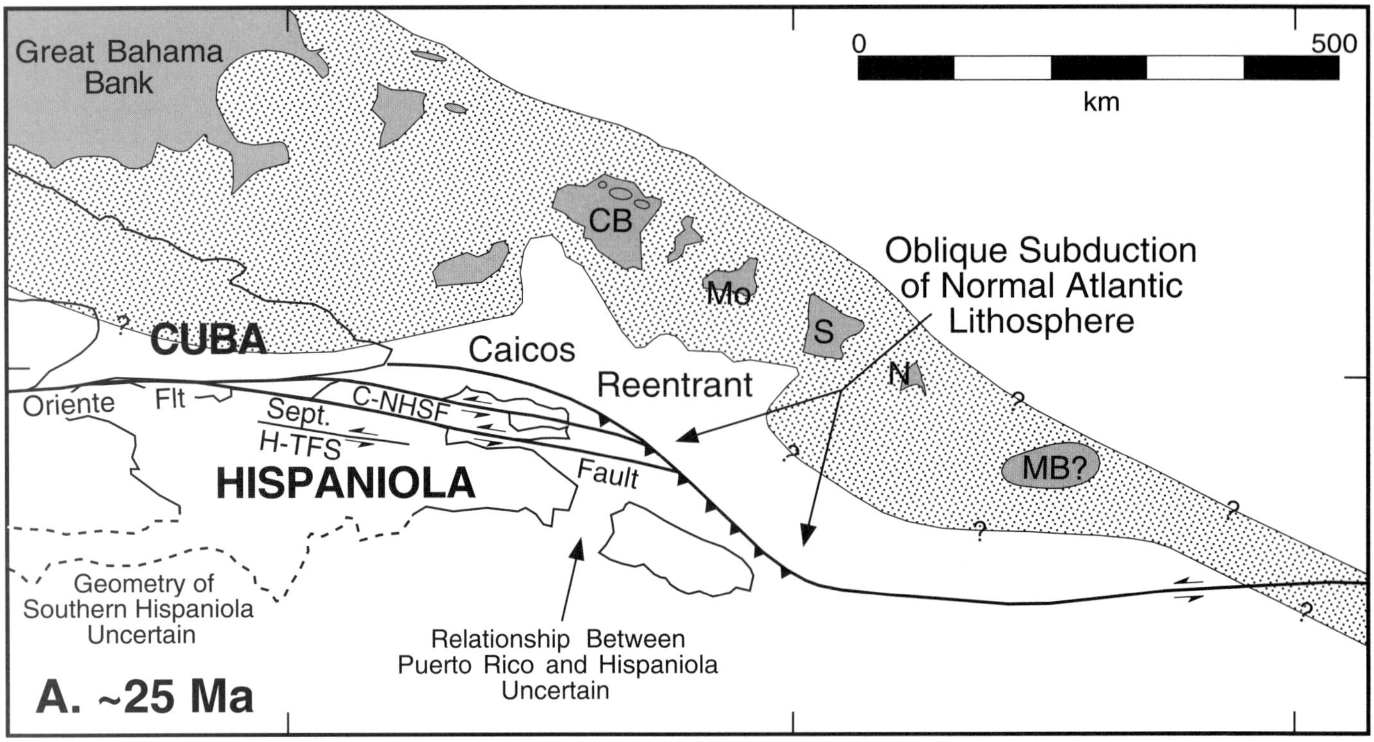

Figure 24 (this and next two pages). Proposed reconstructions of the Neogene-Holocene tectonic evolution of the northern Caribbean plate boundary zone. C–NHSF is Camu–northern Hispaniola slope fault; H–TFS is Hispaniola–Tavera fault system; CB is Caicos bank; Mo is Mouchoir bank; S is Silver bank; N is Navidad bank; MB is Mona bank; Flt is fault; PR is Puerto Rico; EPGFZ is Enriquillo-Plantain Garden fault zone; NHDB is northern Hispaniola deformed belt. (A) Reconstruction for late Oligocene time (~25 Ma). Position of Hispaniola adjacent to Cuban portion of Greater Antilles island arc is based on restoration of ~400 km of left-lateral slip needed to restore pre-strike-slip position of Hispaniola and present position. Most, or all, of this strike-slip motion probably occurred on the Septentrional and Camu fault zones (Pindell and Barrett, 1990; Pindell and Draper, 1991; Erikson et al., 1998). Assuming that all of the ~15 mm/yr Neogene Cayman trough spreading rate (Rosencrantz et al., 1988) has been accommodated along these faults since onset of eastward motion of Hispaniola, then initiation of the system of left-lateral strike-slip faults along the northern part of the plate boundary (including the Septentrional and Camu faults) must have occurred before ~25 Ma. We hypothesize that eastward motion of Hispaniola and Puerto Rico, and initiation of the Septentrional and Camu faults, were facilitated by the presence of "subductible," normal oceanic lithosphere within the proposed Caicos reentrant. Evidence of this period of subduction is provided by deep, near-vertical subducted slab beneath eastern Hispaniola. The position of the early, northwest-trending oblique subduction front northeast of Hispaniola and Puerto Rico shown in figure is based on restoration of the deep, near-vertical Atlantic subducted slab to horizontal. (B) Between late Miocene (~10 Ma) and early Pliocene (~4.5 Ma) time, Puerto Rico rotated 25° counterclockwise (cclw) (Reid et al., 1991). We suggest that this rotation was caused by highly oblique collision of the northwestern corner of Puerto Rico with the southeastern end of the Bahamas ridge (Mona bank). Oblique subduction of normal oceanic lithosphere continued in the Caicos reentrant northeast of Hispaniola. Northern Hispaniola was covered by a carbonate shelf, atop which the shallow-marine Villa Trina Formation was deposited. (C) By early Pliocene time (~4–5 Ma), as Puerto Rico ceased rotating as it began to pass eastward of the oblique Mona bank collision, Hispaniola began to collide with thick, buoyant lithosphere of the Bahamas ridge. The collision initiated uplift of the Villa Trina Formation carbonates in the Cordillera Septentrional of Hispaniola and obliteration of the connection between the originally through-going Camu–northern Hispaniola slope fault zone. As eastward motion of Hispaniola slowed in response to the collision, Mona Passage began to open behind Puerto Rico, which is no longer undergoing active collision and is therefore free to move eastward with the remainder of the Caribbean plate. (D) Hispaniola continues to collide with Silver and Navidad banks, causing the increased mechanical coupling responsible for the nucleation of the 1946 M_s 8.1 northeast Hispaniola earthquake at the collision. A similar zone of enhanced coupling at the active Mona bank collision zone to the southeast caused the 1943 M_s 7.8 north Mona Passage earthquake. Mona Passage continues to open along north-south–trending normal faults in response to collisional slowing of Hispaniola.

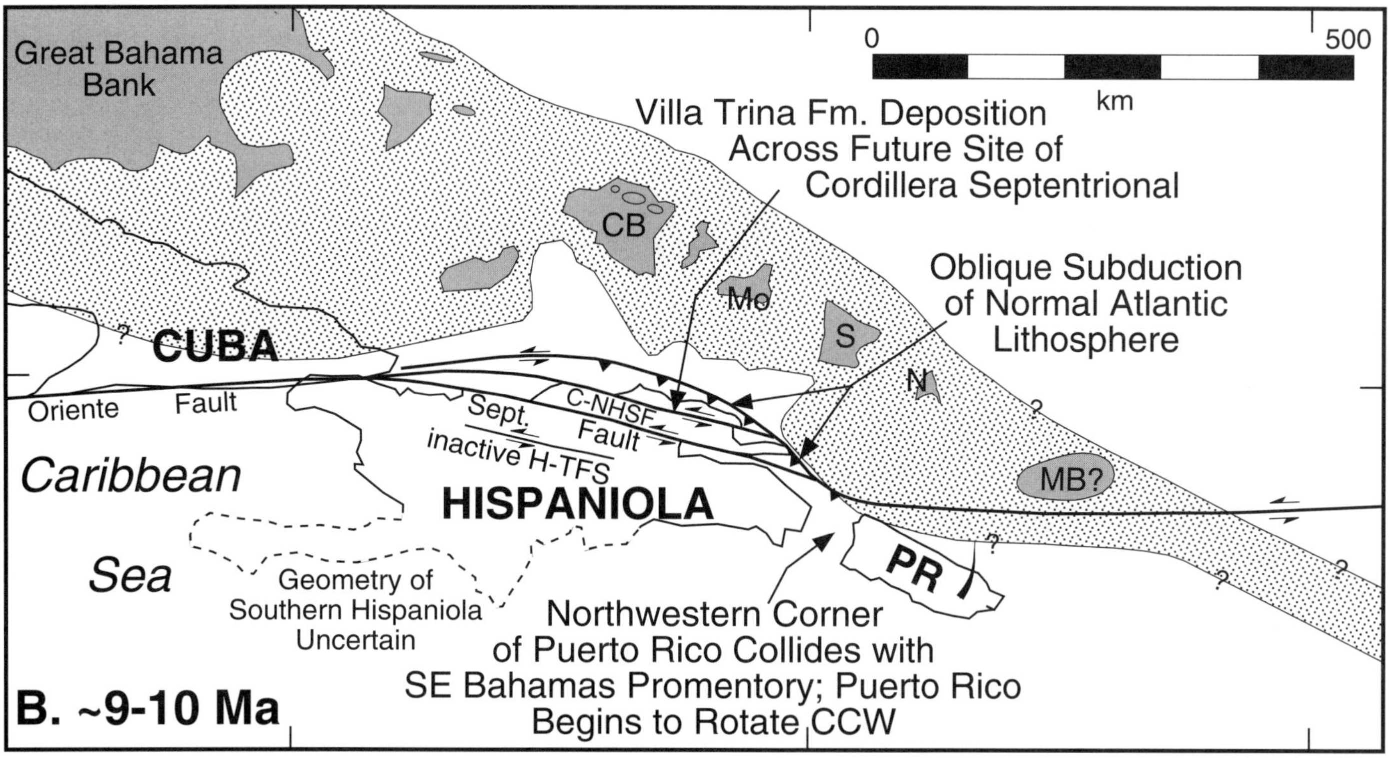
Great Bahama Bank
0 500 km
Villa Trina Fm. Deposition Across Future Site of Cordillera Septentrional
CB
Mo
S
N
Oblique Subduction of Normal Atlantic Lithosphere
CUBA
Oriente Fault
C-NHSF
Sept. Fault
inactive H-TFS
Caribbean
HISPANIOLA
MB?
Sea
Geometry of Southern Hispaniola Uncertain
PR
Northwestern Corner of Puerto Rico Collides with SE Bahamas Promentory; Puerto Rico Begins to Rotate CCW
B. ~9-10 Ma

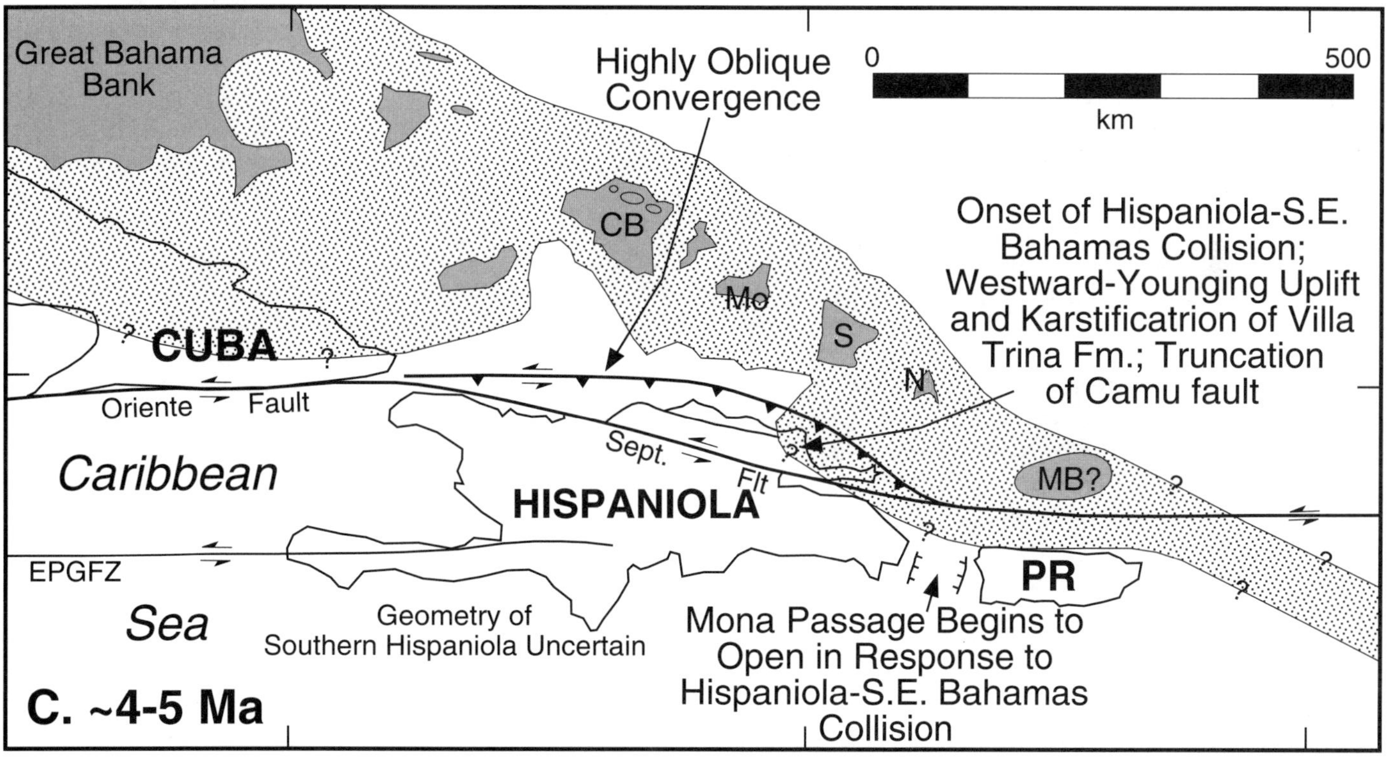
Great Bahama Bank
Highly Oblique Convergence
0 500 km
CB
Mo
S
N
Onset of Hispaniola-S.E. Bahamas Collision; Westward-Younging Uplift and Karstificatrion of Villa Trina Fm.; Truncation of Camu fault
CUBA
Oriente Fault
Sept. Flt
Caribbean
HISPANIOLA
MB?
EPGFZ
PR
Sea
Geometry of Southern Hispaniola Uncertain
Mona Passage Begins to Open in Response to Hispaniola-S.E. Bahamas Collision
C. ~4-5 Ma

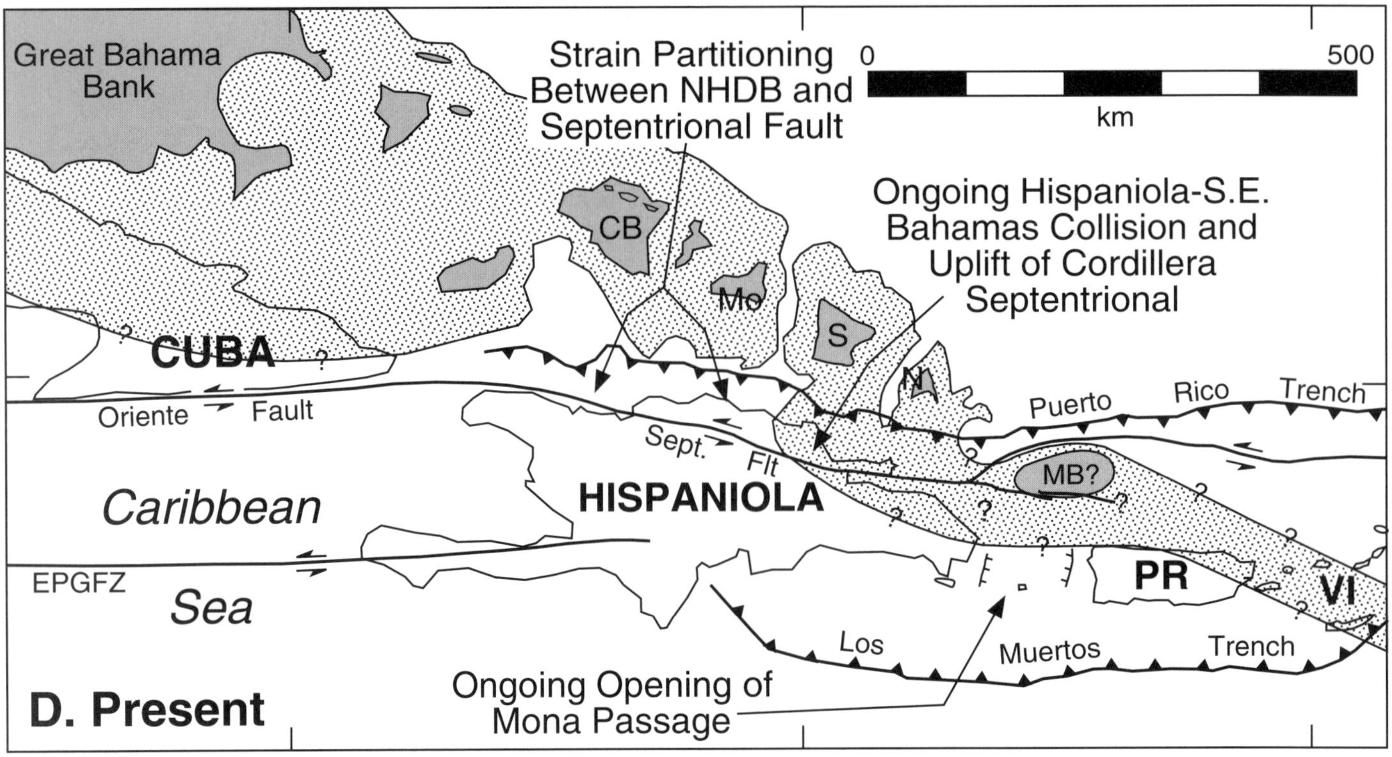

about the tectonic evolution of the north-central Caribbean region, including: (1) How and why did Hispaniola separate from Cuba and move to the east, while Cuba remained behind, firmly welded to the North America plate? (2) Why did the present northern Caribbean plate boundary, and especially the Oriente-Septentrional fault system, become localized where it did? (3) Why is the Hispaniola restraining bend localized where it is? (4) Why did Puerto Rico separate from Hispaniola, opening the Mona Passage in its wake? and (5) Why did Puerto Rico undergo 25° of counterclockwise rotation during Miocene-Pliocene time?

Between Cretaceous and Eocene time, the Greater Antilles island arc is thought to have developed on the leading edge of the Caribbean plate as the plate moved northeastward from the Pacific region along a southwest-dipping subduction zone (Pindell and Dewey, 1982; Burke, 1988; Pindell and Barrett, 1990). East of central Cuba, during Paleocene-Eocene time, southwestward subduction of the North America (Atlantic) plate terminated when buoyant lithosphere of the Bahamas province entered and clogged the subduction zone (Pindell and Barrett, 1990; Mann et al., 1995). This collision resulted in a major shift in plate motion, and since Eocene time the Caribbean plate has moved generally eastward with respect to the North America plate (Pindell and Dewey, 1982; Burke, 1988; Pindell and Barrett, 1990). During Eocene time, most of the resultant left-lateral strike-slip motion was probably accommodated primarily along the southwestern edge of the moribund island arc (Pindell and Barrett, 1990; Dolan et al., 1991). Since Oligocene time, however, most left-lateral motion has occurred along several fault systems that cut across the grain of the extinct island arc in southeastern Cuba and northwestern Hispaniola (Pindell and Barrett, 1990; Dolan et al., 1991; de Zoeten and Mann, 1991; Draper and Barros, 1994). The earliest evidence for this more northerly phase of strike-slip is provided by opening of the Tavera pull-apart basin in northern Hispaniola during early Oligocene time (Dolan et al., 1991). Assuming that all of the ~15 mm/yr Neogene Cayman trough spreading rate (Rosencrantz et al., 1988) has been accommodated along these faults since the onset of eastward motion of Hispaniola, then initiation of the system of left-lateral strike-slip faults along the northern part of the plate boundary (including the Septentrional and Camu faults) must have occurred before ~25 Ma. This Neogene-to-Holocene phase of left-lateral strike-slip has dismembered the island arc, separating Cuba from Hispaniola–Puerto Rico (Fig. 24; Pindell and Barrett, 1990; Erikson et al., 1998). But why did the Oriente-Septentrional fault system, which is responsible for much of this motion, become localized where it did? We suggest that the answer lies in the distribution of normal, neutrally to negatively buoyant Atlantic oceanic lithosphere versus buoyant, less "subductible" Bahamas lithosphere on the subducting plate. This lateral difference in subduction resistance played an important part in controlling the strike-slip dismemberment of the Greater Antilles island arc and the present morphology of Cuba, Hispaniola, and Puerto Rico.

We speculate that the Caicos reentrant was floored by normal Atlantic oceanic lithosphere (Fig. 24A). This oceanic lithosphere was less buoyant than the thick Bahamas basement to the northwest, northeast, and southeast of the reentrant, and was therefore probably much less resistant to oblique subduction. Thus, the

Caicos reentrant represented a "free-face" or "hole" in the thick, laterally extensive Bahamas basement that allowed easy eastward translation of the Hispaniola–Puerto Rico block through a combination of oblique subduction along the northeast edge of Hispaniola and left-lateral strike-slip on the proto-Septentrional fault system. The western limit of the subductible Caicos reentrant therefore controlled the location at which the Hispaniola portion of the Greater Antilles island arc broke away from the Cuba part of the arc, which was pinned in place by the buoyant, Bahamas lithosphere beneath the island. We propose that the Oriente-Septentrional fault system initiated in response to this separation, its location a direct consequence of the distribution of normal oceanic lithosphere beneath the Caicos reentrant.

Eastward motion of the old, cold, northwest-trending Cretaceous to Eocene core of the Greater Antilles island arc resulted in oblique-contractional deformation along the northeastern (leading) edge of the Hispaniola block. Evidence of this eastward motion of Hispaniola is preserved in the deep, near-vertical slab of Atlantic oceanic lithosphere beneath eastern Hispaniola and the Mona Passage (Figs. 20 and 21). The northwest strike of the older, deeper parts of the underthrust slab reflects the original northwest-trending orientation of the subduction zone northeast of Hispaniola (Fig. 24A). By measuring the downdip slab length and restoring it to horizontal, we can reconstruct the approximate position of the original oblique subduction deformation front (Fig. 24A). Restoration of 85 km of left slip on the Septentrional fault (Erikson et al., 1998), 60 km of left slip on the Camu fault (Pindell and Draper, 1991), and the deep slab to its presubduction geographic extent places the old island arc core of Hispaniola adjacent to similar island arc rocks in Cuba (Draper and Barros, 1994).

Over most of the length of the currently active Hispaniola restraining bend, the active plate boundary is roughly parallel to the core of island arc rocks, which appear to have acted as a rigid backstop during oblique subduction in Miocene-Pliocene time. As discussed, the abrupt change in strike of the plate boundary at the northwest end of the restraining bend (Septentrional fault to Oriente fault transition) simply records the point at which the strike-slip fault system that offset the originally contiguous Cuba and Hispaniola parts of the island arc became localized at the western edge of the proposed Caicos reentrant. Thus, the present Hispaniola restraining bend is caused by eastward translation of a piece of the Greater Antilles island arc into the Caicos reentrant.

The origin of the southeastern end of the Hispaniola restraining bend is more complicated. The orientation of all major plate boundary structures, such as the Septentrional fault, the active Silver-Navidad collisional deformation front, and the deep, subducted Atlantic slab, changes abruptly at the southeastern end of the restraining bend. The locus of maximum curvature of these features occurs within the active Bahamas collision zone, between Silver Spur and Mona Passage, suggesting that the change is related to the site of active, oblique underthrusting of the Bahamas ridge. But how does the location of the active Bahamas-Hispaniola collision control the transition from the northwest-trending Hispaniola restraining bend to the east-west plate boundary to the east?

The 1946–53 earthquakes indicate that Hispaniola has collided with buoyant Bahamas ridge material along Silver and Navidad banks (Dolan and Wald, this volume). We suggest that this collision has retarded eastward motion of Hispaniola, and may be responsible for incipient lock up of the oblique subduction zone along the northeastern margin of Hispaniola. Collisional retardation of eastward motion has several important consequences.

The north-south normal faults in Mona Passage (e.g., those along the east side of Mona Canyon) are highly anomalous features, insofar as they are located within an east-trending, transpressional plate boundary. We propose that these faults have formed during relative motion between Hispaniola and Puerto Rico, with Puerto Rico pulling away from Hispaniola, that is, moving eastward at a faster rate relative to the North America plate (Fig. 24). Why did Puerto Rico, which was presumably originally part of the same island arc as Hispaniola, separate from Hispaniola? And why is it now moving eastward at a faster rate than Hispaniola? The most likely answer is that eastward motion of Hispaniola is retarded by the active Bahamas collision, whereas eastward motion of Puerto Rico is not. This line of reasoning suggests that the mechanical transition between "subductible" and "non-subductible" Bahamas ridge lies along the southeastern edge of the proposed Mona bank collision northwest of Puerto Rico (Fig. 24).

Any model of the recent tectonic evolution of Puerto Rico must explain the 25° of counterclockwise rotation of the island during late Miocene-Pliocene time (Reid et al., 1991). We suggest that as the combined Hispaniola–Puerto Rico block moved eastward during late Miocene time, the northwestern corner of Puerto Rico collided with Mona bank, resulting in enhanced interplate mechanical coupling at the point of collision. If our speculation that Mona bank represents the southeastern end of the buoyant Bahamas province is correct, then the remainder of Puerto Rico did not encounter any buoyant material as it moved eastward. Thus, the major lateral differences in the buoyancy at the southeastern end of the Bahamas resulted in significantly different tractions affecting northwestern Puerto Rico relative to the rest of the island. As a result of this collision, Puerto Rico, pinned only at its northwestern corner, rotated counterclockwise (Fig. 24). The 1943 earthquake data (Dolan and Wald, this volume) suggest that the active Mona bank collision now lies to the northwest of Puerto Rico, and that the island has passed to the east of the collision zone since Pliocene time.

Thus, Puerto Rico is not actively colliding with any buoyant lithosphere and is free to move eastward, in contrast to northern Hispaniola, which remains partially pinned in place by the active Silver-Navidad bank collision. The southeastern end of the Hispaniola restraining bend therefore became localized along the eastern margin of the currently active Hispaniola-Bahamas collision, in much the same way that the northwestern end of the restraining bend became localized at the earlier Cuba-Bahamas collision during Paleogene time. In summary, near-purely eastward motion of Puerto Rico, and the consequent east-west orientation of the plate boundary east of the Hispaniola restraining bend, resulted from the northwest-to-southeast decrease in the buoyancy and subductibility of the Bahamas ridge.

CONCLUSIONS

Marine geophysical and seismicity data indicate that the southeastern Bahamas are actively colliding with Hispaniola along a 350-km-long segment of the northern Caribbean plate boundary zone. This oblique collision is responsible for many of the structural and seismological features of the northern Caribbean, including localization of the large thrust earthquakes at sites of enhanced collisional coupling, Early to mid-Pliocene initiation of uplift in the Cordillera Septentrional of northern Hispaniola, reorganization of the Camu fault system, and the location of the Hispaniola restraining bend.

Highly oblique subduction of Atlantic oceanic lithosphere along the deformation front north of Hispaniola and Puerto Rico is partitioned onto two major structures: (1) the left-lateral Septentrional-northern Puerto Rico slope strike-slip fault system; and (2) the oblique left-lateral thrust fault along the south to south-southwest–dipping interface along the top of the underthrust Atlantic slab. This strain partitioning indicates that the part of Hispaniola north of the Septentrional fault constitutes a small microplate caught up within the plate boundary that is moving eastward relative to the North America plate at a much slower rate than the bulk of Hispaniola to the south. The Septentrional fault probably merges into the southwest-dipping decollement along the top of the subducted slab at a depth of ~15 to 30 km.

Seismicity data indicate that the Atlantic slab, which is obliquely subducted along the deformation front north of Hispaniola and Puerto Rico, and the Caribbean slab, which is subducted along the Los Muertos trough south of the islands, are colliding beneath southeastern Hispaniola and the Mona Passage. This upper mantle collision results in roll back and steepening of the Atlantic slab to near-vertical as the Caribbean slab overrides it.

In contrast to the Cretaceous oceanic lithosphere of the Atlantic slab, the southeastern Bahamas block is a buoyant feature that will resist subduction. We propose that a major reentrant in the southern edge of the Bahamas—the Caicos reentrant—controlled the location at which Hispaniola detached from the Cuba portion of the originally contiguous Greater Antilles island arc massif. The Cuba portion of the arc was welded in place by the presence of buoyant, subducted Bahamas lithosphere beneath the island. The reentrant to the east, however, was floored by subductible, normal oceanic lithosphere. The presence of this patch of normal oceanic lithosphere represented a free-face, allowing the Hispaniola–Puerto Rico portion of the arc to move eastward, separating from Cuba. Eastward motion of Hispaniola via oblique subduction of Atlantic lithosphere continued until eventual collision with the banks of the southeastern Bahamas during Pliocene time. Thus, the location of the Septentrional-Oriente fault zone and the Hispaniola restraining bend was ultimately controlled by the distribution of buoyant Bahamas lithosphere relative to subductible, normal oceanic lithosphere. The active oblique Hispaniola-Bahamas collision retards continued eastward motion of Hispaniola, and allows Puerto Rico, which is not currently undergoing collision, to pull away from Hispaniola. The consequent relative eastward motion of Puerto Rico results in active east-west extension in the Mona Passage between Hispaniola and Puerto Rico.

ACKNOWLEDGMENTS

This study was supported by the National Science Foundation (OCE 8812082). We would like to thank Eric Calais, Bill Dillon, Grenville Draper, Dave Kemp, Paul Mann, Bill McCann, and Fred Taylor for many helpful comments and suggestions. Thanks also to Eli Silver and Don Reed for helpful advice about side-scan sonar acquisition and interpretation, to Sandy Shor and the scientific staff and crew of the R/V *Moana Wave* for an enjoyable and productive cruise, and to the staff of the Lamont-Doherty Earth Observatory seismic reflection processing group. We also greatly appreciate thoughtful reviews of the manuscripts by Gren Draper, Eric Calais, and Paul Mann. We are indebted to Larry Feldman and Bill McCann for sharing their unpublished catalog of historical earthquakes. Allan Tucker and Jeff Beard provided much-appreciated help with the illustrations.

REFERENCES CITED

Abe, K., 1981, Magnitudes of large shallow earthquakes from 1904–1980: Physics of the Earth and Planetary Interiors, v. 27, p. 72–92.

Abrams, L. J., Detrick, R. S., and Fox, P. J., 1988, Morphology and crustal structure of the Kane fracture zone transverse ridge: Journal of Geophysical Research, v. 93, p. 3195–3210.

Austin, J. A., 1983, Overthrusting in a deep-water carbonate terrane, *in* Balley, A. W., ed., Seismic expression of structural styles: Tulsa, Oklahoma, American Association of Petroleum Geologists Studies in Geology 15, p. 3.4.2-167–3.4.2-172.

Birch, F. S., 1986, Isostatic, thermal, and flexural models of the subsidence of the north coast of Puerto Rico: Geology, v. 14, p. 427–429.

Bouyesse, P., and Westercamp, D., 1990, Subduction of Atlantic aseismic ridges and late Cenozoic evolution of the Lesser Antilles island arc: Tectonophysics, v. 175, p. 349–380.

Bowin, C. O., Nalwalk, A. J., and Hersey, J. B., 1966, Serpentinized peridotite from the north wall of the Puerto Rico trench: Geological Society of America Bulletin, v. 77, p. 257–270.

Bracey, D. W., and Vogt, P. R., 1970, Plate tectonics in the Hispaniola area: Geological Society of America Bulletin, v. 81, p. 2855–2860.

Burke, K., 1988, Tectonic evolution of the Caribbean: Annual Review of Earth and Planetary Sciences, v. 16, p. 201–230.

Byrne, D. B., Suarez, G., and McCann, W. R., 1985, Muertos Trough subduction—Microplate tectonics in the northern Caribbean: Nature, v. 317, p. 420–421.

Calais, E., and Mercier de Lépinay, B., 1995, Strike-slip tectonic processes in the northern Caribbean between Cuba and Hispaniola (Windward Passage): Marine Geophysical Research Letters, v. 17, p. 63–95.

Calais, E., Mercier de Lépinay, B., Renard, V., and Tardy, M., 1989, Geometrie et regime tectonique le long d'une limite de plaques en coulissage: La frontiere nord-Caraibe de Cuba a Hispaniola, Grandes Antilles (Geometry and tectonic regime along a strike-slip plate boundary: the northern

Caribbean plate boundary from Cuba to Hispaniola, Greater Antilles), Compte Rendu de l'Académie Science Paris, v. 308, p. 131–135.

Calais, E., Bethoux, N., and Mercier de Lépinay, B., 1992, From transcurrent faulting to frontal subduction: A seismotectonic study of the northern Caribbean plate boundary from Cuba to Puerto Rico: Tectonics, v. 11, p. 114–123.

Carson, B., Holmes, M. L., Umstattd, K., Strasser, J. C., and Johnson, H. P., 1991, Fluid expulsion from the Cascadia accretionary prism: Evidence from porosity distribution, direct measurements, and GLORIA imagery: Philosophical Transactions of the Royal Society of London, v. 335, p. 331–340.

Carson, B., Seke, E., Paskevich, V., and Holmes, M. L., 1994, Fluid expulsion sites on the Cascadia accretionary prism: Mapping diagenetic deposits with processed GLORIA imagery: Journal of Geophysical research, v. 99, p. 11959–11969.

Case, J., and Holcombe, T., 1980, Geologic-tectonic map of the Caribbean region: U.S. Geological Survey Miscellaneous Investigations Series, Map I–1100.

Case, J. E., Mac Donald, W. D., and Fox, P. J., 1990, Caribbean crustal provinces; seismic and gravity evidence, *in* Dengo, G., and Case, J. E., eds., The Caribbean region: Boulder, Colorado, Geological Society of America, The Geology of North America, v. H., p. 15–36.

Chalas-Jimenez, J. A., 1989, Probabilidad de un sismo catastrófico en la República Dominicana: Historia sísmica, predicción, y magnitudes esperadas (Probability of a catastrophic earthquake in the Dominican Republic: Earthquake history, prediction, and expected magnitudes): Revista Geofisica (Instituto de Geografia e Historia), v. 31, p. 185–203.

Collot, J-Y, and Fisher, M. A., 1991, The collision zone between the north d'Entrecasteaux Ridge and the New Hebrides island arc: 1. Sea Beam morphology and shallow structure: Journal of Geophysical Research, v. 896, p. 4457–4478.

DeMets, C., 1993, Earthquake slip vectors and present-day plate motions: Journal of Geophysical Research, v. 98, p.6703–6714.

DeMets, C., Gordon, R. G., Argus, D. F., and Stein, S., 1990, Current plate motions: Geophysical Journal International, v. 101, p. 425–478.

Deng, J., and Sykes, L. R., 1995, Determination of Euler pole for contemporary relative motion of Caribbean and North American plates using slip vectors of interplate earthquakes: Tectonics, v. 14, p. 39–53.

de Zoeten, R., and Mann, P., 1991, Structural geology and Cenozoic tectonic history of the central Cordillera Septentrional, Dominican Republic, *in* Mann, P., Draper, G., and Lewis, J. F., eds., Geologic and tectonic development of the North America–Caribbean plate boundary: Geological Society of America Special Paper 262, p. 265–279.

de Zoeten, R., Draper, G., and Mann, P., 1991, Geologic map of the northern Dominican Republic, *in* Mann, P., Draper, G., and Lewis, J., eds., Geologic and tectonic development of the North America-Caribbean plate boundary in Hispaniola, Geological Society of America Special Paper 262, Plate 1, scale 1:100,000.

Dillon, W., Austin, J. A., Scanlon, K. M., Edgar, N. T., and Parson, L. M., 1992, Structure and development of the insular margin north of western Hispaniola: A tectonic accretionary wedge on the Caribbean plate boundary, Marine and Petroleum Geology, v. 9, p. 70–88.

Dillon, W. P., Edgar, N. T., Scanlon, K. M. and Coleman, D. F., 1994, A review of the tectonic problems of the strike-slip northern boundary of the Caribbean Plate and examination by GLORIA, *in* Gardner, J. V., Field, M. E., and Twichell, D. C., eds., Geology of the United States' Seafloor: The View from Gloria: United Kingdom, Cambridge University Press, p. 135–164.

Ditty, P. S., Harmon, C. J., Pilkey, O. H., Ball, M. M., and Richardson, E. S., 1977, Mixed terrigenous-carbonate sedimentation in the Hispaniola-Caicos turbidite basin: Marine Geology, v. 24, p. 1–20.

Dolan, J. F., and Wald, D. J., 1997, The 1946 Hispaniola earthquake and the tectonics of the North America–Caribbean plate boundary zone, northeastern Hispaniola: Comment: Journal of Geophysical Research, v. 102, p. 785–792.

Dolan, J. F., Mann, P., de Zoeten, R., Heubeck, C., Shiroma, J., and Monechi, S., 1991, Sedimentologic, stratigraphic, and tectonic synthesis of Eocene-Miocene sedimentary basins, Hispaniola and Puerto Rico, *in* Mann, P., Draper, G., and Lewis, J. F., eds., Geologic and tectonic development of the North America–Caribbean plate boundary, Geological Society of America Special Paper 262, p. 217–263.

Draper, G., and Barros, J. A., 1994, Cuba, *in* Donovan, S. K., and Jackson, T. A., eds., Caribbean Geology: An Introduction: Kingston, Jamaica, University of West Indies Publication Association, p. 65–86.

Draper, G., and Nagle, F., 1991, Geology, structure, and tectonic development of the Rio San Juan Complex, northern Dominican Republic, *in* Mann, P., Draper, G., and Lewis, J. F., eds., Geologic and tectonic development of the North America–Caribbean plate boundary in Hispaniola: Geological Society of America Special Paper 262, p. 77–95.

Eberle, W., Hirdes, W., Muff, R., and Palaez, M., 1982, The geology of the Cordillera Septentrional (Dominican Republic); *in* Transactions of the 9th Caribbean Geological Conference, 1980: Santo Domingo, Dominican Republic, p. 619–632.

Eberli, G. P., and Ginsburg, R. N., 1987, Segmentation and coalescence of Cenozoic carbonate platforms in northwestern Great Bahama Bank: Geology, v. 15, p. 75–79.

Eberli, G. P., and Ginsburg, R. N., 1989, Cenozoic progradation of northwestern Great Bahama Bank, a record of lateral platform growth and sea-level fluctuations, *in* Crevallo, P. D., Wilson, J. L., Sarg, J. F., and Read, J. F., eds., Controls on carbonate platform and basin development: Society of Economic Paleontologists and Minerologists Special Publication 44, p. 339–351.

Edgar, N. T., 1991, Structure and geologic development of the Cibao Valley, *in* Mann, P., Draper, G., and Lewis, J. F., eds., Geologic and tectonic development of the North America–Caribbean plate boundary in Hispaniola: Geological Society of America Special Paper 262, p. 281–299.

Edgar, N. T., Ewing, J. I., and Hennion, J., 1971, Seismic refraction and reflection in the Caribbean Sea: American Association of Petroleum Geologists Bulletin, v. 55, p. 833–870.

EEZ SCAN 85 Scientific Staff, 1987, Atlas of the U.S. Exclusive Economic Zone, eastern Caribbean: Reston, Virginia, U. S. Geological Survey Miscellaneous Investigations Series I–1864-B, 58 pp.

Erikson, J. P., Pindell, J. L., Karner, G., Sonder, L. J., Dent, L., and Fuller, W. E., 1998, Neogene sedimentation and tectonics in the Cibao basin and northern Hispaniola: Journal of Geology (in press).

Fischer, K., M., and McCann, W. R., 1984, Velocity modelling and earthquake relocation in the North East Caribbean: Seismological Society of America Bulletin, v. 74, p. 1249–1262.

Fox, P. J., and Heezen, B. C., 1975, Geology of the Caribbean crust, *in* Nairn, A. E. M., and Stehli, F. G., eds., The Ocean Basins and Margins, Volume 3, The Gulf of Mexico and the Caribbean: New York, Plenum Press, p. 421–466.

Hagen, R. A., and Moberly, R., 1994, Tectonic effects of a subducting aseismic ridge: The subduction of the Nazca Ridge at the Peru Trench: Marine Geophysical Researches, v. 16, p. 145–161.

Haq, B. U., Hardenbol, J., and Vail, P. R., 1987, Chronology of fluctuating sea levels since the Triassic: Science, v. 235, p. 1156–166.

Heezen, B. C., Rawson, M., Lynde, R. P., and Nesterhoff, W. D., 1976, In situ submersible observations in the western Puerto Rico Trench: Geological Society of America Abstracts with Programs, v. 6, p. 911–912.

Heezen, B. C., Nesteroff, W. D., Rawson, M., and Freeman-Lynde, R. P., 1985, Visual evidence for subduction in the western Puerto Rico trench: Geodynamique des Caribes, Symposium Proceedings, Editions Technip, February 27, 1985, Paris, p. 287–304.

International Seismological Summary of 1953, 1961, prepared and edited by the International Union of Geodesy and Geophysics, The British Association Seismological Committee: Richmond, Surrey, U.K., Kew Observatory, p. 823.

Jordan, T. H., 1975, The present-day motions of the Caribbean plate: Journal of Geophysical Research, v. 80, p. 4433–4449.

Joyce, J., 1991, Blueschist metamorphism and deformation on the Samana Peninsula; a record of subduction and collision in the Greater Antilles, *in* Mann, P., Draper, G., and Lewis, J. F., eds., Geologic and tectonic development of the North America–Caribbean plate boundary in Hispaniola: Geological Society of America Special Paper 262, p. 47–76.

Kafka, A. L., and Weidner, T. J., 1979, The focal mechanisms and depths of small earthquakes as determined from Rayleigh wave radiation patterns: Bulletin of the Seismological Society of America, v. 69, p. 1379–1390.

Kelleher, J., Sykes, L., and Oliver, J., 1973, Possible criteria for predicting earthquake locations and their application for major plate boundaries of the Pacific and the Caribbean: Journal of Geophysical Research, v. 78, p. 2547–2585.

Klitgord, K. D. and Schouten, H., 1986, Plate kinematics of the central Atlantic: *in* Vogt, P. R., and Tucholke, B. E., eds., The Western North Atlantic Region, Boulder, Colorado, Geological Society of America, The Geology of North America, v. M, p. 351–378.

Ladd, J. W., and Sheridan, R. E., 1987, Seismic stratigraphy of the Bahamas: American Association of Petroleum Geologists Bulletin, v. 71, p. 719–736.

Ladd, J. W., and Watkins, J. S., 1978, Tectonic development of trench-arc complexes on the northern and southern margins of the Venezuela Basin, *in* Watkins, J. S., Montadert, L., and Dickerson, P., eds., Geophysical and geological invetigations of continental margins: American Association of Petroleum Geologists Memoir 29, p. 363–372.

Ladd, J. W., Holcombe, T. L., Westbrook, G. K., and Edgar, N. T., 1990, Caribbean marine geology; Active margins of the plate boundary, *in* Dengo, G., and Case, J. E., eds., The Caribbean region: Boulder, Colorado, Geological Society of America, The Geology of North America, v. H., p. 261–290.

Lallemand, S., and Le Pinchon, X., 1987, Coulomb wedge model applied to the subduction of seamounts in the Japan Trench: Geology, v. 15, p. 1065–1069.

Larson, R. L., 1991, Latest pulse of the Earth: Evidence for a mid-Cretaceous super plume: Geology, v. 19, p. 547–550.

Larue, D., and Ryan, H., 1990, Extensional tectonism in the Mona Passage, Puerto Rico and Hispaniola: A preliminary study: Transactions of the 12th Caribbean Geological Conference, St. Croix, Virgin Islands: Miami, Florida, Geological Society, p. 223–230.

Larue, D. K., and Ryan, H. F., 1998, Seismic reflection profiles of the Puerto Rico trench: Shortening between the North America and Caribbean Plates, *in* Lidiak, E., and Larue, D., eds., Tectonics and Geochemistry of the Northeastern Caribbean: Geological Society of America Special Paper 322, p. 193–210.

Larue, D. K., Joyce, J., and Ryan, H. F., 1990, Neotectonics of the Puerto Rico Trench: Extensional tectonism and forearc subsidence: Transactions of the 12th Caribbean Geological Conference, St. Croix, Virgin Islands: Miami, Florida, Miami Geological Society, p. 231–247.

Lewis, J. F., Draper, G., Bourdon, C., Bowin, C., Mattson, P., Maurrasse, F., Nagle, F., and Pardo, G., 1990, Geology and tectonic evolution of the northern Caribbean region, *in* Dengo, G., and Case, J. E., eds., The Caribbean region: Boulder, Colorado, Geological Society of America, The Geology of North America, v. H., p. 77–140.

Mann, P., and Burke, K., 1984, Neotectonics of the Caribbean: Reviews of Geophysics and Space Physics, v. 22, p. 309–362.

Mann, P., Burke, K., and Matumoto, T., 1984, Neotectonics of Hispaniola: Plate motion, sedimentation, and seismicity at a restraining bend: Earth and Planetary Science Letters, v. 70, p. 311–324.

Mann, P., Schubert, C., and Burke, K., 1990, Review of Caribbean neotectonics, *in* Dengo, G., and Case, J. E., eds., The Caribbean region: Boulder, Colorado, Geological Society of America, The Geology of North America, v. H., p. 405–432.

Mann, P., Taylor, F., Edwards, R. L., and Ku, T., 1995, Actively evolving microplate formation by oblique collision and sideways motion along strike-slip faults: An example from the northeastern margin of the Caribbean plate: Tectonophysics, v. 246, p. 1–69.

Masson, D. G., and Scanlon, K. M., 1991, The neotectonic setting of Puerto Rico: Geological Society of America Bulletin, v. 103, p. 144–154.

McCann, W. R., and Pennington, W. D., 1990, Seismicity, large earthquakes, and the margin of the Caribbean Plate, *in* Dengo, G., and Case, J. E., eds., The Caribbean Region: Boulder, Colorado, Geological Society of America, The Geology of North America, v. H, p. 291–306.

McCann, W. R., and Sykes, L. R., 1984, Subduction of aseismic ridges beneath the Caribbean Plate: Implications for the tectonics and seismic potential of the northeastern Caribbean: Journal of Geophysical Research, v. 89, p. 4493–4519.

Minster, J. B., and Jordan, T. H., 1978, Present-day plate motions: Journal of Geophysical Research, v. 83, p. 5331–5354.

Molnar, P., and Sykes, L. R., 1969, Tectonics of the Caribbean and Middle America regions from focal mechanisms and seismicity: Geological Society of America Bulletin, v. 80, p. 1639–1684.

Moussa, M. T., Sieglie, G. A., Meyerhoff, A. A., and Tanner, I., 1987, The Quebradillas Limestone (Miocene-Pliocene), northern Puerto Rico, and tectonics of the northeastern Caribbean margin: Geological Society of America Bulletin, v. 99, p. 427–439.

Muller, R. D., and Smith, W. H. F., 1993, Deformation of the oceanic crust between the North American and South American plates: Journal of Geophysical Research, v. 98, p. 8275–8291.

Mullins, H. T., and Cook, H. F., 1986, Carbonate apron models: Alternative to the submarine fan model for paleoenvironmental analysis and hydrocarbon exploration: Sedimentary Geology, v. 48, p. 37–79.

Mullins, H. T., and Hine, A., 1989, Scalloped bank margins: Beginning of the end for carbonate platforms?: Geology, v. 17, p. 30–33.

Mullins, H. T., and Lynts, G. W., 1977, Origin of northwestern Bahama Platform: Review and reinterpretation: Geological Society of America Bulletin, v. 88, p. 1447–1461.

Mullins, H. T., Dolan, J., Breen, N., Andersen, B., Gaylord, M., Petruccione, J. L., Wellner, R. W., Melillo, A. J., and Jurgens, A. D., 1991, Retreat of carbonate platforms: Response to tectonic processes: Geology, v. 19, p. 1089–1092.

Mullins, H. T., Breen, N., Dolan, J., Wellner, R. W., Petruccione, J. L., Gaylord, M., Anderses, B., Melillo, A. J., Jurgens, A. D., and Orange, D., 1992, Carbonate platforms along the southeast Bahamas-Hispaniola collision zone: Marine Geology, v. 105, p. 169–209.

Nalwalk, A. J., 1969, Geology of a portion of the north wall of the Puerto Rico trench: Tectonophysics, v. 8, p. 403–425.

Orange, D. L., and Breen, N., 1992, Effects of fluid escape on accretionary wedges 2. Seepage force, slope failure, headless submarine canyons, and vents: Journal of Geophysical Research, v. 97, p. 9277–9295.

Pacheco, J. F., and Sykes, L. R., 1992, Sesimic moment catalog of large, shallow earthquakes, 1900–1989: Bulletin of the Seismological Society of America, v. 82, p. 1306–1349.

Perfit, M. R., Heezen, B. C., Rawson, M., and Donnelly, T. W., 1980, Chemistry, origin, and tectonic significance from the Puerto Rico Trench: Marine Geology, v. 34, p. 125–156.

Pierson, B. J., 1982, Cyclic sedimentation, limestone diagenesis, and dolomitization in Upper Cenozoic carbonates of the southeast Bahamas [Ph.D. thesis]: Miami, Florida, University of Miami, 294 p.

Pindell, J. L., and Barrett, S. F., 1990, Geologic evolution of the Caribbean region; a plate-tectonic perspective, *in* Dengo, G., and Case, J. E., eds., The Caribbean region: Boulder, Colorado, Geological Society of America, The Geology of North America, v. H., p. 405–432.

Pindell, J. L., and Dewey, J. F., 1982, Permo-Triassic reconstruction of western Pangea and the evolution of the Gulf of Mexico/Caribbean region: Tectonics, v. 1, p. 179–211.

Pindell, J., and Draper, G., 1991, Stratigraphy and geological history of the Puerto Plata area, northern Dominican Republic, *in* Mann, P., Draper, G., and Lewis, J. F., eds., Geologic and tectonic development of the North America–Caribbean plate boundary in Hispaniola: Geological Society of America Special Paper 262, p. 97–114.

Pockalny, R. A., Gente, P., and Buck, R., 1996, Oceanic transverse ridges: A flexural response to fracture-zone-normal extension: Geology, v. 24, p. 71–74.

Prentice, C. S., Mann, P., Taylor, F. W., Burr, G., and Valastro, S., 1993, Paleoseismicity of the North America–Caribbean plate boundary (Septentrional fault), Dominican Republic: Geology, v. 21, p. 49–52.

Reid, J. A., Plumley, P. W., and Schellekens, J. H., 1991, Paleomagnetic evidence for late Miocene counterclockwise rotation of north coast carbonate sequence, Puerto Rico: Geophysical Research Letters, v. 18, p. 565–568.

Rosencrantz, E., and Mann, P., 1991, SeaMARC II mapping of transform faults in the Cayman Trough, Caribbean Sea: Geology, v. 19, p. 690–693.

Rosencrantz, E., Ross, M., and Sclater, J. G., 1988, Age and spreading history of the Cayman Trough as determined from depth, heat flow, and magnetic anomalies: Journal of Geophysical Research, v. 93, p. 2141–2157.

Russo, R. M., and Villaseñor, A., 1995, The 1946 Hispaniola earthquake and the tectonics of the North America–Caribbean plate boundary zone, northeastern Hispaniola: Journal of Geophysical Research, v. 100, p. 6265–6280.

Russo, R. M., and Villaseñor, A., 1997, The 1946 Hispaniola earthquake and the tectonics of the North America–Caribbean plate boundary zone, northeastern Hispaniola, Reply: Journal of Geophysical Research, v. 102, p. 793–802.

Schell, B. A., and Tarr, A. C., 1978, Plate tectonics of the northeastern Caribbean Sea region: Geologie En Mijnbouw, v. 57, p. 319–324.

Sheridan, R. E., Mullins, H. T., Austin, J. A., Ball, M. M., and Ladd, J. W., 1988, Geology and geophysics of the Bahamas: *in* Sheridan, J. E., and Grow, J. A., eds., The Atlantic continental margin, U.S., Boulder, Colorado, Geological Society of America, v. I–2, p. 329–362.

Sieglie, G. A., Froelich, P. N., and Pilkey, O. H., 1976, Deep-sea sediments of Navidad Basin: Corellation of sand layers: Deep-Sea Research, v. 23, p. 89–101.

Speed, R. C., and Larue, D. K., 1991, Extension and transtension in the plate boundary zone of the northeastern Caribbean: Geophysical Research Letters, v. 18, p. 573–576.

Stein, S., DeMets, C., Gordon, R. G., Brodholt, J., Argus, D., Engeln, J. F., Lundgren, P., Stein, C., Wiens, D. A., and Woods, D. F., 1988, A test of alternative Caribbean Plate relative motion models: Journal of Geophysical Research, v. 93, p. 3041–3050.

Sykes, L. R., and Ewing, M., 1965, The seismicity of the Caribbean region: Journal of Geophysical Research, v. 70, p. 5065–5074.

Sykes, L. R., and McCann, W. R., and Kafka, A. L., 1982, Motion of the Caribbean plate during the last 7 million years and implications for earlier Cenozoic movements: Journal of Geophysical Research, v. 87, p. 10656–10676.

Uchupi, E., Milliman, J. D., Luyendyk, B. P., Bowin, C. O., and Emery, K. O., 1971, Structure and origin of the southeastern Bahamas: American Association of Petroleum Geologists Bulletin, v. 55, p. 687–704.

Wanless, 1979, H. R., Tedesco, L. P., Rossinsky, V., and Dravis, J. J., 1989, Carbonate environments and sequences of the Caicos Platform: International Geological Congress, 28th, Field Trip Guidebook T374: Washington, D.C., American Geophysical Union, 75 p.

Weaver, J. D., Smith, A. L., and Sieglie, G. A., 1975, Geology and tectonics of Mona Passage: Eos (Transactions, American Geophysical Union), v. 56, p. 451–452.

Manuscript Accepted by the Society January 9, 1998

Printed in U.S.A.

Geological Society of America
Special Paper 326
1998

Tectonic geomorphology and paleoseismology of the Septentrional fault system, Dominican Republic

Paul Mann*
Institute for Geophysics, University of Texas, 4412 Spicewood Springs Road, Bldg. 600, Austin, Texas 78759-8500
Carol S. Prentice
U.S. Geological Survey, MS 977, 345 Middlefield Road, Menlo Park, California 94025
G. Burr
National Science Foundation Accelerator Facility, PAS Building #81, University of Arizona, Tucson, Arizona 85721
Luis R. Peña
Avenida Cuesta Colorado Esquina Calle no. 9, Las Colinas, Santiago, Dominican Republic
F. W. Taylor
Institute for Geophysics, University of Texas, 4412 Spicewood Springs Road, Bldg. 600, Austin, Texas 78759-8500

ABSTRACT

The Septentrional fault system accommodates part of the active left-lateral strike-slip motion between the North America and Caribbean plates in northern Hispaniola (Dominican Republic). The Septentrional fault system and the Motagua fault zone of northern Central America to the west are the largest subaerial exposures of the 3,200-km-long North America–Caribbean plate boundary zone. We have delineated the main active trace of the Septentrional fault system in the Cibao Valley of the Dominican Republic based on geomorphic expression of the fault interpreted from a SEASAT radar image, aerial photographs, and field reconnaissance. For descriptive purposes, we subdivide the Septentrional fault system into eastern, central, and western map areas. In the eastern area along the southeastern Samana Peninsula, the submarine extension of the Septentrional fault zone lies about 2 km south of a prominent mountain-front fault along the highlands of the Samana Peninsula. This submarine fault aligns with a 30-km-long lineament on SEASAT radar imagery in the alluvial area west of the Samana Peninsula. We presume that this lineament and its mapped offshore extension represent the active trace of the fault zone.

In the central area the 100-km-long fault trace is marked by a well-defined linear scarp in alluvium that ranges in relief from 1.1 to 11.3 m and changes facing directions at intervals of 4–10 km. This trace is located about 2–3 km south of the topographic mountain front of the Cordillera Septentrional that we interpret as an inactive trace of the strike-slip plate boundary. Geomorphic fault-related features along the active fault zone in the valley include left-laterally deflected streams and stream terrace risers, scarps in alluvium, and sag ponds. At two sites, measured lateral offsets on stream terrace risers range from 40 ± 10 m at Rio Juan Lopez to 55 ± 10 m at Rio Licey and allow calculation of two independent, maximum slip rates for the Septentrional fault zone of 13 ± 4 and 23 ± 7 mm/yr, respectively.

*email:paulm@utig.ig.utexas.edu

Mann, P., Prentice, C. S., Burr, G., Peña, L. R., and Taylor, F. W., 1998, Tectonic geomorphology and paleoseismology of the Septentrional fault system, Dominican Republic, *in* Dolan, J. F., and Mann, P., eds., Active Strike-Slip and Collisional Tectonics of the Northern Caribbean Plate Boundary Zone: Boulder, Colorado, Geological Society of America Special Paper 326.

In the western area the Septentrional fault system exhibits two traces: (1) The 90-km-long southern branch (named here the Septentrional fault zone) trends 100° across low-lying areas of the west-central Cibao Valley; this fault is largely obscured because of recent fluvial sedimentation and erosion, but we propose that it represents the active, western continuation of the fault zone; (2) the 70-km-long northern branches (named here the mountain front, Villa Vasquez, and Monte Cristi faults) strike 110° and are both well exposed and obscured by mountain-front alluvial fans and pediment surfaces. The Monte Cristi and Villa Vasquez faults are well exposed in the westernmost area, but exhibit no evidence for late Quaternary activity. Geomorphic fault-related features, which are particularly well displayed along the Villa Vasquez and Monte Cristi fault zones, include a 12-km-long push-up at a 1.5-km compressional stepover in the fault trace, prominent scarps juxtaposing different sedimentary rock types, and offset streams.

Trenches excavated in late Holocene alluvial fan sediments at two sites, 91-1 and 91-2, on the central segment of the Septentrional fault zone revealed faulted and folded deposits that provide evidence for the nature of faulting and the timing of the most recent ground-rupturing earthquake at these localities. Trench 91-2 is sited on the fault where it is completely buried by young, well-stratified sediments of the Cenoví alluvial fan. This trench provides critical data used to constrain the minimum age of the most recent surface rupture. Radiocarbon analyses of 32 samples collected from these sediments in trench 91-2 show that this most recent surface rupture occurred before A.D. 1260 (i.e., there has been no surface faulting at site 91-2 since about A.D. 1260, more than 730 yr ago). Trench 91-1 is sited on a low scarp across the fan and provided no information to constrain the age of the last groundbreaking event. Both trenches revealed vertical and horizontal offset of gravel units within finer grained stratified units. The absence of deformation in either trench correlatable with historical earthquakes felt in the Cibao Valley in 1564, 1783, and 1842 suggests that some or all of these events may have originated on either offshore thrust or strike-slip faults of the north Hispaniola deformed belt, or the western extension of Septentrional fault zone in the Cibao Valley and offshore along the northern coast of Haiti (Calais and Mercier de Lépinay, 1995).

We estimate that the Septentrional fault rate is at least 5 mm/year, based on the lowest estimate of total plate-boundary slip rate of 9 mm/year by DeMets et al. (1990) and the assumption that about half of the interplate motion could be accommodated by other subparallel strike-slip faults in Hispaniola like the Enriquillo-Plantain Garden fault zone and faults within the north Hispaniola deformed belt. Based on these two assumptions, a slip deficit of more than 3.6 m of potential slip has accumulated at the 5 mm/yr rate in the more than 730 yr since the last ground-rupturing earthquake. Because this estimate is based on a minimum estimated slip rate, the amount of accumulated slip and the implied seismic hazard may be significantly higher. These paleoseismic data, along with the stream and terrace offset data, support the earlier finding of Prentice et al. (1993) that the Septentrional fault zone is a source of high seismic potential in this densely populated region.

1. INTRODUCTION

Seismic hazards of the North America–Caribbean plate boundary zone

Geologic and marine geophysical studies have established that the northern boundary of the Caribbean plate is a seismically active zone of left-lateral strike-slip faults and thrust faults that extends westward over 3,200 km from the Lesser Antilles island arc to the Middle America arc (Fig. 1A). The North America plate slides westward past the much smaller Caribbean plate at a rate of about 20 mm/yr (cf. Mann et al., 1990, for review of rates and offsets of all strike-slip faults within the plate boundary zone) (Fig. 1B). Only two subaerial exposures of this

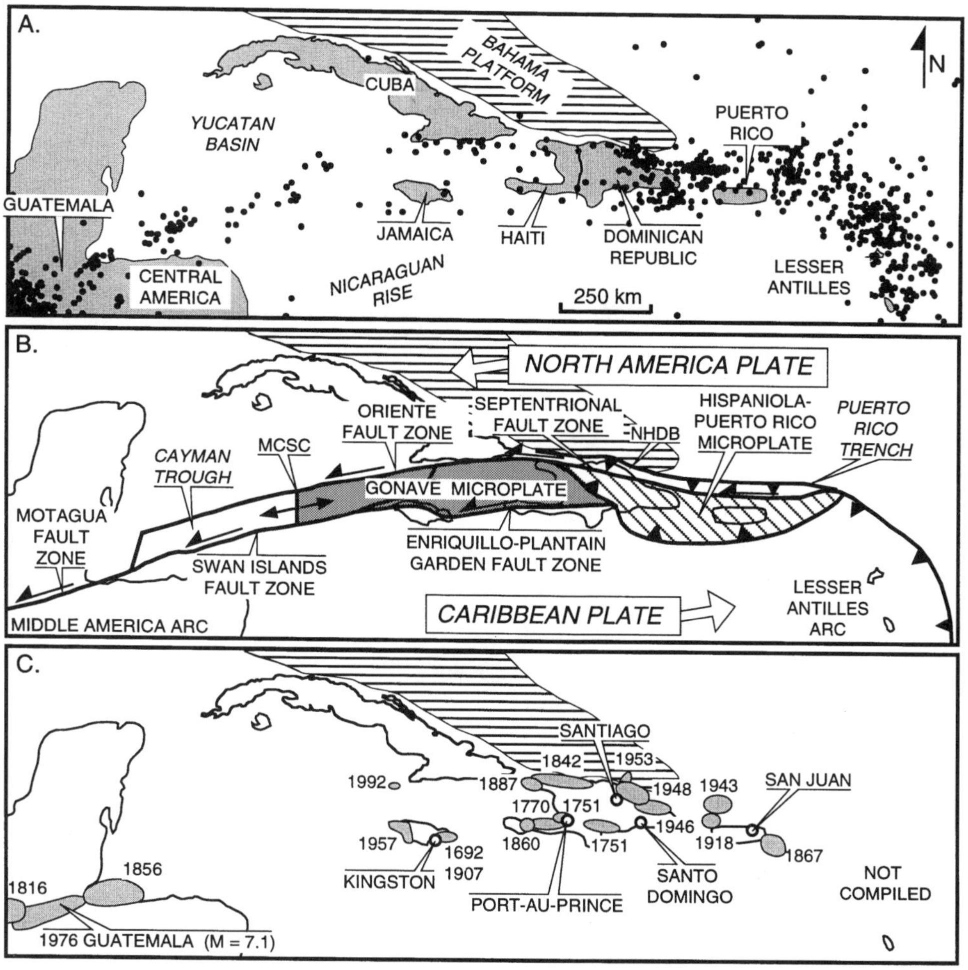

Figure 1. (A) Location map and seismicity (1963–1992, M > 4.5, source: International Seismic Commission) of the North America–Caribbean plate boundary zone. (B) Tectonic map modified from Mann et al. (1995) of the northern Caribbean showing microplates, major strike-slip fault zones, and directions of North America and Caribbean plates. MCSC = Mid-Cayman spreading center; NHDB = northern Hispaniola deformed belt. (C) Felt zones of large northern Caribbean historical earthquakes modified from Kelleher et al. (1973); Robinson et al. (1958); Plafker (1976); Osiecki (1981); and White (1985). Note dual zone of historical earthquakes in the northeastern Caribbean that corresponds to vague belts of earthquake epicenters in A. The 1976 Guatemala earthquake has been the most devastating historical rupture of the northern Caribbean plate boundary zone in the 20th century and resulted in the loss of 22,780 lives.

strike-slip fault system occur along this mainly submarine, strike-slip plate boundary zone: the Motagua fault zone of northern Central America (Guatemala) and the Septentrional fault system of northern Hispaniola (Dominican Republic) (Fig. 1B).

Relevance of 1976 earthquake on the Motagua fault, Guatemala, to this study. The Motagua fault zone is a better studied segment of the North America–Caribbean plate boundary zone than the Septentrional fault system because a 230-km-long segment of the Motagua fault zone ruptured in 1976 with about 1 m of left-lateral slip. This catastrophic M 7.1 earthquake took the lives of 22,780 Guatemalans and left more than one million homeless in a country with a total 1976 population of about 5.5 million people (Plafker, 1976; Fig. 1C). Because of the lack of geologic and paleoseismologic information for the Motagua fault zone, government and disaster relief officials were unprepared for a large earthquake on this fault. Geologists who had studied the Mesozoic and early Cenozoic geologic and tectonic history of Guatemala viewed the prominent valley of the Motagua fault zone as an extinct convergent suture zone inactive since Late Cretaceous time (e.g., Donnelly et al., 1968). For this reason, the low relief (<2 m high), late Quaternary fault scarp within late Holocene alluvial and fluvial deposits of the Motagua Valley was not recognized as a major fault within the plate boundary zone or studied in detail until after the 1976 disaster had occurred (Plafker, 1976; Bucknam et al., 1978; Schwartz et al., 1979).

Seismic potential of the strike-slip plate boundary in Hispaniola and Puerto Rico. As was the case in Guatemala, a large earthquake generated by displacement on the Septentrional fault system of the North America–Caribbean strike-slip plate boundary in the Puerto Rico–Hispaniola area would severely affect this rapidly developing, densely populated region (population density of the Dominican Republic is 151 people/km^2; Puerto Rico is 392 people/km^2) (Fig. 1B). The city of Santiago, Dominican Republic (present population: 705,000), is situated 10 km south of the Septentrional fault zone, and the coastal city of San Juan, Puerto Rico (population: 437,000), is within 50–125 km south of known and inferred active strike-slip faults within this plate boundary zone. One of the known strike-slip faults northwest of Puerto Rico includes the eastern, submarine extension of the Septentrional fault zone mapped by Dolan et al. (this volume; Grindlay et al., 1997). This fault has been studied in the offshore Puerto Rico area by Grindlay et al. (1997) where they have named it the South Puerto Rico Slope fault and have shown that this fault appears active.

Regions of strong shaking associated with historical earthquakes in A.D. 1564, 1783, 1842, 1887, and 1897 in Hispaniola coincide closely with the Septentrional–South Puerto Rico Slope fault system and suggest that some of these events may be related to movements along these faults (Fig. 1C). For example, the 1842 event along the north coast of Haiti and Dominican Republic (Scherer, 1912) was possibly related to movement on the Septentrional fault zone in the Cibao Valley and its westward offshore extension. In addition to seismic shaking, destructive tsunamis were associated with several historical earthquakes that affected the northeastern Caribbean. A tsunami associated with the 1842 northern Hispaniola earthquake (Scherer, 1912) was possibly caused by movement along the western, offshore extension of the Septentrional fault zone mapped by Calais and Mercier de Lépinay (1995). Tsunamis were also associated with the 1887 (Scherer, 1912) and 1946 (Dolan and Wald, this volume) earthquakes along the north coast of Hispaniola as well as the 1867 earthquake in the Puerto Rico area (McCann, 1985) (Fig. 1C).

Thrust and strike-slip partitioning of plate motion in Hispaniola

A complete understanding of the seismic hazard and active tectonics of the Septentrional fault requires study of how oblique-slip plate motion along this part of the North America–Caribbean plate boundary may be partitioned into thrust and strike-slip faulting. Teleseismically recorded earthquakes (Fig. 1A) and the felt areas of historical earthquakes (Kelleher et al., 1973; McCann et al., 1979; Fig. 1C) suggest that the plate boundary zone in the northeastern Caribbean region is wider than the seismogenic zone to the west near Central America. Mapping in Hispaniola has shown that late Neogene thrust faults, strike-slip faults, and even normal faults have accommodated this plate motion across a diffuse zone (Mann et al., 1991). Sykes et al. (1982) and Mann et al. (1995) attributed the wider plate boundary zone in the northeastern Caribbean region to the existence of two, subparallel zones of active strike-slip faulting separated by about 125 km (Fig. 1B).

The northern, or Septentrional fault zone, is flanked to the north by the north Hispaniola deformed belt, an active fold-thrust belt (Fig. 1B). The mainly onshore strike-slip faults of the Septentrional fault zone accommodate the purely strike-slip component of deformation while offshore faults of the north Hispaniola deformed belt mainly accommodate the convergent part of the transpressional North America–Caribbean plate motion (Calais and Mercier de Lépinay, 1995; Dolan et al., this volume). Because both strike-slip and thrust zones are active, subparallel, and generally separated by less than 100 km, it is difficult to distinguish large, preinstrumental earthquakes on the Septentrional fault system from earthquakes generated by the north Hispaniola deformed belt (Fig. 1C).

The association of the 1946 earthquake in northern Hispaniola with a tsunami (Lynch and Bodle, 1948; Dolan et al., this volume) supports the idea that some large historical earthquakes in this region are produced by offshore faulting in the north Hispaniola deformed belt. Russo and Villaseñor (1995) have challenged the view that the large earthquake series in the 1940s represents slip on shallowly dipping, offshore thrusts and propose an alternative interpretation that the earthquakes represent left-lateral strike-slip displacement attributed to deep crustal motion on the subvertical Septentrional fault zone near Samana Bay. Detailed, paleoseismic study of the subaerial Septentrional fault zone provides a direct test of this hypothesis. For example, a long period of quiescence on the Septentrional fault zone would indicate that the historical seismic events are related to other, possibly offshore, faults.

Objectives and significance of this study

The objective of this chapter is to present a detailed description of the tectonic and geologic setting and tectonic geomorphology and paleoseismology of the left-lateral Septentrional fault system exposed on northern Hispaniola (Fig. 2). Because there is virtually no previous literature on this important plate boundary structure, we include abundant documentation of the Quaternary fault traces using maps, aerial photographs, and satellite imagery. This documentation may prove important to future workers as geomorphic fault features may become altered and obscured by human activities in this densely populated and rapidly urbanizing area.

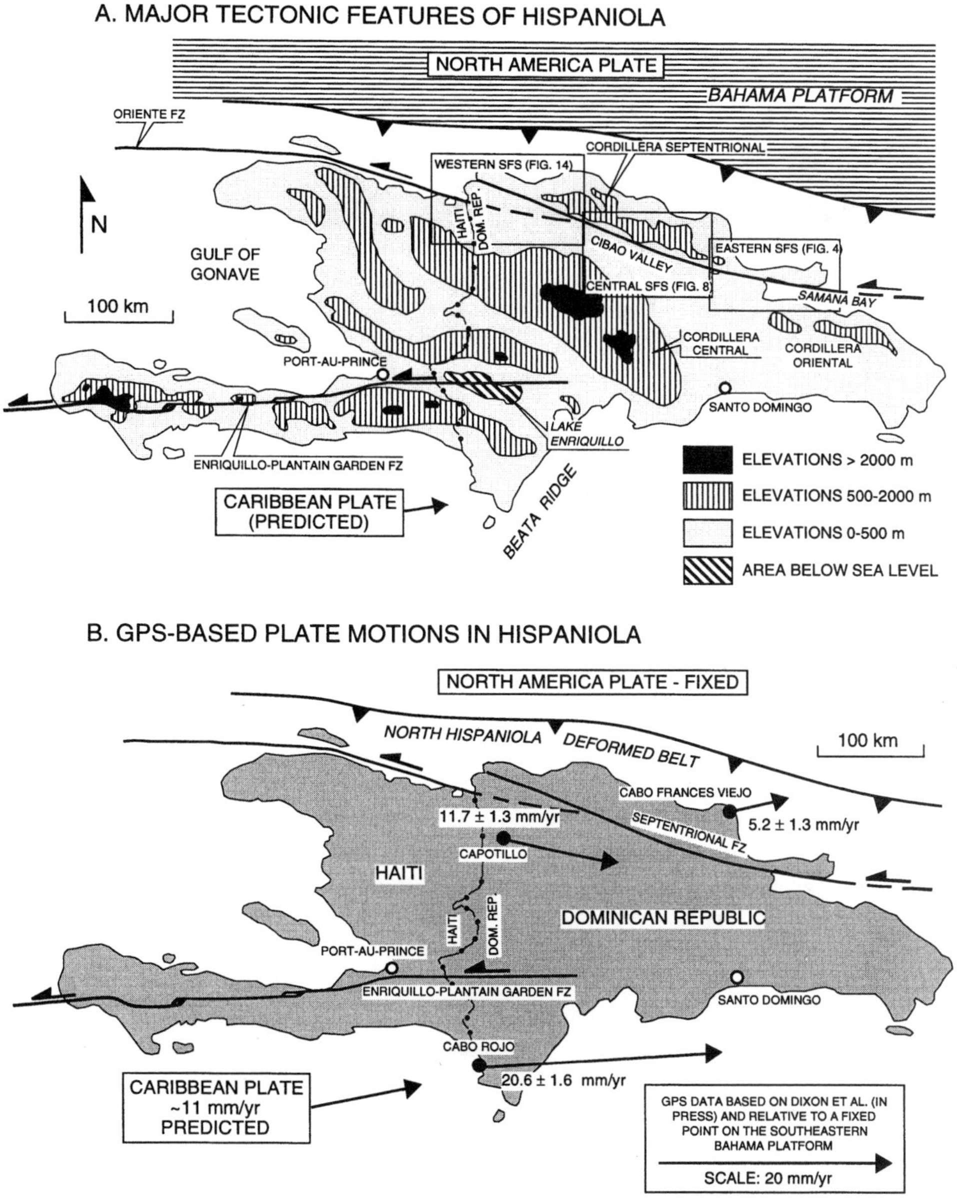

Figure 2. (A) Map showing major tectonic features of the island of Hispaniola (Dominican Republic and Haiti) and topographic expression of two zones of interplate strike-slip faults. The northern strike-slip fault, the Septentrional fault system (SFS), is the subject of this chapter. Boxes indicate locations of more detailed maps of the eastern, central, and western Septentrional fault system shown in Figures 4, 8, and 14. (B) The approximately east-west direction of North America–Caribbean relative plate motion predicted by the Nuvel plate model of DeMets et al. (1990) (shown in box) is slower and more eastward than GPS-based plate rates based on the velocities of the three sites shown relative to a fixed point on the North America plate on the southeastern Bahamas platform schematically shown by horizontally lined area in A (Dixon et al., 1998).

Description of the tectonic geomorphology of the subaerial part of the fault system is intended to complement descriptions of the subparallel, submarine fault system along the North Hispaniola deformed belt by Dolan et al. (this volume) (Fig. 2). In addition, Dolan et al. (this volume) and Grindlay et al. (1997) describe the morphology of the eastern, offshore continuation of the Septentrional fault zone in the area of the northern Mona Passage between the Dominican Republic and Puerto Rico and along the southern slope of the Puerto Rico trench. Calais et al. (this volume) and Dolan and Wald (this volume) describe the seismology of both the on- and offshore parts of the Septentrional fault system.

The submarine North Hispaniola deformed belt and the eastern and western submarine extensions of the Septentrional fault system are inaccessible for trenching, radiocarbon dating, and other paleoseismic methods developed in recent years for the study of subaerial faults (e.g., Sieh, 1981; Sieh and Jahns, 1984) (Fig. 1B). Field-based studies of the tectonic geomorphology and paleoseismology of subaerial faults of the North America–Caribbean plate boundary in northern Hispaniola are therefore critical to evaluation of seismic hazards associated with individual faults within this zone. Questions related to the tectonic geomorphology and paleoseismology of the Septentrional fault system that we address in this chapter include the following:

(1) What is the effect of wet-to-dry microclimatic variation in the Cibao Valley on geomorphic fault features found along the Septentrional fault system? Can variations in fault geomorphology for different sections of the fault be attributed to weathering variations related to contrasting microclimates, or are these geomorphic differences related to more fundamental differences in fault displacement history or the types of rocks juxtaposed by faulting?

(2) Why does the central part of the Septentrional fault system form such a straight and continuous fault scarp? Other active strike-slip faults, such as the San Andreas fault of California (Jennings, 1975) and the North Anatolian fault of Turkey (Barka and Kadinsky-Cade, 1988) exhibit more complex, less continuous, and multiple fault traces.

(3) Does the Septentrional fault system exhibit any geomorphic features indicative of segmented strike-slip boundaries, or does the entire length of the onland fault rupture during large earthquakes? Are any fault segment boundaries present that might determine points where fault rupture could start or stop during large earthquakes? Such structures in the Cibao Valley might include fault branches characterized by bifurcation and major strike changes, minor fault bends along continuous sections of the fault, transverse structures, and the hydrologic environment of the fault trace. This question of segmentation bears on the questions of "characteristic earthquakes," estimates of maximum magnitudes, stress drops, and prediction of strong ground motions (Schwartz and Coppersmith, 1984).

(4) What is the sense of movement, slip amount of major earthquakes, and recurrence interval between them for the Septentrional fault system? What are implications of the prehistoric record of earthquakes on the Septentrional fault system for the rate of strain accumulation and frequency of future earthquakes on the fault?

(5) How much contemporary North America–Caribbean plate motion is occurring on the Septentrional fault system? The total plate motion across this complex plate boundary zone integrates motion across the Septentrional fault; motion on active submarine thrust and possibly strike-slip faults within the North Hispaniola deformed belt offshore (Mullins et al., 1992; Dolan et al., this volume); and motion of the parallel, left-lateral Enriquillo-Plantain Garden fault zone to the south (Mann et al., 1995) (Fig. 2A).

(6) Are the historical earthquakes of A.D. 1564, 1783, 1842, 1887, and 1897 that produced strong shaking in the Cibao Valley (Scherer, 1912) related to strike-slip movement on the Septentrional fault system, or are they related to the thrust or strike-slip displacement on submarine faults in the North Hispaniola deformed belt? Historical reports of felt areas in coastal zones like those shown for Cibao Valley earthquakes in Figure 1C bias the inferred rupture area onshore when it could possibly be centered offshore.

2. ORGANIZATION OF THIS CHAPTER

This chapter is divided into 10 main sections. Sections 3 and 4 provide the geologic, tectonic, and climatic setting of the Septentrional fault zone. Section 5 provides a general description of the tectonic geomorphology of the entire fault trace using radar imagery and montages of aerial photographs. The overall description of the fault trace is divided into an arbitrarily defined eastern, central, and western area with the subdivision shown in Figure 2. Section 6 provides data documentation for recent movement (scarps, terrace riser, and stream offsets) from the central area using ground photographs and aerial photographs at scales ranging from 1:20,000 to 1:60,000 and presented as mosaics and stereopairs. Aerial photographs vary widely in quality of original 8 × 8 inch prints and vintage (1958 to 1983). Aerial photography is complemented by oblique views of the fault from a chartered, single-engine plane taken in January, 1993, and by ground views of the fault taken in January and February, 1991, and January and Februrary, 1993. Rod and transit profiles across the fault scarp were measured at 10 localities along the central Septentrional fault and three localities along the northern branch of the western Septentrional fault to quantify and compare slope angles. Section 7 presents evidence for the lack of late Quaternary movement on the northern branch of the western segment of the fault. Section 8 presents the results of a fault trench study carried out to determine the timing and size of the most recent earthquake along the central Septentrional fault zone. Preliminary results of this study were published by Prentice et al. (1993). All of the results presented in this chapter support the main conclusions of the Prentice et al. (1993) study. Section 9 presents a discussion of the results based on questions posed in section 1 and section 10 summarizes the main conclusions of the study.

3. GEOLOGIC AND TECTONIC SETTING OF STRIKE-SLIP FAULTS IN HISPANIOLA

Geologic setting of strike-slip faults in Hispaniola

Physiography of Hispaniola. The physiography of Hispaniola consists of northwest-trending, en echelon fault-bounded mountain ranges up to 3,000 m in elevation (Fig. 2A). Previous geologic mapping of Cretaceous and Cenozoic rocks (see Mann et al., 1991, and accompanying chapters) has shown that the more northwest-trending mountain fronts are bounded by reverse and thrust faults, the more east-trending mountain fronts are bounded by strike-slip faults, and the mountains themselves are large anticlinoriums exposing Mesozoic and lower Cenozoic rocks. The intervening valleys separating the ranges are mainly filled with thick deposits of clastic detritus produced during Neogene uplift and erosion of the ranges. Regional northeast-to-southwest shortening across the island has deformed the basins and their clastic sedimentary fill into large synclines bounded on their edges by reverse or thrust faults. These tectonically depressed basins are much lower than the surrounding mountain ranges. In the case of the southernmost valley, the Enriquillo, the surface of the valley floor is as much as 100 m below sea level while the surrounding mountains are more than 2,000 m in elevation (Mann et al., 1991) (Fig. 2).

The Cibao Valley occupies a synclinal lowland separating the Cordillera Septentrional ("Northern Range") from the Cordillera Central ("Central Range") in the Dominican Republic (Fig. 2A). Modern drainage from both the Cordillera Central in the south and Cordillera Septentrional in the north is longitudinal along the axis of the Cibao Valley. The eastern axial river (Rio Camú–Rio Yuna) empties into Samana Bay in the east while the western axial river (Rio Yaqué del Norte) empties into the Atlantic Ocean near the Haitian border.

Structure of the Cibao Valley. The Cibao Valley is underlain by a broad synclinal structure filled with 3,000–5,000 m of mainly Miocene to lower Pliocene sedimentary clastic rocks derived from erosion of Eocene and older island arc rocks and younger sedimentary cover. These older eroded sections are exposed in the Cordillera Septentrional and Cordillera Central ranges (Edgar, 1991; de Zoeten and Mann, 1991). A shelf ramp formed from carbonate-terrigenous sediments flanked the Cordillera Central along the southern edge of the Cibao basin during late Miocene and early Pliocene time (Evans, 1986). The former margin is well preserved and forms a prominent present-day plateau along the southern edge of the Cibao Valley (Saunders et al., 1986). Based on regional studies of micro- and macrofossils in basinal sedimentary rocks, Saunders et al. (1986) inferred that the Cibao basin was uplifted above sea level to form the present-day Cibao Valley in middle to late Pliocene time. Gently folded to flat-lying shallow-water, massive limestone of late Miocene–early Pliocene age has been uplifted to elevations ranging from 500 to 1,200 m on either side of the Cibao Valley. This uplift supports the idea of regionally extensive uplift beginning in middle to late Pliocene time (de Zoeten and Mann, 1991).

Septentrional fault zone. Previous mapping has shown that the most conspicuous strands of the Septentrional fault system form semicontinuous, linear faults that crop out over a distance of 320 km and typically form an abrupt break in slope between the highlands of the Cordillera Septentrional and the lowlands of the Cibao Valley (de Zoeten and Mann, 1991; Winslow et al., 1991; Calais et al., 1992b) (Fig. 2A). We suggest that this fault accommodates much of the interplate strike-slip motion in the area of northern Hispaniola because: (1) it forms the most prominent and most continuous lineament relative to nearby faults (Fig. 2); (2) it forms a scarp in Quaternary alluvium of the Cibao Valley over much of the length of the mountain front (Winslow et al., 1991; Calais et al., 1992b; Prentice et al., 1993); and (3) it is continuous with active strike-slip faults studied to the east in the Puerto Rico trench (Masson and Scanlon, 1991; Grindlay et al., 1997; Dolan et al., this volume) and to the west along the northern coast of Haiti (Calais and Mercier de Lépinay, 1995). Recent marine studies of the North Hispaniola deformed belt have shown evidence for active folds, but not long and continuous strike-slip faults like those of the Septentrional fault system or the Enriquillo-Plantain Garden fault zone (Dillon et al., 1992; Mullins et al., 1992; Dolan et al., this volume) (Fig. 2A).

Enriquillo-Plantain Garden fault zone. The Enriquillo-Plantain Garden fault zone is a second zone of active strike-slip faults that extends across the southern margin of Hispaniola and has been proposed as the boundary of a recently detached, elongate microplate within the North America–Caribbean interplate zone (Mann et al., 1991, 1995; Rosencrantz and Mann, 1991) (Fig. 1C). The eastern edge of the microplate between the two strike-slip fault systems in Hispaniola has been topographically uplifted and undergone complex internal deformation related to its east-west transpression in Neogene time. Near Lake Enriquillo, recent lateral offset along the Enriquillo-Plantain fault zone has been negligible, based on a well-dated Holocene coral reef that has been tectonically uplifted, but not laterally offset (Mann et al., 1995). One explanation for the lack of Holocene lateral offset on the fault at Lake Enriquillo is its proximity to the apparent termination of the fault about 50 km to the east (Fig. 2A).

Tectonic setting of strike-slip faults in Hispaniola

Plate motions derived from Global Positioning System (GPS) measurements determined by Dixon et al. (1998) at three sites in the Dominican Republic and one site assumed to represent North America on the southeastern Bahamas platform confirm roughly east-west (to east-northeast), strike-slip plate motions between the North America and Caribbean plates in Hispaniola at rates up to 20 mm/yr (Fig. 2B). Measurements at some sites were begun as early as 1986 (Dixon et al., 1991); other sites have been surveyed only in 1991 and 1994 or 1995. GPS-based velocity data in this and other areas are not comparable to geologically determined late Quaternary slip rates across individual strike-slip faults. Rather, GPS-based velocities provide a measure of the regional strain field produced by all active faults in the area. Elastic models of the strain fields around specific strike-slip faults can be modeled given certain assumptions concerning the locking depth of faults and the rake of movement along these fault surfaces (Dixon et al., 1998).

One site on the southern coast of Hispaniola south of the Enriquillo-Plantain Garden fault zone is moving about 20.6 ± 1.6 mm/yr relative to a fixed North America plate and is presumed by Dixon et al. (1998) to reflect the integrated motion of all faults in Hispaniola including the Enriquillo-Plantain Garden and Septentrional fault zones. The 20.6 ± 1.2 mm/yr rate and the direction of the southernmost site relative to North America are parallel to but almost twice as fast as that predicted by the Nuvel-1A North America–Caribbean plate motion model of DeMets et al. (1990) based on global analysis of plate pairs (Fig. 2B). Two sites within 50 km to the north and south of the Septentrional fault zone, respectively, are moving at rates of 5.2 ± 1.3 and 11.7 ± 1.3 mm/yr relative to North America and therefore measure only part of the total motion occurring between the North America and Caribbean plates.

4. CLIMATIC SETTING OF THE SEPTENTRIONAL FAULT SYSTEM

In Hispaniola diverse microclimates are produced by the interplay of topography and prevailing trade winds (García, 1976). Most moisture transported by northeasterly trade winds from the Atlantic Ocean falls as rain on the 1,000–3,000-m-high Cordillera Oriental, Cordillera Septentrional, and Cordillera Central (Fig. 3).

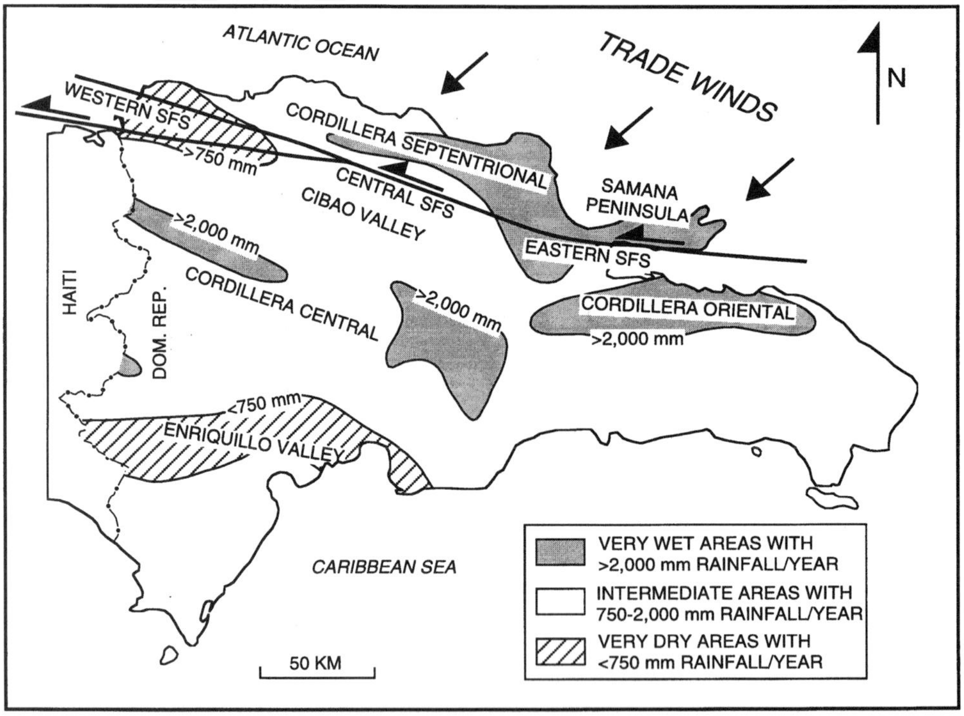

Figure 3. Amount of annual rainfall in the Dominican Republic (modified from figure 297 of Garcia, 1976). Northeasterly tradewinds off the Atlantic Ocean bring rainfall to the northeastern slopes of the Cordillera Septentrional, Cordillera Oriental, and Cordillera Central. Rain shadows form in adjacent Cibao and Enriquillo Valleys southwest of these mountains. Note that the western, central, and eastern Septentrional fault system (SFS) in the Cibao Valley occupies dry, intermediate, and very wet climatic zones.

The Cordillera Septentrional and Cordillera Central act as moisture barriers that produce rain shadows or arid microclimates in the Cibao and Enriquillo Valleys, respectively (Fig. 3). These two areas receive an average of ~750 mm of rain per year with most falling in the month of June (García, 1976). In contrast, the mountainous areas and the eastern Cibao Valley receive almost three times as much rain distributed more evenly throughout the year.

The effects of these microclimates are readily apparent as one moves from tropical flora in the area of the eastern Septentrional fault zone to an area of desert vegetation in the western Cibao Valley. An intermediate climate in the central Cibao Valley gives rise to moderate rainfall. These widely varying microclimates strongly influence the preservation of tectonic geomorphic features found along the length of the Septentrional fault system.

5. GENERAL DESCRIPTION OF THE SEPTENTRIONAL FAULT TRACE

Samana Bay and Peninsula

Geologic setting. The alluvium-covered eastern Cibao Valley (elevation 0–5 m) and shallow-marine Samana Bay (average water depth 6–10 m b.s.l.) forms a graben-like, 10–17-km-wide depression flanked by highlands of the Samana Peninsula to the north and highlands of the Cordillera Oriental to the south (Fig. 4). The highest peaks in both ranges are capped by uplifted and tilted to gently folded, 250-m-thick, upper Miocene–lower Pliocene shallow-water limestone of the Cañitas Formation (Joyce, 1991). Lithologic and age equivalent rocks to the Cañitas Formation are also found in the Cordillera Oriental (Cevicos Formation) and Cordillera Septentrional (Villa Trina Formation) (de Zoeten and Mann, 1991). These carbonate lithologies tend to be more resistant to weathering than the underlying clastic and metamorphic rocks (de Zoeten and Mann, 1991). Along the southern edge of the Samana Peninsula, the Cañitas Formation reaches elevations of 500 m, dips 10–15° to the north, and is heavily karstified (Joyce, 1991) (Fig. 5). In the Cordillera Oriental, the limestone occurs at lower elevations up to 230 m, has subhorizontal dips, and is also heavily karstified (Fig. 4).

GPS-based crustal motions. Recent GPS measurements made on the Cabo Frances Viejo Peninsula (Fig. 2B) indicate that northeastern Hispaniola within the plate boundary zone moves east-northeastward at a rate of 5.2 ± 1.3 mm/yr relative to a site fixed on the stable North America plate in the southeastern Bahamas north of Hispaniola (Dixon et al., 1998). This plate motion is slightly oblique to the Septentrional fault system suggesting transpressional deformation in the region of roughly west-northwesterly striking faults of the eastern and central Cibao Valley.

Offshore fault zones in Samana Bay. Edgar (1991) mapped or inferred four high-angle fault zones in Samana Bay using widely spaced single-channel seismic lines and deeper penetration, multichannel seismic lines. Because none of these offshore faults has been previously named, we have named the four fault zones he defined and show them on Figure 4.

Extension of mountain-front fault beneath Samana Bay. This fault is the northernmost fault inferred by Edgar (1991). It is about 3 km long and located ~1–2 km southeast of the town of Samana (Fig. 4). Although Edgar (1991) obtained seismic lines in this area, he states that this fault is not imaged on the seismic data but is inferred from the juxtaposition of elevated Cretaceous metamorphic rocks on Samana Peninsula (Joyce, 1991) with sediments overlying the inferred, subsided metamorphic basement beneath Samana Bay. We project the submarine extension of the mountain-front fault shown by Edgar (1991) to the area 1 km south of the town of Samana. In this area, we infer that the upthrown, northern side of the fault is marked by a string of small islets exposing southeast-dipping beds of Oligocene conglomerate composed mostly of metamorphic clasts (Joyce, 1991) (Fig. 4).

Our projection of the mountain-front fault intersects the coast about 3 km west of the city of Samana and is colinear with a prominent scarp separating coastal alluvium to the south from elevated Mio-Pliocene limestone and underlying Oligocene conglomerate and Cretaceous metamorphic rocks to the north (Fig. 5). Bedding in limestone and conglomerate dips about 50° northward away from the fault scarp. To the west of the area shown in Figure 5, the scarp separates only limestone from alluvium. The relief along the scarp is more subdued in the metamorphic and conglomeratic areas than in the limestone areas and varies from 150 to 200 m. The Samana Peninsula is one of the highest rainfall areas in the Dominican Republic (Fig. 3), and tropical weathering may contribute to topographic relief across this scarp.

Septentrional fault zone in Samana Bay. As in the case of the mountain-front fault zone, this fault, named here the Septentrional fault zone, was inferred by Edgar (1991) based on juxtaposition of metamorphic rocks onshore and younger, clastic sedimentary strata of presumed Oliogocene to Holocene age overlying acoustic basement beneath Samana Bay. The Septentrional fault zone south of the town of Samana is inferred to be downthrown to the north because of the presence of an elongate island, Cayo Levantado or "Elevated Key," south of its projected trace. As in the case of the mountain-front fault zone, the inferred offshore fault is colinear with an onshore lineament in the eastern Cibao Valley.

We propose that this lineament represents the active trace of the fault zone in this area of the eastern Cibao Valley (Dixon and Daily, 1981) (Fig. 6). It is interesting to note that this relationship between an older, prominent mountain-front fault and a younger, more subtle trace about 1 km to the south in Quaternary sediments is identical to the relationship we describe in a following subsection for the better exposed and studied fault traces of the mountain-front and Septentrional fault zones in the

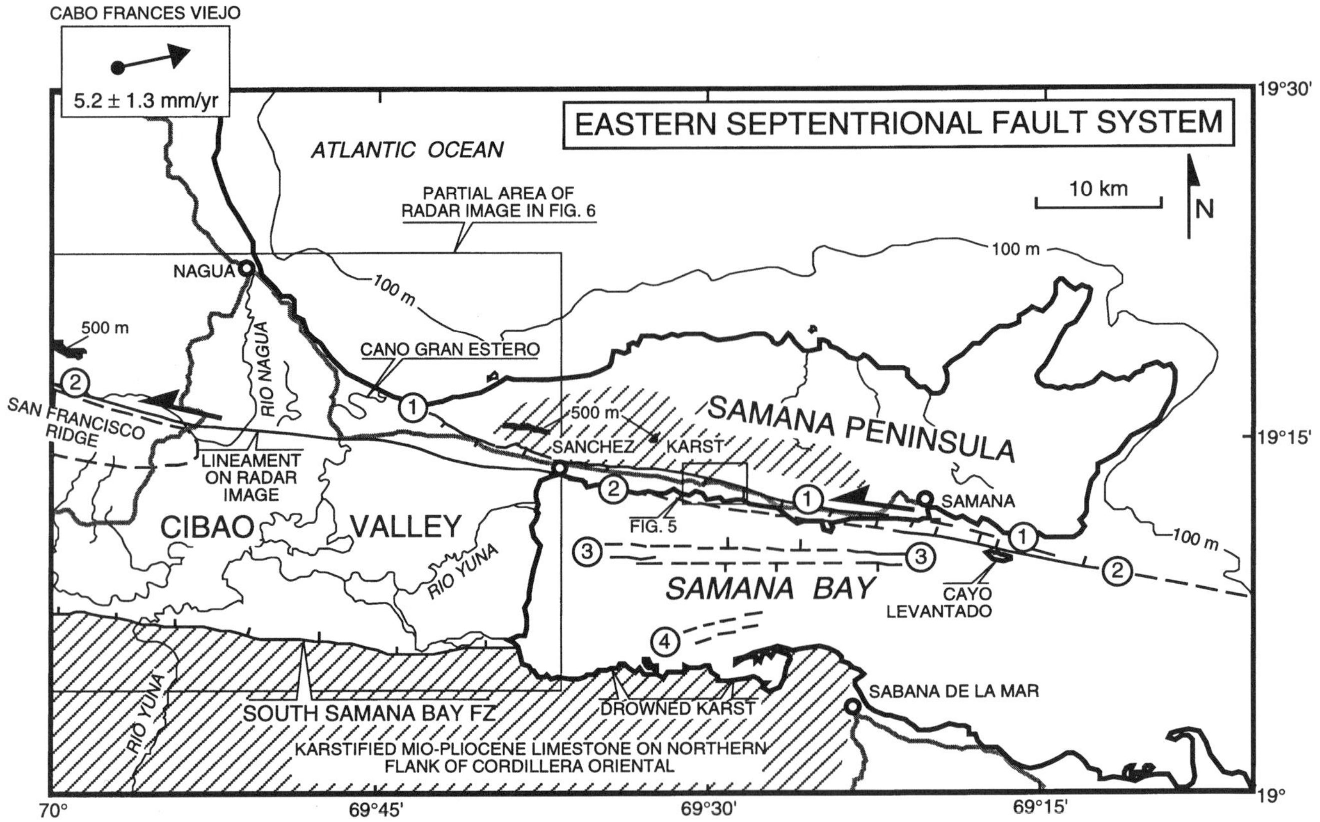

Figure 4. Map of late Quaternary fault scarps in the area of the eastern Septentrional fault system based on interpretation of SEASAT radar imagery, conventional aerial photographs taken in 1967 (1:20,000 scale), and field observations. Boxes show locations of montage of aerial photographs in Figure 5 and SEASAT radar image shown in Figure 6. Diagonal lines indicate outcrop area of karstified Mio-Pliocene shallow water limestone now at elevations up to 500 m a.s.l. (gray areas). Submarine fault zones 1–4 shown in Samana Bay and named in this chapter are taken from Edgar et al. (1991): 1 = Extension of mountain-front fault; 2 = Septentrional fault zone; 3 = Central Samana Bay fault zone; 4 = South-central Samana Bay fault zone. Onland scarps of the mountain-front fault and Septentrional fault zones are taken from radar imagery. The Septentrional fault zone, considered here to be an active fault trace, is more subtle and defines the coastline and forms a faint lineament. This same relationship between an inactive mountain-front fault and a more active fault within the lower lying basinal area is seen in the central Cibao Valley. Gray lines represent paved two-lane roads. Inset shows GPS-based rate and direction of relative motion of a point north of the eastern Septentrional fault system relative to a fixed point on the North American plate on the southeastern Bahamas platform (Dixon et al., in press, cf. Fig. 2B).

central Cibao Valley. We are unable to find any onshore, geomorphic expression of the active trace of the Septentrional fault zone in the coastal alluvium, where the fault projects onshore for about 5 km west of the town of Samana (Fig. 4). Dolan et al. (this volume; Grindlay et al., 1997) have mapped the colinear, eastern extension of the offshore active fault trace using marine geophysical techniques in the deeper water area to the northwest of Puerto Rico.

Central Samana Bay fault zone. This fault was inferred by Winslow et al. (1991) to project westward to form the southern edge of a right-stepping, compressional push-up known as San Francisco ridge (Fig. 6). However, the single-channel seismic lines of Edgar (1991) indicate that the Central Samana Bay fault zone does not reach the upper 20 ms (two way time) of reflectors (about 16 m) that he infers to be Holocene in age. Therefore, this fault may not have been active for the last 10,000 yr.

Edgar (1991) presents single-channel seismic reflection records showing folds in well-layered reflectors beneath the undisturbed Holocene? horizon. He considers these folded rocks to be of Plio-Pleistocene age over most of his survey area in Samana Bay. Some of this folding is possibly related to motion along the central Samana Bay fault zone and may be coeval to late Neogene push-up–related deformation proposed by Winslow et al. (1991) for the San Francisco ridge.

South-central Samana Bay fault zone. This fault near the southern edge of Samana Bay has a more northeasterly strike than the strikes of faults to the north (Edgar, 1991). This zone may merge with the scarp of the south Samana Bay fault zone

Figure 5. Montage of two 1967 aerial photographs showing the mountain front north of Samana Bay (aerial photographs: DR-B67-V65-19-14, January, 1968, nos. 3795 and 3796; original scale, 1:20,000). We interpret the prominent, embayed scarp as an inactive mountain-front fault lying about 2 km north of the active submarine Septentrional fault inferred from previous marine surveys (Edgar, 1991). A similar relationship between an inactive mountain-front fault and an active basinward scarp is observed along the central Septentrional fault. The scarp bounding late Miocene–early Pliocene shallow-water limestone is higher and sharper than the scarp bounding the more easily eroded Oligocene conglomerate.

A.

B.

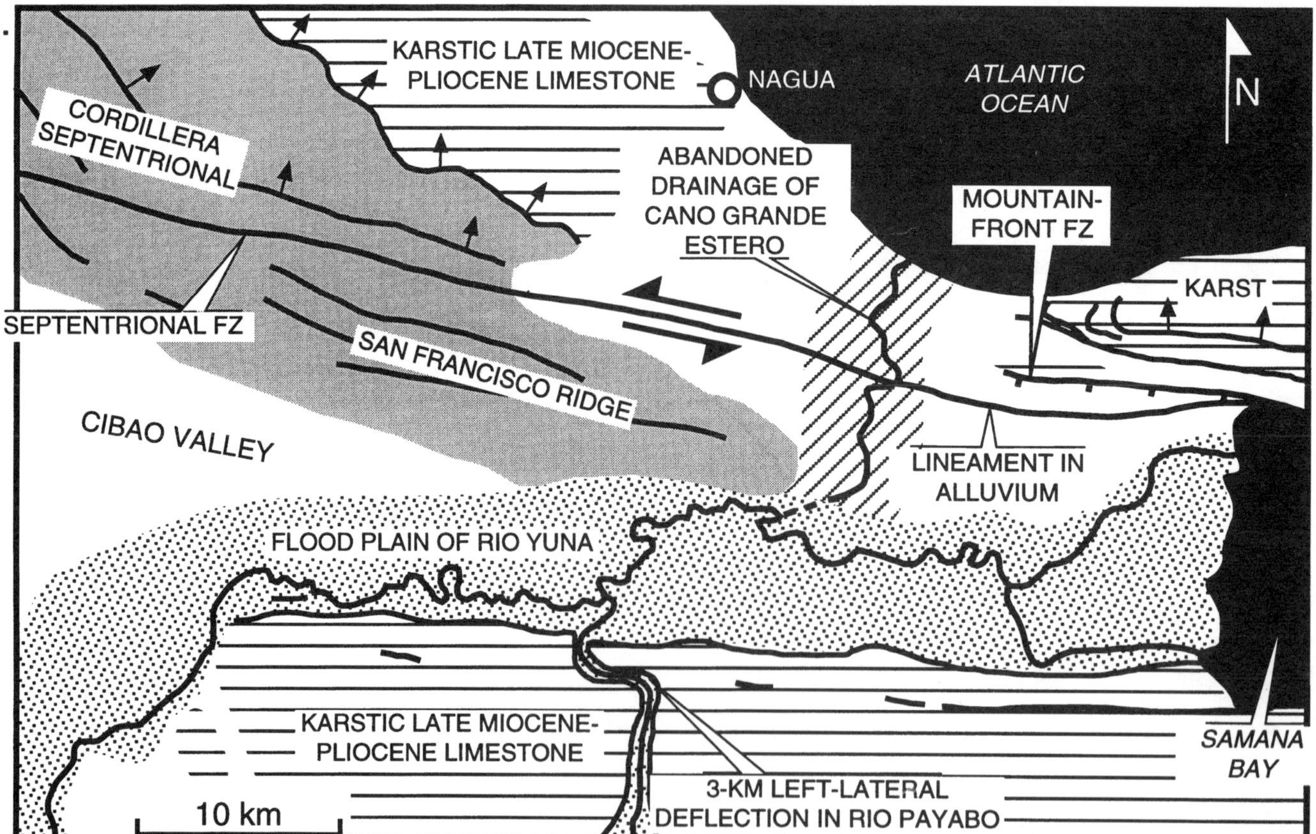
KARSTIC LATE MIOCENE-PLIOCENE LIMESTONE
NAGUA
ATLANTIC OCEAN
N
CORDILLERA SEPTENTRIONAL
ABANDONED DRAINAGE OF CANO GRANDE ESTERO
MOUNTAIN-FRONT FZ
KARST
SEPTENTRIONAL FZ
SAN FRANCISCO RIDGE
CIBAO VALLEY
LINEAMENT IN ALLUVIUM
FLOOD PLAIN OF RIO YUNA
KARSTIC LATE MIOCENE-PLIOCENE LIMESTONE
SAMANA BAY
3-KM LEFT-LATERAL DEFLECTION IN RIO PAYABO
10 km

that defines the edge of karstified Mio-Pliocene limestone along the southern edge of Samana Bay. East of Samana Bay, the south Samana Bay fault zone is covered by Quaternary alluvium in the Cibao Valley and is therefore believed to be inactive.

Onshore strike-slip faults in the eastern Cibao Valley. SEASAT radar imagery reveals a faint lineament in the area of low-lying alluvium separating the Samana Peninsula from the Cordillera Septentrional (Fig. 6). The lineament is slightly sinusoidal with a wavelength of about 20 km. The lineament projects eastward to the Septentrional fault zone in Samana Bay and westward to the sharp fault valley of the Septentrional fault in the area between the Cordillera Septentrional and the San Francisco ridge (Fig. 4). The Rio Nagua parallels the fault trace north of San Francisco ridge and accentuates its physiographic expression on the radar image (Fig. 6).

The low-lying area where the lineament is observed between the Cordillera Septentrional and Samana Peninsula is characterized by level, marshy ground that is a few meters above sea level. Three drainages are present in this area: (1) The Rio Nagua drains through the western part of the area near the San Francisco ridge and empties into the Atlantic Ocean near Nagua; (2) The active Rio Yuna, which is the major river system of the eastern Cibao Valley and currently drains into Samana Bay; it probably once flowed through the Caño Gran Estero to the Atlantic Ocean; (3) an isolated channel (Caño Gran Estero—"Stream of the Big Prairie") about 5 km long and east of the present channel of the Rio Nagua may represent an abandoned channel of the Rio Yuna (Fig. 4). The Caño Gran Estero now consists of elongated ponds and swamps in the topographically-lowest part of the former stream channel and is visible at the scale of the radar image (Fig. 6).

We carried out field checks to determine whether the lineament corresponds to a scarp and to investigate whether the abandoned, north-south–trending channel of the Caño Gran Estero was laterally offset by the east-west–striking active trace of the Septentrional fault (Fig. 4). No fault scarps were found along the lineament. Many of the areas where the scarp is present on the 1978 radar image are now cultivated so it is possible that a scarp could have been obscured by the extensive cultivation and canal building in this region. Abandoned, meandering river channels such as Caño Gran Estero shown on Figure 4 show no offsets but intriguing truncations and deflections are seen on aerial photographs near the projected trace of the lineament. Although we considered exploratory trenching at this site, we concluded that trenching would not be feasible because the water table is at or near the ground surface.

Onland south Samana Bay fault zone. The radar image reveals a prominent 3-km left-lateral deflection of the Rio Yuna 1 km south of the main trace of the South Samana Bay fault zone along the south side of the eastern Cibao Valley (Dixon and Daily, 1981) (Fig. 6). As described in a later section, the western projection of the South Samana Bay fault zone across the central Cibao Valley is onlapped by Quaternary alluvium. We consider the onlapped western extension inactive in Holocene time. The South Samana Bay fault zone in the Rio Yuna area deserves more detailed field studies to determine its activity.

Historical maps of the Samana area. A compendium of European historical maps of Hispaniola compiled by Rodríguez-Demorizi (1979) includes maps of the Samana area for the period 1509 through the 19th century. Maps dated 1750 (by French cartographer M. Bellin) and 1787 (by French cartographer Chastenet Puisegur) show a marine strait separating the area of the present-day Samana Peninsula from the mainland (Fig. 7A–E). Maps before and after this period show the Samana Peninsula.

It is possible that the marine strait shown on the 1750 and 1787 maps corresponds to the now-abandoned Caño Gran Estero. The relative youth of the Caño Gran Estero channel, if formed during the late 18th century, would make the possibility of any perceptible offset by the Septentrional fault zone minimal.

Central area of the Cibao Valley

Two parallel scarps. As on the south of Samana Peninsula, two parallel fault scarps separated by about 1 km are present along the mountain front of the Cordillera Septentrional in the central area of the Cibao Valley (Fig. 8). The more northerly, more topographically prominent, inactive fault separates Cretaceous–lower Cenozoic bedrock of the Cordillera Septentrional from Neogene sedimentary rocks of the Cibao Valley. The southern, less prominent scarp displaces alluvium in the Cibao Valley. Both scarps are remarkably straight in the central area of the Cibao Valley. The southern scarp exhibits only one major discontinuity, a 1-km-wide extensional left-step north of the city of San Francisco de Macorís (Fig. 8). We here name the northern fault the "mountain-front fault" and the southern fault the "Septentrional fault zone." Collectively, both faults form the Septentrional fault system of the central Cibao Valley.

Plate motions. Recent GPS measurements made in the Cordillera Central near the Dominican-Haitian border (Fig. 2B) indicate that central Hispaniola within the plate boundary zone moves east-southeastward at a rate of 11.7 ± 1.3 mm/yr relative to a site fixed on the stable North America plate in the southeastern Bahamas north of Hispaniola (Dixon et al., 1998). This

Figure 6. (A) SEASAT radar image of the eastern Septentrional fault system. (B) Interpretation of SEASAT image in A. A faint lineament in alluvium separates the Samana Peninsula in top right from the San Francisco ridge and Cordillera Septentrional in top left. The present drainage in the eastern Cibao Valley flows into Samana Bay and follows the southern margin of the valley. A north-trending corridor of abandoned river meanders leading toward the Atlantic Ocean indicates a previous drainage pattern northward across the Septentrional fault zone. Studies of aerial photographs and field observations did not reveal any displacement on these late Holocene channels at the point where they are intersected by the lineament.

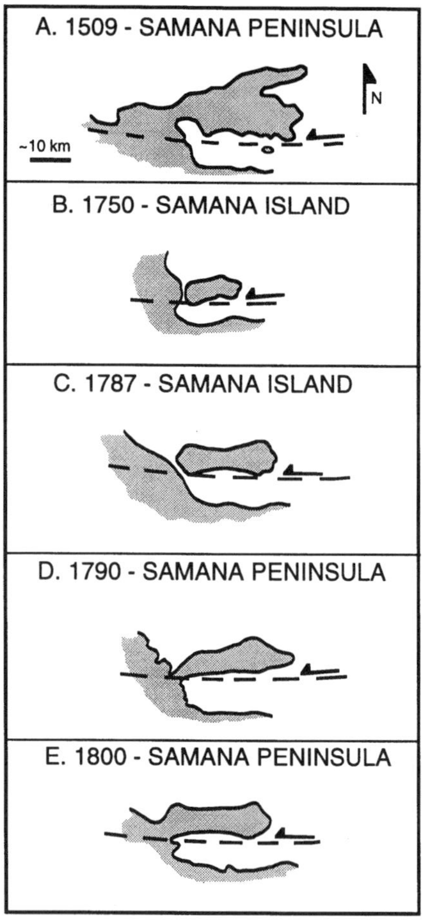

Figure 7. Compilation of historical maps for the period from 1509 to 1800 suggesting a short-lived seaway isolated the present-day Samana Peninsula from Hispaniola (from Rodríguez-Demorizi, 1979). Dashed line represents the approximate trace of the Septentrional fault zone. (A) Map by Andrés de Morales (1509) compiled by Rodríguez-Demorizi (1979) showing the Samana Peninsula. (B) Map by M. Bellin (1750) compiled by Rodríguez-Demorizi (1979) showing Samana Island. (C) Map by Chastenet Puisegur (1787) compiled by Rodríguez-Demorizi (1979) showing Samana Island. (D). Map by Logroño (1790) compiled by Rodríguez-Demorizi (1979) showing Samana Peninsula. (E) Map by J. Poirson (1800) compiled by Rodríguez-Demorizi (1979) showing Samana Peninsula. The lack of strike-slip offset on fluvial channels between Samana Peninsula and the mainland by the Septentrional fault zone may reflect their recent formation during the late 18th century.

plate motion is slightly oblique to the Septentrional fault system suggesting transpressional deformation in the region of roughly west-northwesterly striking faults of the central and western Cibao Valley.

Relation of fault scarps to restraining bend and uplift history. In contrast to the roughly east-west strike (098°) of both the mountain front and valley scarps in the Samana Bay area (Fig. 4), dual scarps in the central area curve markedly to the northwest in the area 10 km east of the town of Tenares where both scarps assume a strike of 111° (Fig. 8). Associated with this change of fault strikes in the central area is an abrupt increase in the elevation of the Cordillera Septentrional to elevations more than 1,000 m with an accompanying thickening up to 2,500 m in the adjacent late Neogene clastic basin fill of the Cibao Valley (Edgar et al., 1991). Mann et al. (1984), de Zoeten and Mann (1984), and J. Erikson (personal communication, 1995) attributed the increase in elevation of the fault-parallel Cordillera Septentrional and the deepening in the adjacent basin to the restraining bend character of this central, curved segment of the Septentrional fault zone. The Septentrional fault zone curves in a direction to place the adjacent crust under increased interplate convergence based on the east-southeast to east northeast–trending GPS-derived plate vectors (Dixon et al., 1998) (Fig. 2B).

Post–early Pliocene uplift at the restraining bend is documented by de Zoeten and Mann (1991) by regional variations in the elevation of shallow-water late Miocene–Pliocene limestone in the Cordillera Septentrional. Limestone elevations range from near sea level out of the bend area in the eastern Cibao Valley and northernmost Cordillera Septentrional to elevations more than 1,000 m within the area of maximum curvature of the Septentrional fault zone in the south-central Cordillera Septentrional (Fig. 8). Erosion of the limestone in the high rainfall area of the uplifted bend area in the Cordillera Septentrional (Fig. 3) has left isolated and conspicuous remnants of the carbonate cap along the topographic crest of the Cordillera Septentrional. These remnants of a once more extensive limestone cap along with northeast-dipping strata in the Cordillera Septentrional are apparent on the radar image shown in Figure 9. The relationships between the uplifted and deformed limestone cap, the older scarp bounding the mountain front, and the young scarp affecting Quaternary alluvium is well exposed when viewed from the south in the city of Santiago (Fig. 10). Southwestward dips of the limestone cap toward the Cibao Valley are the result of gentle folding of the carbonate cap in the post–early Pliocene restraining bend setting (de Zoeten and Mann, 1991).

Mountain-front fault. Although straight in overall morphology (Figs. 8 and 9), the mountain-front scarp of the Septentrional fault zone is strongly embayed and dissected at the scales of 1:20,000 and 1:40,000 aerial photographs. In the area northwest of Santiago, de Zoeten and Mann (1991) found the mountain-front fault to form a highly sheared contact between Paleogene sedimentary rocks of the Cordillera Septentrional and late Neogene sedimentary rocks of the Cibao Valley. In the area north of Moca and San Francisco de Macorís, the mountain-front fault exhibits a similar relationship to northeast-dipping Mio-Pliocene limestone (Fig. 9). In both the central and eastern areas, the more recent scarp is located 1–2 km to the south in late Quaternary sediments of the Cibao Valley. During this study, we found no evidence for late Holocene offset of the mountain-front scarp, but abundant evidence for late Holocene movement on the valley scarp.

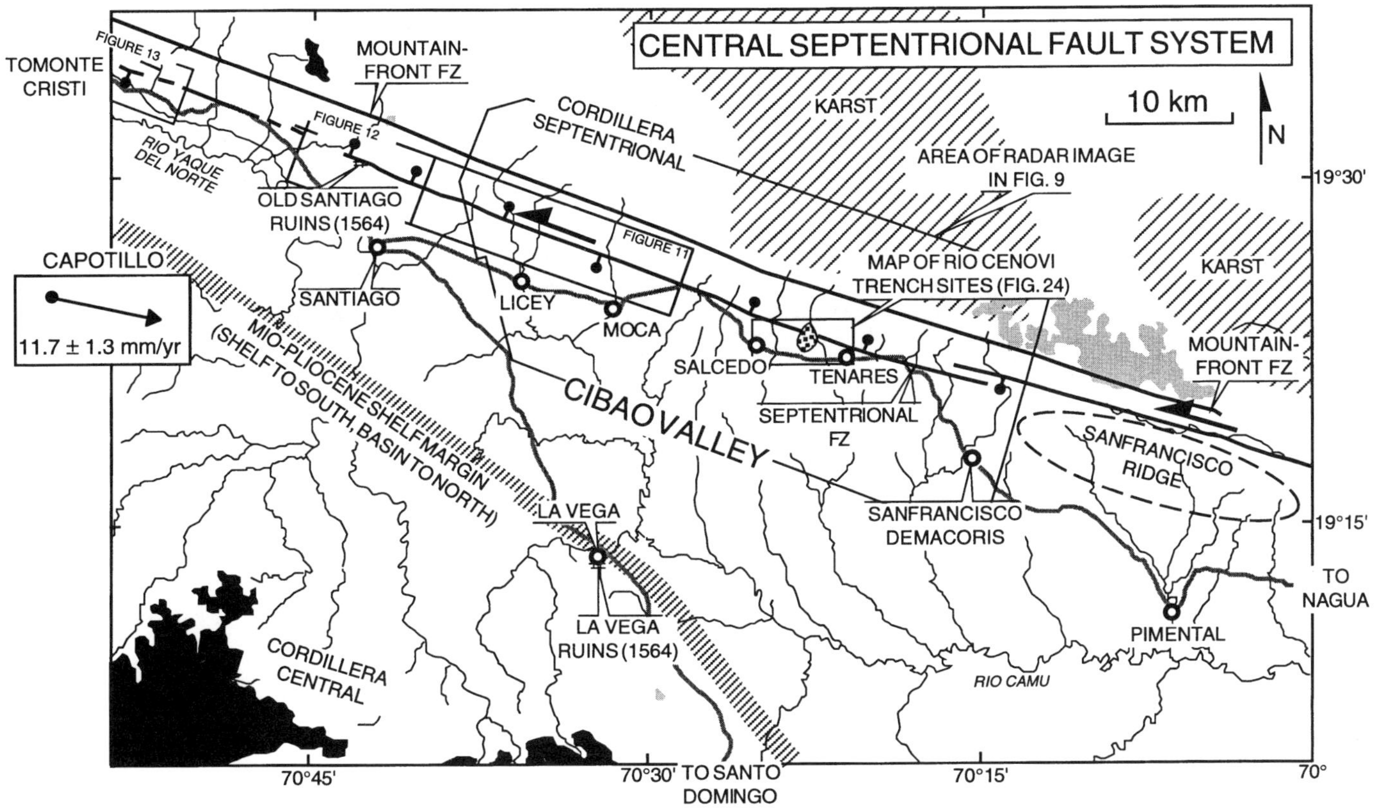

Figure 8. Map of late Quaternary fault traces along the central part of the Septentrional fault system based on interpretation of SEASAT radar imagery, conventional aerial photographs taken in 1967 (1:20,000 scale) and 1983 (1:40,000 scale), and field observations in 1991 and 1993 (cf. Fig. 2A for location). Gray areas represent topographic elevations between 500 m and 1,000 m; black areas are above 1,000 m. Note offset and deflected drainages near the fault scarp and tectonically controlled topographic fronts. Ball and ticks indicate downthrown side of Septentrional scarp. The towns of Old Santiago and La Vega were destroyed in a large earthquake in 1564. Boxes show locations of SEASAT radar image shown in Figure 9 and montages of aerial photographs shown in Figures 11 and 12. Gray lines represent paved roads. Inset shows GPS-based rate and direction of relative motion of a point at Capotillo in the western Dominican Republic relative to a point on the North American plate on the southeastern Bahamas platform (Dixon et al., 1998, cf. Fig. 2B).

Scarp in Quaternary alluvium of the Cibao Valley. The Septentrional fault zone in the central Cibao Valley is expressed in most areas as a north- or south-facing scarp in alluvium varying from about 1–11 m in height. A single, 1-km-wide, left-stepping extensional stepover in the fault trace is present 5 km east of Tenares, but this stepover lacks a sag pond (Fig. 8). In contrast to its simple and straight surface expression, multichannel seismic lines collected across the Septentrional fault zone in the Cibao Valley by the Weeks Petroleum Company, and published by Edgar (1991), reveal a wide zone of complex, subsurface flower structures affecting a 10-km-wide belt of rocks of late Neogene age.

Drainages in the Cibao Valley cross the fault at right angles (Fig. 8) in contrast to the area between the San Francisco ridge and the Cordillera Septentrional where major rivers and streams parallel the fault (Figs. 4, 8). Left-lateral deflections of these drainages are evident on SEASAT radar images of the central valley area (Fig. 9).

Rio Moca–Rio Licey area of the central Cibao Valley

Fault-related features interpreted from 1:40,000 scale aerial photographs of the inactive mountain-front fault and active Septentrional fault zone in the Rio Moca–Rio Licey area are shown on Figure 11. The main features visible at this scale are summarized on the inset map of Figure 11 and include the linear, inactive fault defining the topographic mountain front of the Cordillera Septentrional and the active trace of the Septentrional fault zone in late Quaternary alluvium of the Cibao Valley. Fault features are variably preserved in this rapidly developing area of small urban centers, small farms, and livestock pastures.

Drainages originating in the highlands of the Cordillera Septentrional show little or no deflection as expected at the inactive mountain-front fault, but deflections ranging from 0.1 to 0.5 km occur along the active fault valley. The scarp in the area of Figure 11 changes from north-facing in the east, to

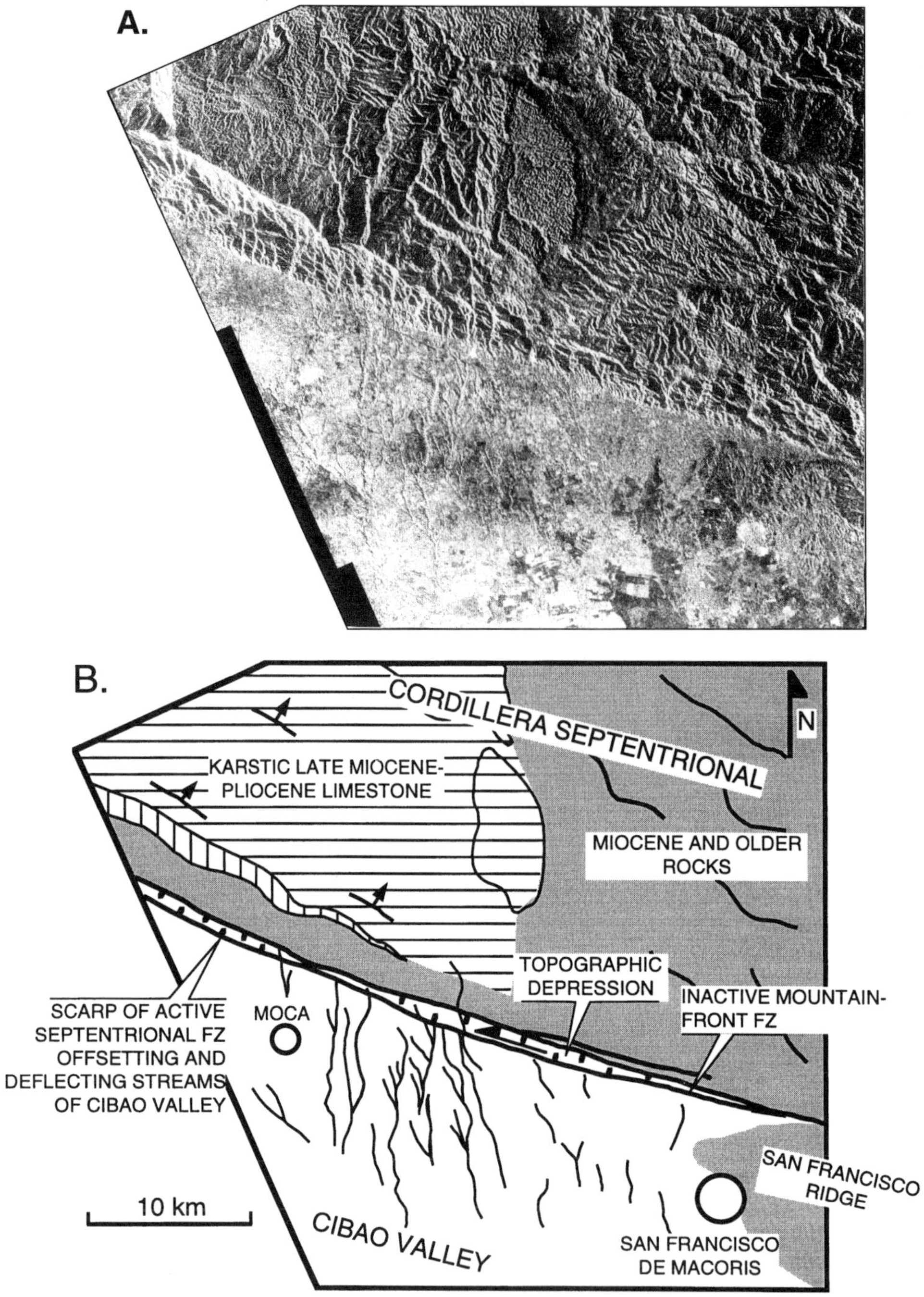

Figure 9. (A) SEASAT radar image of the central Septentrional fault system. (B) Interpretation of SEASAT image in A. The prominent mountain-front fault is an inactive strike-slip fault; the active trace of the Septentrional fault is the faint lineament about 2 km to the south of the mountain-front fault. See Figure 11 for a more detailed map of river offsets and deflections along the lineament. The intervening area between the two faults commonly forms an elongate, topographic depression. Note karstified Mio-Pliocene limestone capping the highest area of the Cordillera Septentrional and dipping to the northeast.

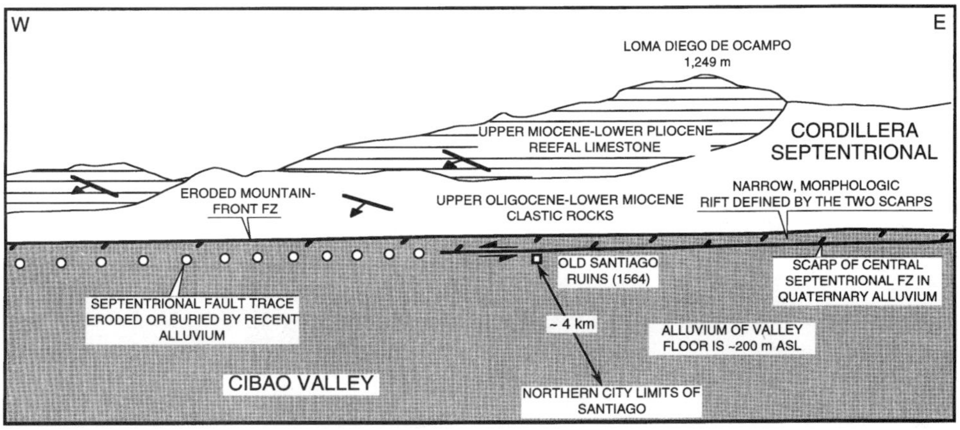

Figure 10. View taken in 1993 from Monument Hill in Santiago looking toward the Cordillera Septentrional. The floor of the Cibao Valley is about 200 m in elevation. The highest peak in the Cordillera Septentrional, Loma Diego de Ocampo, reaches 1,249 m and is capped by upper Miocene-lower Pliocene limestone. The mountain front is bounded by the inactive mountain-front fault; the active trace of the Septentrional fault zone is parallel and about 2 km to the south in alluvial sediments of the floor of the Cibao Valley. The active, north-facing scarp has less than 5 m of relief. A narrow topographic depression is defined by the north-facing scarp of the Septentrional fault zone and the south-facing scarp of the mountain-front fault zone. This depression becomes less distinct to the west.

south-facing in the central 10-km-long segment, to north-facing in the west. The transition between the north-facing eastern segment and the south-facing central segment is marked by a small, 200-m-long and 20-m-wide pull-apart basin occupied by a small sag pond. The change in scarp facing between the central and western segments is not as abrupt, but occurs near the Rio Licey. Southside-up vertical movement on the eastern fault segment has caused erosion within late Quaternary sediments and scarp retreat on the upthrown side of the fault. Offset features and locations of scarp profiles are indicated on Figure 11 and are discussed in greater detail in section 6.

Jacagua area of the central Cibao Valley

Fault-related features visible on 1:40,000 scale aerial photographs of the Jacagua area are shown on the photomosaic and inset map of Figure 12, and are similar to those interpreted on Figure 11 in the Rio Moca–Rio Licey area to the east. The main features visible at this scale include the inactive fault defining the topographic mountain front of the Cordillera Septentrional and the active trace of the Septentrional fault zone in late Quaternary alluvium of the Cibao Valley.

Scarp. In the Jacagua area, the scarp in alluvium ranges from 1 to 5 m high and is uniformly downthrown to the north over a distance of 13 km. Fault features are well preserved in this slowly developing, rural area dominated by livestock pastures and small farms. This scarp becomes indistinct in the western part of the area, probably because of recent alluvial deposition by the Rio Quinigua (Fig. 12).

Southside-up vertical movement creates a topographic depression between the active Septentrional fault zone in the Cibao Valley and the inactive mountain front that widens slightly to the west and varies in width from 1 to 2 km (Fig. 12). A low, swampy area marked by a vegetation contrast within this tectonically controlled valley is apparent north of the ruins of Old Santiago and is perhaps one reason that 16th century Spanish colonists selected this site for their farming and ranching community. Southside-up vertical movement on the fault between Old Santiago and Jacagua has caused headward erosion within late Quaternary sediments south of the fault.

Villa Vasquez–Navarrete area of the central Cibao Valley

The tectonic geomorphology of the area between the towns of Navarrete and Villa Vasquez is controlled by an increasingly divergent angle and distance between the west-northwest–striking inactive mountain-front fault and the west-southwest–striking, north-facing active valley scarp (Fig. 13). This area marks a general change to a drier microclimate from the wetter climate affecting the Jacagua–Rio Moca–Rio Licey areas to the east (Fig. 3). Fault features are generally not well preserved in the area shown in the photomosaic of Figure 13 because of intensive agricultural cultivation and related building of concrete-lined irrigation canals. The widening angle between the two faults is accompanied by widening (1–7 km) and deepening of the basin formed between the two faults (Fig. 13). The floor of the basin is at an elevation between 120 and 140 m a.s.l., whereas the southern ridge in Mio-Pliocene rocks south of

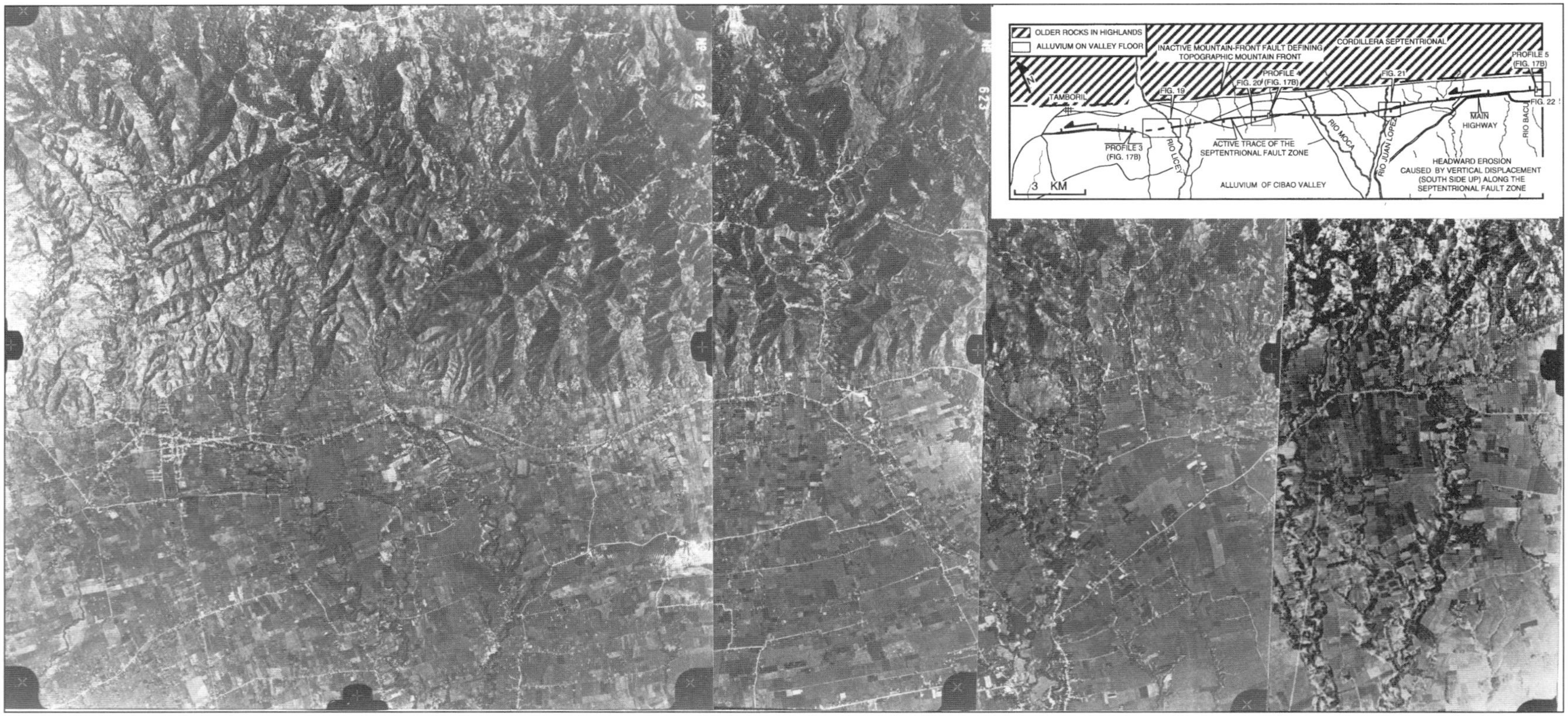

Figure 11. Montage of four 1983 aerial photographs showing the scarp of the central Septentrional fault zone between Tamboril and the area northwest of Salcedo (aerial photographs: roll 7, flight 9, February 16, 1983, nos. 622, 623, 624 and 625; original scale, 1:40,000). Inset map shows main tectonic features discussed in the text. The late Holocene scarp in alluvium of the valley floor is about 2 km to the south of the inactive mountain front. Southward drainage of the Cordillera Septentrional flows freely across the south-facing scarp in the center of the map but is impeded by the northward-facing scarp at the eastern and western edges of the map. This area is probably the mostly densely populated area along the Septentrional fault system, and, therefore, there has been considerable cultural modification of the fault scarp.

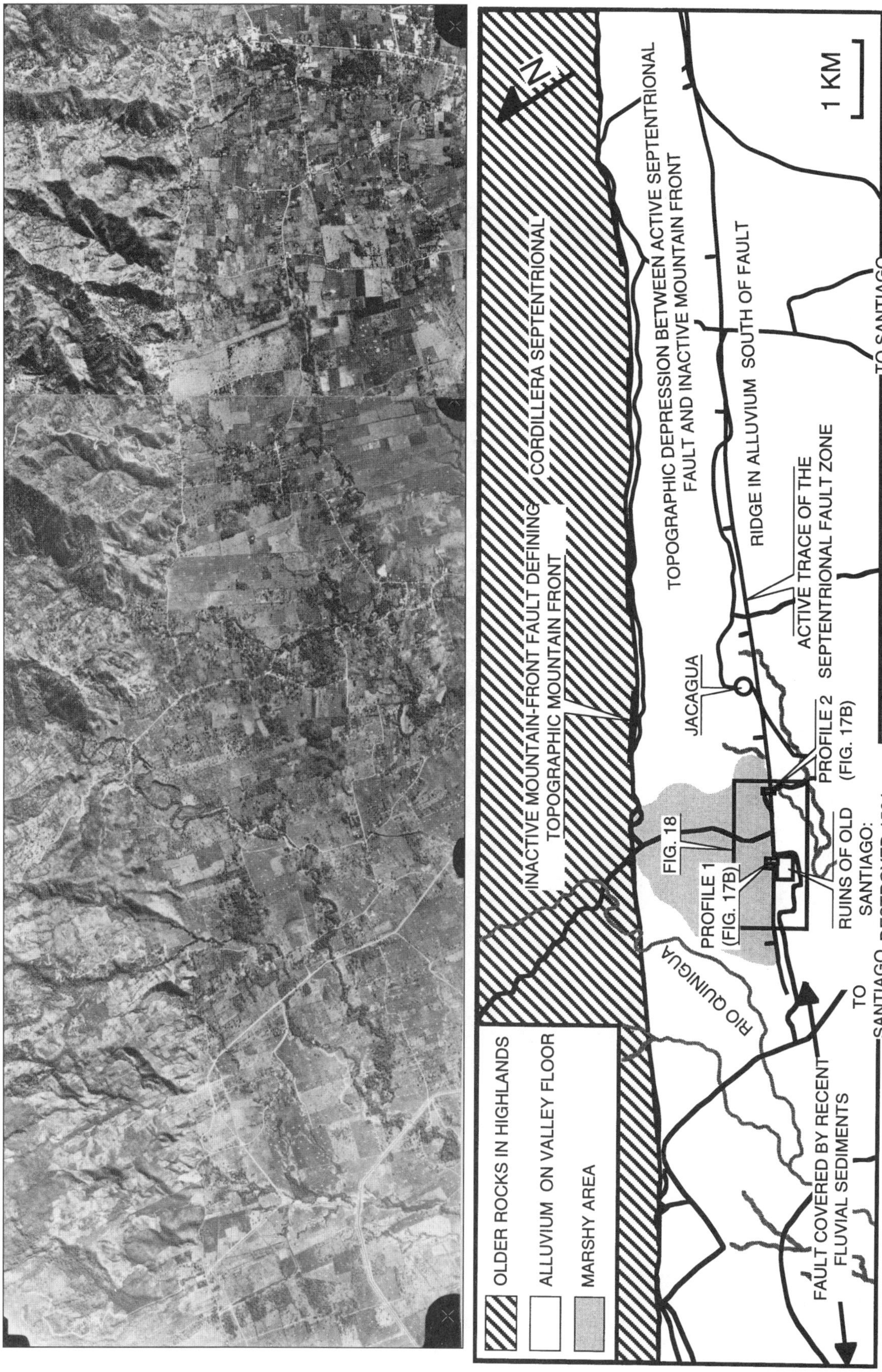

Figure 12. Montage of two 1983 aerial photographs showing the scarp of the central Septentrional fault zone between the ruins of Old Santiago and the area north of Santiago (aerial photographs: roll 7, flight 9, February 16, 1983, nos. 618 and 619; original scale, 1:40,000). Map shows main tectonic features discussed in the text. The active scarp in alluvium of the valley floor is about 2 km to the south of the inactive mountain front. The southward drainage of the Cordillera Septentrional flows freely downslope in the western part of the map, where the scarp is apparently buried, but drainage is impeded by the northward- and uphill-facing scarp in the central and eastern parts of the map. The ruins of Old Santiago are located from 0 to 200 m south of the scarp. A marshy area (shown in gray) formed by blocked drainages north of the fault is found immediately to the north of the ruins of old Santiago.

Figure 13 (this and opposite page). Montage of three 1958 aerial photographs showing late Quaternary faults in the bifurcation area of the northern and southern branches of the western Septentrional fault zone (aerial photographs: ICM-1(1958)-13-V28-12-29, December 1958, nos. 1117, 1119, and 1121; original scale, 1:60,000). Inset map shows main tectonic features discussed in the text. The late Holocene scarp in alluvium of the valley floor is about 2–8 km to the south of the inactive mountain front, which is marked by inactive, eroding alluvial fans. Drainage is impeded by the northward- and uphill-facing scarp of the southern active fault in the valley, and a topographic, alluvium-filled depression has formed between the two diverging faults. The strike of the northern branch is 110 and the strike of the southern branch is 100. Southwestward shifting of the trace of the southern branch across the Cibao Valley occurs at poorly defined pull-apart basins marked by topographic depressions that concentrate groundwater and produce vegetation contrasts to the surrounding drier areas.

the fault is ~3–5 m higher. The crest of the Cordillera Septentrional fault north of this area is at an elevation of ~1,000 m (Fig. 8).

Scarp. The floor of the westward-widening basin is low lying and flat bottomed and ideal for heavily irrigated crops such as rice. Up-to-the-south vertical motion on the southern Septentrional fault has produced a low ridge in Mio-Pliocene sedimentary rocks of the Cibao Valley that dip southward from 10 to 15° (Fig. 13). This fault-parallel ridge is onlapped to the south by the flood plain of the Rio Yaqué del Norte. To the east, the ridge narrows to less than 1 km wide south of Navarrete. West of Navarrete, a 3-km-long and 1-km-deep embayment in the northern edge of the ridge suggests scarp retreat related to localized erosion in this area. The break in slope at the northern edge of the ridge is closely followed by a large, 3-m-deep, concrete-lined irrigation canal whose construction parallel to the base of the scarp has modified the expression of the scarp.

Pull-apart basins. A 2-km left-step in the trace of the active fault zone north of Navarrete has produced the Navarrete pull-apart basin. We define this basin on the basis of the topographic depression and the elongate, isolated pattern of vegetation contrasts north of Navarrete that are shown on Figure 13. Several old, abandoned alluvial fans occur along the inactive mountain-front fault. These inactive fans are larger in the Navarrete pull-apart basin area and smaller to the west along the mountain front. This changing fan-size relationship suggests larger, long-term vertical motions at the mountain-front fault in the pull-apart

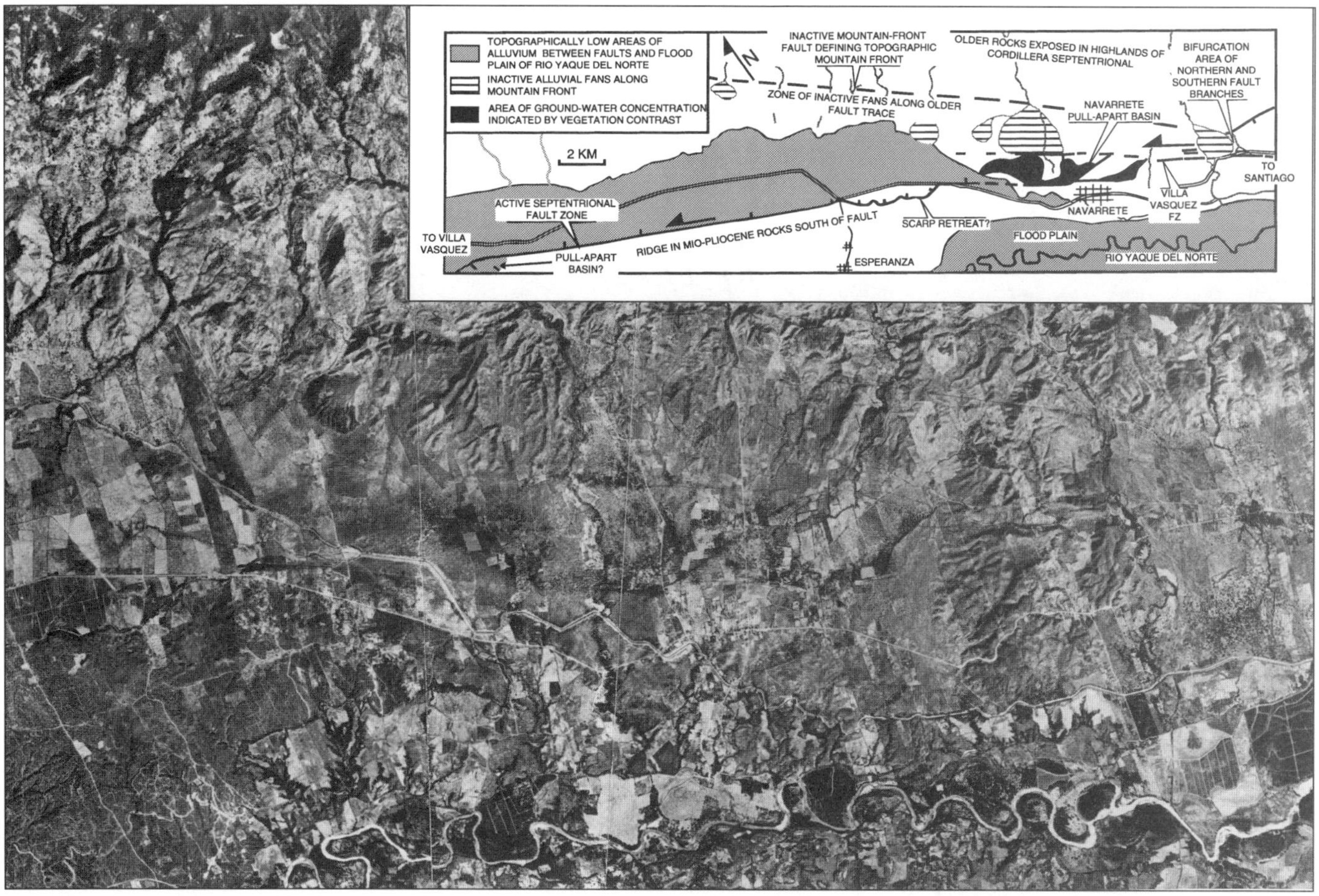

area. We infer a second, smaller pull-apart basin at the western edge of the area shown in Figure 13. This basin is marked by a stepover in the fault trace and an area of swampy ground and vegetation contrast. This basin marks the first in a series of three stepover basins to the west along the western extension of the Septentrional fault zone that are shown on Figure 14.

Northern and southern branches of the Septentrional fault system in the western Cibao Valley

Fault nomenclature. In the western Cibao Valley, four distinct strands of the Septentrional fault are named here—the mountain-front fault, the Monte Cristi fault, the Villa Vasquez fault, and the Septentrional fault—and are identified on Figure 14. The three northern faults constitute a geomorphically prominent 5–8-km-wide zone, which we collectively refer to in this chapter as the "northern branch of the Septentrional fault system." The Villa Vasquez push-up structure, a 12-km-long area of folded Neogene rocks located 5 km north of Villa Vasquez, is the result of a 1.5-km-wide step between the Villa Vasquez and Monte Cristi faults. We propose that the active trace of the Septentrional fault system trends more westerly (100°), closely following the Rio Yaqué del Norte near the city of Guayabín, continuing west to the city of Pepillo Salcedo, and connecting to the offshore faults mapped by Calais and Mercier de Lépinay (1995) (Fig. 14). Left-stepping pull-apart basins occur along the Septentrional fault zone in the west-central Cibao Valley east of the town of Guayabín (Fig. 14). These pull-apart basins range in length from 5 to 8 km and in width from 1 to 2 km and appear to influence southward jogs in the direction of the main channel of the Rio Yaqué del Norte and a former channel to the north near the easternmost pull-apart basin (Rio Viejo) (Fig. 14).

Correlation with offshore faults. The projected trace of the Septentrional fault zone connects with the active, offshore fault zone mapped or inferred along the northern coast of Haiti by other workers from sidescan and seismic-profiling data (Edgar, 1991; Calais and Mercier de Lépinay, 1995). Edgar (1991) found no offshore continuation of the mountain-front, Villa Vasquez, and Monte Cristi fault zones on roughly north-south seismic-reflection lines crossing the shelf between the Siete Hermanos Islands and the city of Monte Cristi (Fig. 14). However, vertical faults cutting the massive carbonate bank occupying this shelf may not be readily apparent on the profiles owing to the lack of distinctive seismic reflectors.

Southern Cibao basin margin. We are confident that

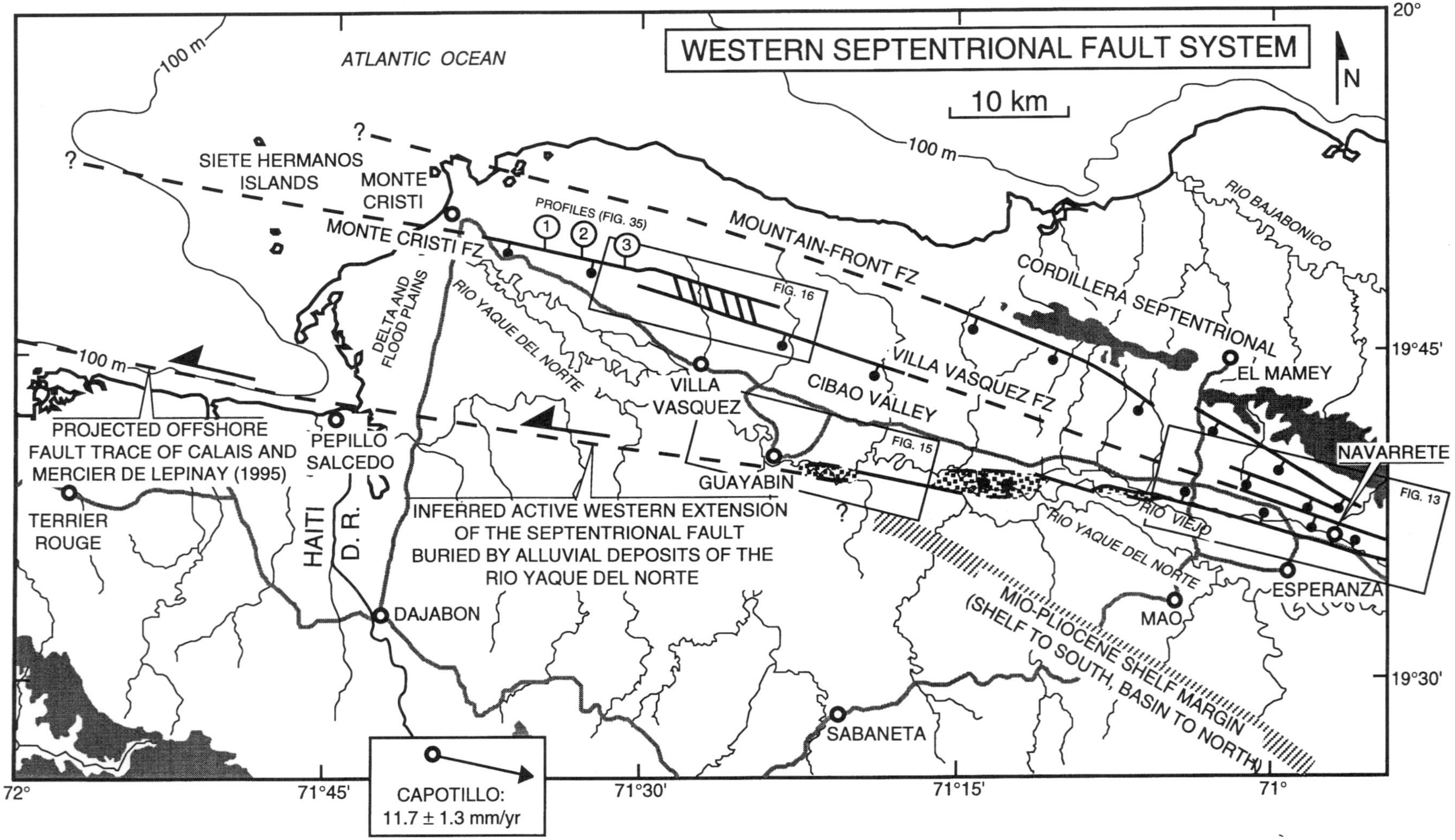

Figure 14. Map of Quaternary fault scarps in the area of the western Septentrional fault system based on interpretation of aerial photographs taken in 1967 (1:20,000 scale) and 1983 (1:40,000 scale) and field observations in 1991 and 1993. Gray areas represent topographic elevations about 500 m. Boxes show locations of montages of aerial photographs shown in Figures 13, 15, and 16. Based on evidence presented here, we propose that the mountain-front, Villa Vasquez, and Monte Cristi faults are inactive in late Quaternary time and that late Holocene motion is occurring on the Septentrional fault that is largely buried under alluvium of the Rio Yaqué del Norte. Note that no strike-slip faults can pass to the south of the town of Guayabín because there are no offsets of the continuous and geomorphically prominent Mio-Pliocene shelf margin along the southern edge of the Cibao basin. Inset shows GPS-based rate and direction of relative motion of a point at Capotillo relative to a point on the North American plate on the southeastern Bahamas platform (Dixon et al., 1998; cf. Fig. 2B).

there are no additional strike-slip fault strands accommodating North America–Caribbean plate motion south of the Septentrional fault in the Cibao Valley because the Mio-Pliocene shelf margin along the west-southwesterly edge of the valley forms a remarkably undeformed, linear feature that is not offset by faults of any orientation (Evans, 1986) (Fig. 14). Evans (1986) demonstrated that gentle dips of carbonate and clastic shelf platform and margin units are depositional, and not tectonic, in origin.

Plate motions. Recent GPS measurements at the town of Capotillo near the Dominican-Haitian border (Fig. 2B) indicate that the motion of this part of central Hispaniola on the Caribbean plate is east-southeastward at a rate of 11.7 ± 1.3 mm/yr relative to the stable North America plate in the southeastern Bahamas north of Hispaniola (Dixon et al., 1998) (Fig. 2B). This plate motion predicts a strike-slip motion along the western Septentrional fault system striking 100° (Fig. 14). The presence of pull-apart basins supports the existence of a local transtensional environment where the western, 100°-striking Septentrional fault zone bifurcates in the Navarrete area and steps left across the floor of the Cibao Valley (Fig. 14).

Climate, flood plains, and scarp preservation. Fault scarps along the three strands of the western Septentrional fault system are generally more prominent than scarps of the Septentrional fault zone in the eastern and central Cibao Valley because there is (1) less precipitation (Fig. 3); (2) there is lower population density and less cultural modification; and (3) scarps occur in indurated, late Neogene sandstone in the case of the mountain-front, Villa Vasquez, and Monte Cristi fault zones.

One exception is the inferred, buried extension of the Septentrional fault zone between the towns of Guayabín and Pepillo Salcedo. This flat-lying region in the center of the western Cibao Valley occupies an area of rapid late Quaternary sedimentation on the flood plain of the Rio Yaqué del Norte (Fig. 14). In this area, we found no surface expression of the Septentrional fault zone, although we inferred that it is present as a continuous, late Holocene structure in the subsurface. Unpublished gravity maps suggest a basement structure in the subsurface of the western Cibao Valley parallel to the projected late Holocene trace of the Septentrional fault zone (W. McCann, personal communication, 1995).

General features of the Septentrional fault zone in the Guayabín area

Features visible on 1:40,000 scale aerial photographs along the southern branch of the Septentrional fault system are shown on the inset of the photomosaic of Figure 15. The main tectonic features include a fault-controlled valley in late Neogene sedimentary rocks defining a low valley 15 km northeast of Guayabín. This fault, along with another fault south of Guayabín, forms an inferred pull-apart basin that is occupied by an east-west–trending segment of the Rio Yaqué del Norte. It is unlikely that the eastern strike-slip fault trace continues on trend to the west because the large anticline developed in Mio-Pliocene sedimentary rocks north of Guayabín is not offset (Fig. 15). The southern strike-slip fault trace shown on the inset map may account for the abrupt termination and possible westward translation of the northwest-trending Mio-Pliocene shelf margin southeast of Guayabín (Evans, 1986) (Fig. 14). We attribute the lack of strike-slip geomorphic features to rapid fluvial sedimentation along the Rio Yaqué del Norte.

General features of the northern branch of the Septentrional fault system in the Villa Vasquez–Monte Cristi area

Mountain-front fault. This fault forms a prominent topographic break between the highlands of the Cordillera Septentrional near the town of El Mamey and the low-lying foothills formed by folded late Neogene sedimentary rocks (Calais et al., 1992b) (Fig. 14). The resistant lithology of the cap of late Miocene–early Pliocene shallow-water limestone north of the fault may contribute to the height of the south-facing mountain front scarp in the eastern area of the fault near El Mamey (de Zoeten and Mann, 1991). To the west toward the city of Monte Cristi, the mountain-front fault becomes a progressively less distinct lineament with elevations less than 200 m (Fig. 14).

Monte Cristi and Villa Vasquez faults. Together these two faults are 35 km long and are the most geomorphically prominent of the entire Septentrional fault system (Fig. 16). These two faults display many geomorphic features typical of a recently active fault such as scarps without embayments and left-lateral stream offsets. The prominent scarps associated with these two faults are related to lithologic contrast across the fault (more resistant sandstones north of the fault, and less resistant sandstone and shale south of the fault) and slow rates of erosion in the dry climate in this region (Fig. 3). Vertical movements associated with the formation of the 12-km-long, 1.5-km-wide Villa Vasquez push-up at a right-step between the Monte Cristi and Villa Vasquez fault zones may have accentuated the badlands topography within and immediately adjacent to the push-up structure (Fig. 16).

6. EVIDENCE FOR RECENT MOVEMENT ON THE CENTRAL SEPTENTRIONAL FAULT ZONE

Scarps in Quaternary alluvium

Evidence for recent fault movement on the central part of the Septentrional fault zone includes the presence of scarps in Quaternary alluvium and strike-slip offsets of youthful streams and terrace risers. The preservation of both types of features in spite of their rapid erosion in the high to moderate rainfall area of the central Cibao Valley (Fig. 3) requires recent, left-lateral strike-slip movement consistent with the tectonic setting of the fault zone.

The most abundant scarps in Quaternary alluvium along the Septentrional fault system occur in the central area of the fault (Fig. 8). To document these scarps, we measured fault scarp profiles at 10 representative localities along the central

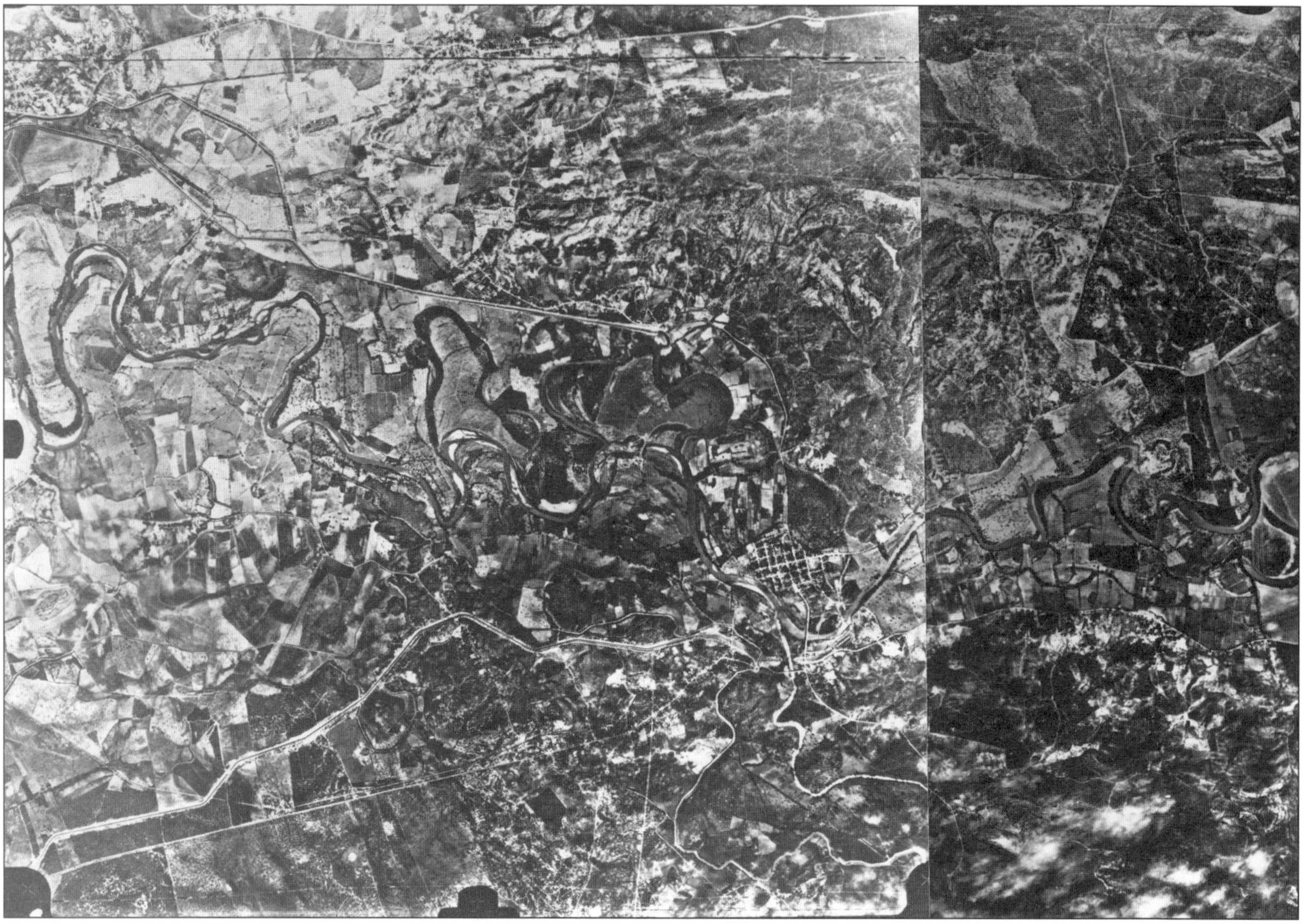

Figure 15 (this and opposite page). Montage of three 1983 aerial photographs showing the Septentrional fault zone in the Guayabín area (aerial photographs: roll 14, strip 11, January 11, 1983, nos. 1823, 1825, and 1827; original scale, 1:40,000). Inset map shows main tectonic features discussed in the text. We infer that the western continuation of the scarp prominently displayed north of Esperanza (cf. inset of Fig. 13) continues via a series 100°-trending scarps connected by pull-apart basins. In this area the fault is proposed to step over across a pull-apart basin (shown by the black dots) that controls but is obscured by the flood plain of the Rio Yaqué del Norte. The presence of an anticline showing no offset in Mio-Pliocene sedimentary rocks north of Guayabín and a Mio-Pliocene shelf margin to the southeast of Guayabín only allow the presence of a 100°-trending fault zone in this zone. Our preferred location of the buried trace west of Guayabín is along the southern margin of the flood plain because this position aligns the fault with the 100°-trending offshore fault mapped along the margin of Haiti (Calais and Mercier de Lépinay, 1995 - Fig. 14) and is consistent with the pull-apart interpretation shown.

Septentrional fault zone using a transit and rod (Fig. 17A). The scarps change from north-facing to south-facing four times at intervals ranging from 4 to 10 km along the 50-km-long fault segment between San Francisco de Macorís and Old Santiago.

Scarp near Old Santiago. Profile 1 at Old Santiago shows a 1.1-m-high, north-facing scarp (Fig. 17B) exposed within the ruins of the Spanish colonial town destroyed by the earthquake of 1564. The 1564 earthquake also destroyed the colonial town of La Vega in the south-central Cibao Valley (Scherer, 1912) (Fig. 8). Because of almost complete destruction, the colonists rebuilt the town about 10 km southward to the present site of the modern city of Santiago (Fig. 10).

The fault scarp at Old Santiago appears as a lineament on the stereopair shown in Figure 18. The scarp affects undated alluvium and forms a gentle mountain-facing slope (Fig. 17B).

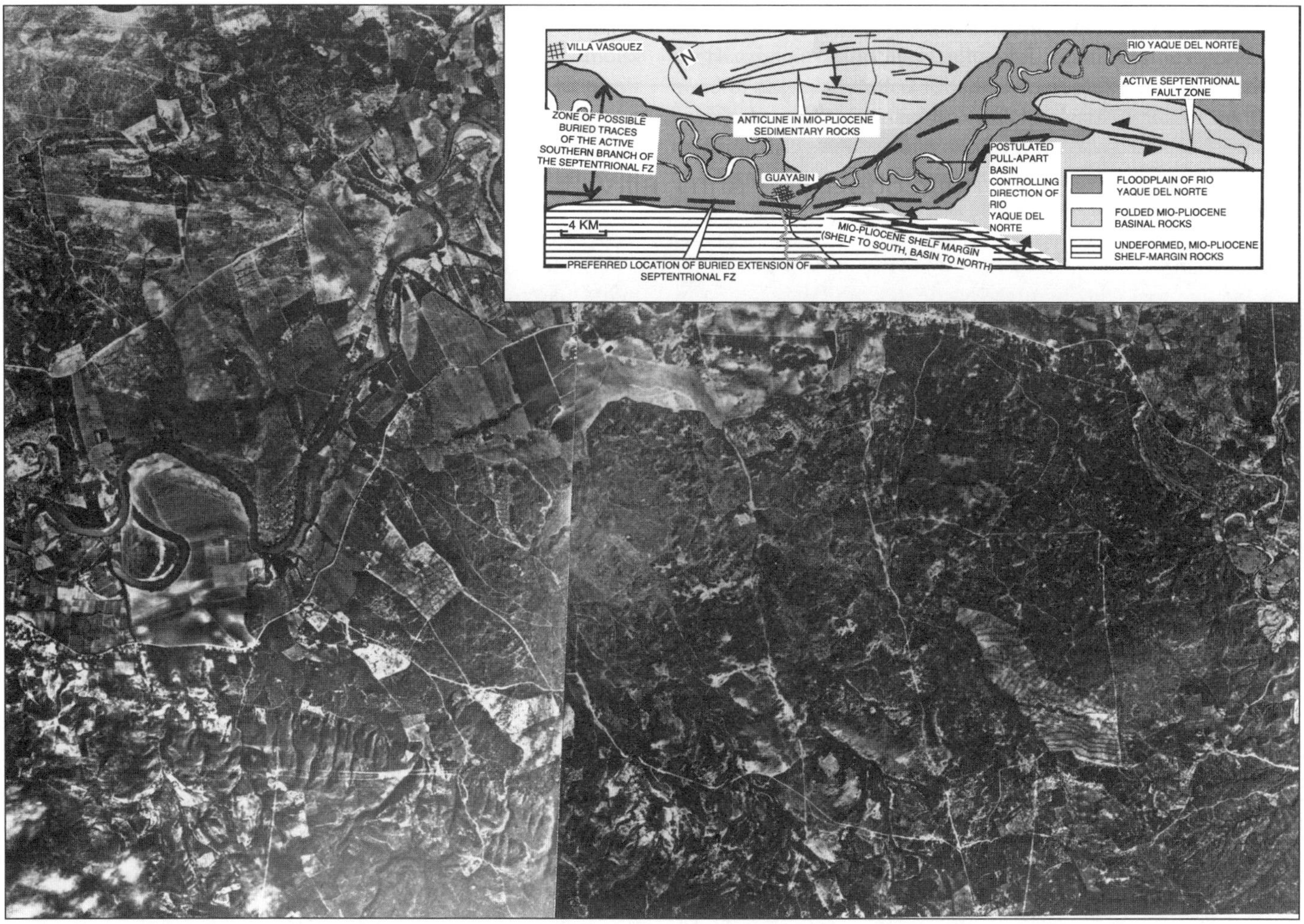

The low-lying marshy area between the scarp and the south-facing mountain front of the Cordillera Septentrional is the result of ground water impounded behind the scarp (Fig. 18). Stone and brick ruins are the only remains of the 16th century buildings of Old Santiago on the uplifted southern side of the fault. The scarp passes along the northern edge of the Old Santiago ruins and within 1 m of the foundation and the 1-m-thick walls of a church that was destroyed during the 1564 event and partially reconstructed during the 19th century. The scarp is discontinuous and disappears to the west of Old Santiago under alluvium deposited by modern, south-flowing rivers, including the Rio Quinigua, that head in the Cordillera Septentrional (Fig. 1).

We were unable to find any historical information to support or refute the possibility that formation or reactivation of the scarp that passes through Old Santiago (Fig. 18) accompanied the 1564 earthquake, but this question remains an intriguing target for future paleoseismic and historical research. It is likely that the scarp near Old Santiago was a preexisting feature of the topography at the time of the Spanish settlement of the Cibao Valley. Groundwater, impounded north of the scarp, may have made this area desirable for farming and ranching. The elevated southern side of the fault provided a drier area for constructing the church and other brick and stone buildings (Fig. 12).

Scarp near Jacagua. Profile 2 across the north-facing scarp south of the village of Jacagua east of Old Santiago shows a 5.0-m-high scarp (Fig. 17B). The north-facing scarp, which is more conspicuous on aerial photographs than its western continuation at Old Santiago, has impounded a 10-m-wide marshy area north of the scarp (Fig. 18).

Scarp near La Ermita. Profile 3 near La Ermita shows a 4.4-m-high, north-facing scarp in alluvium that dams small, south-flowing streams and forms an 8-m-wide swampy area north of the scarp (Fig. 17B). The scarp is slightly arcuate in this area with a southward curve near the Rio Licey (Fig. 19). Erosion along stream channels running along the base of the scarp has added about 3 m of relief to the fault scarp (pre-erosional level approximated on the La Ermita profile in Fig. 17B).

The fault scarp is separated from the north-facing Old Santiago profile to the west by a 5-km-long, south-facing scarp across which no profiles were measured (Fig. 17A).

Scarp near Paso de Moca. Profile 4 near Paso de Moca shows a 5.3-m-high, south-facing scarp in alluvium (Fig. 17B). The aerial photograph stereopair shown in Figure 20 reveals the sharp contrast between the older, eroded mountain-front scarp that is so prominent on the SEASAT image shown in Figure 9 and the younger scarp in alluvium of the Cibao Valley. Agriculture and livestock grazing in the patchwork quilt of small farms shown on the stereopair in Figure 20 may have caused modification of the fault scarp.

Scarp near Rio Juan Lopez and change in scarp-facing direction. The abrupt transition in scarp-facing direction between the south-facing scarp measured on profile 4 near Paso de Moca and the north-facing scarp measured on profile 5 near Rio Bacú (Fig. 17B) occurs at a left-stepping, 200-m-long, 20-m-wide pull-apart basin occupied by a shallow pond 200 m east of Rio Juan Lopez (Fig. 21). The strike of the western, south-facing fault scarp also changes slightly to a more north-northwesterly strike at the basin.

Scarp near Rio Bacú. Profile 5, measured 300 m east of Rio Bacú (Fig. 17B), shows a 4.1-m-high, north-facing scarp in late Quaternary alluvium (Fig. 22). Agriculture may have altered the natural slope of the scarp. Two left-laterally offset stream channels east of the measured profile are indicated by numbers 1 and 2 on Figure 22.

Scarp near Monte Adentro. The 11.3-m-high scarp meas-

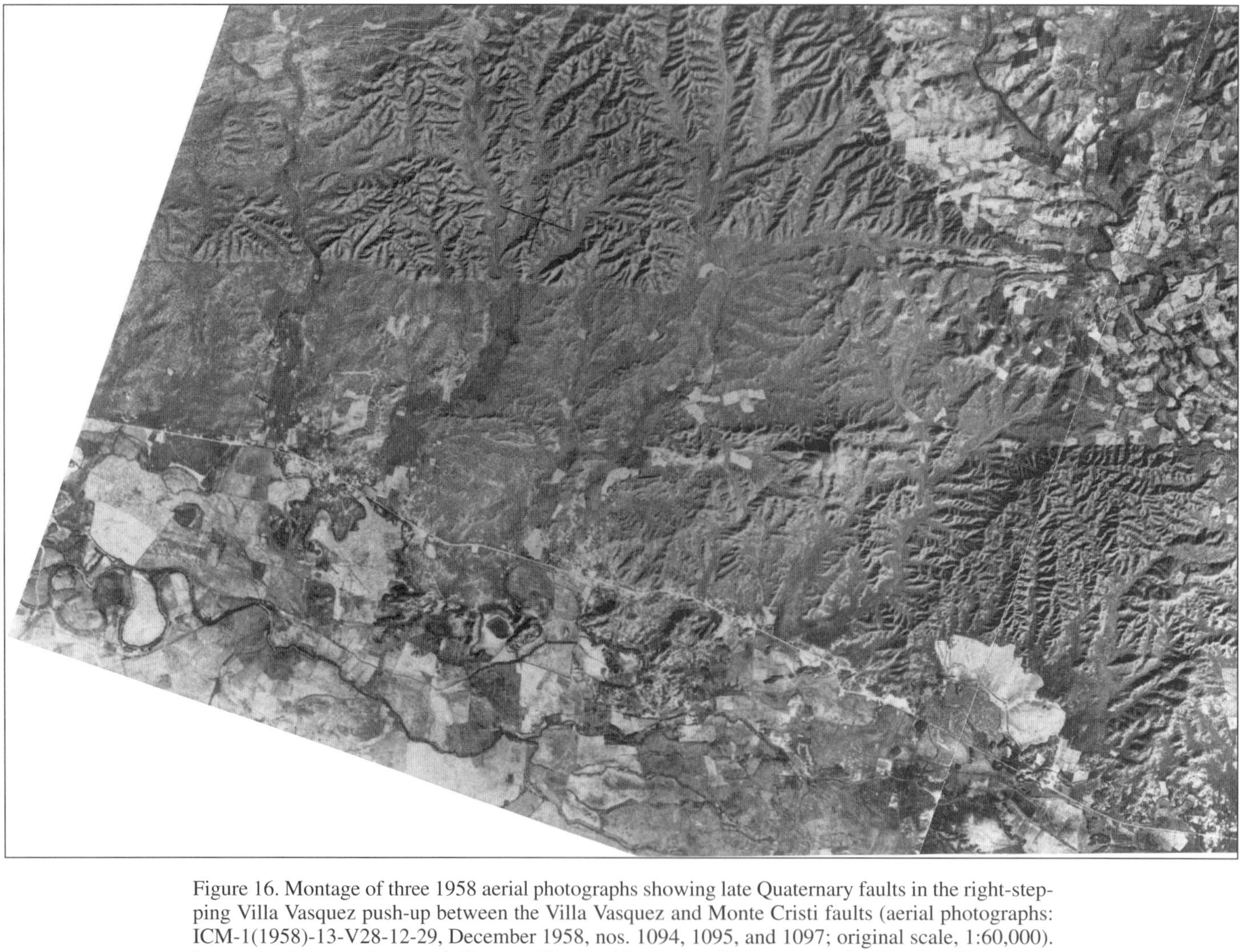

Figure 16. Montage of three 1958 aerial photographs showing late Quaternary faults in the right-stepping Villa Vasquez push-up between the Villa Vasquez and Monte Cristi faults (aerial photographs: ICM-1(1958)-13-V28-12-29, December 1958, nos. 1094, 1095, and 1097; original scale, 1:60,000). Inset map shows main tectonic features discussed in the text. The gray area is the zone of most pronounced topography and most prominently folded and tilted Miocene-Pliocene clastic sedimentary rocks within the push-up block. Rocks within the push-up are thickly bedded, indurated sandstone, which are more erosionally resistant facies than found outside the push-up. Rocks outside the push-up are more easily erodible shales and fine-grained sandstones that form a characteristic badlands topography in areas adjacent to the push-up.

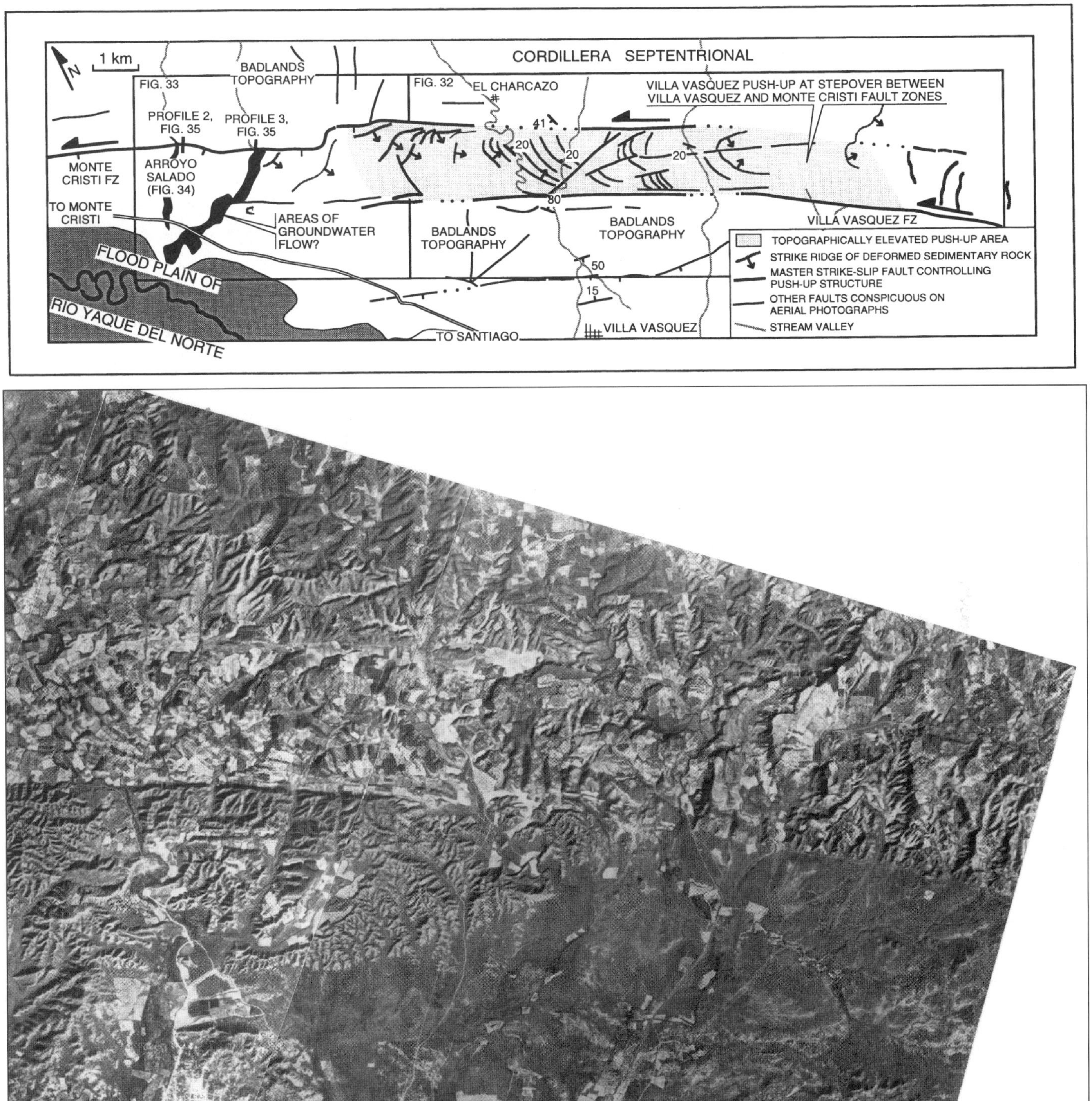
N
1 km
CORDILLERA SEPTENTRIONAL
BADLANDS TOPOGRAPHY
FIG. 33
FIG. 32
EL CHARCAZO
VILLA VASQUEZ PUSH-UP AT STEPOVER BETWEEN VILLA VASQUEZ AND MONTE CRISTI FAULT ZONES
PROFILE 2, FIG. 35
PROFILE 3, FIG. 35
41
20
20
20
80
MONTE CRISTI FZ
ARROYO SALADO (FIG. 34)
TO MONTE CRISTI
AREAS OF GROUNDWATER FLOW?
BADLANDS TOPOGRAPHY
BADLANDS TOPOGRAPHY
VILLA VASQUEZ FZ
FLOOD PLAIN OF
RIO YAQUE DEL NORTE
50
15
TO SANTIAGO
VILLA VASQUEZ
TOPOGRAPHICALLY ELEVATED PUSH-UP AREA
STRIKE RIDGE OF DEFORMED SEDIMENTARY ROCK
MASTER STRIKE-SLIP FAULT CONTROLLING PUSH-UP STRUCTURE
OTHER FAULTS CONSPICUOUS ON AERIAL PHOTOGRAPHS
STREAM VALLEY

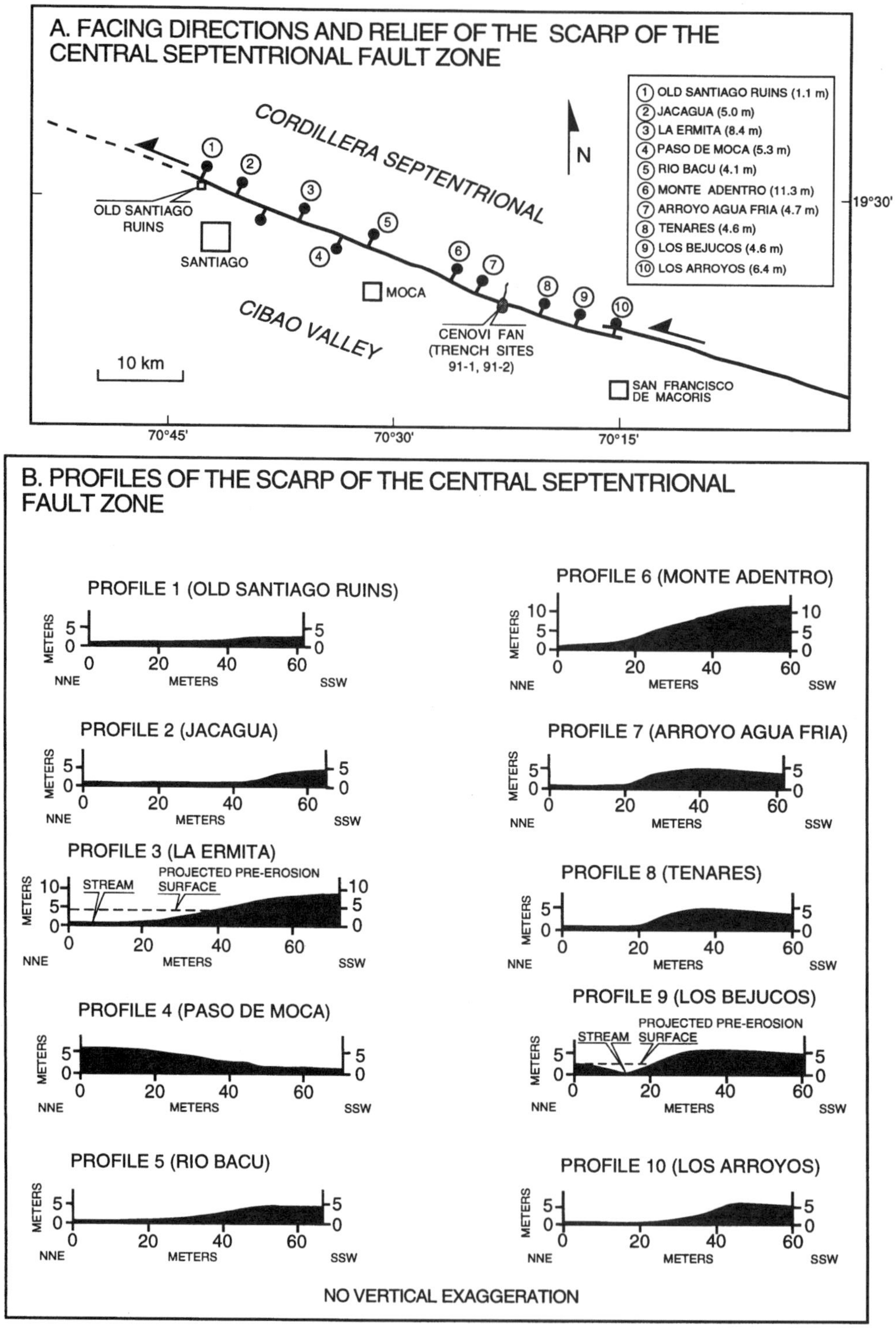

Figure 17. (A) Map of the late Holocene trace of the central Septentrional fault zone showing facing directions of the scarp. Inset box gives locality names and maximum height of the scarp in meters at each locality. Note the change in slope along strike at intervals of 4–10 km. Locations of Cenoví fan and 1991 trench sites are indicated. (B) Leveled topographic profiles of the scarp of the central Septentrional fault zone at the localites shown in A. See text for discussion.

Figure 18. Stereopair of 1983 aerial photographs showing scarp in the area of the ruins of Old Santiago (S on photo) and north of the modern city of Santiago (cf. Fig. 12 inset for location). The fault scarp (indicated by arrows) is uphill-facing (downthrown to the north) and tends to impound drainages in the depression between the fault and the inactive mountain front fault. A topographic profile measured at the ruins of Old Santiago is shown in Figure 17B. Aerial photographs: roll 7, flight 9, February 16, 1983, nos. 618 and 619; original scale, 1:40,000.

Figure 19. Stereopair of 1967 aerial photographs showing scarp (indicated by large arrows) in the area of La Ermita and Rio Licey (cf. Fig. 11 inset for location). Two thin arrows indicate left-laterally offset fluvial terrace riser of terrace T1 along Rio Licey (see Fig. 29). Left-lateral offset is 55 ± 10 m. Aerial photographs: DR-B67-V19-22-4, February 25, 1968, nos. 4571 and 4572; original scale 1:20,000.

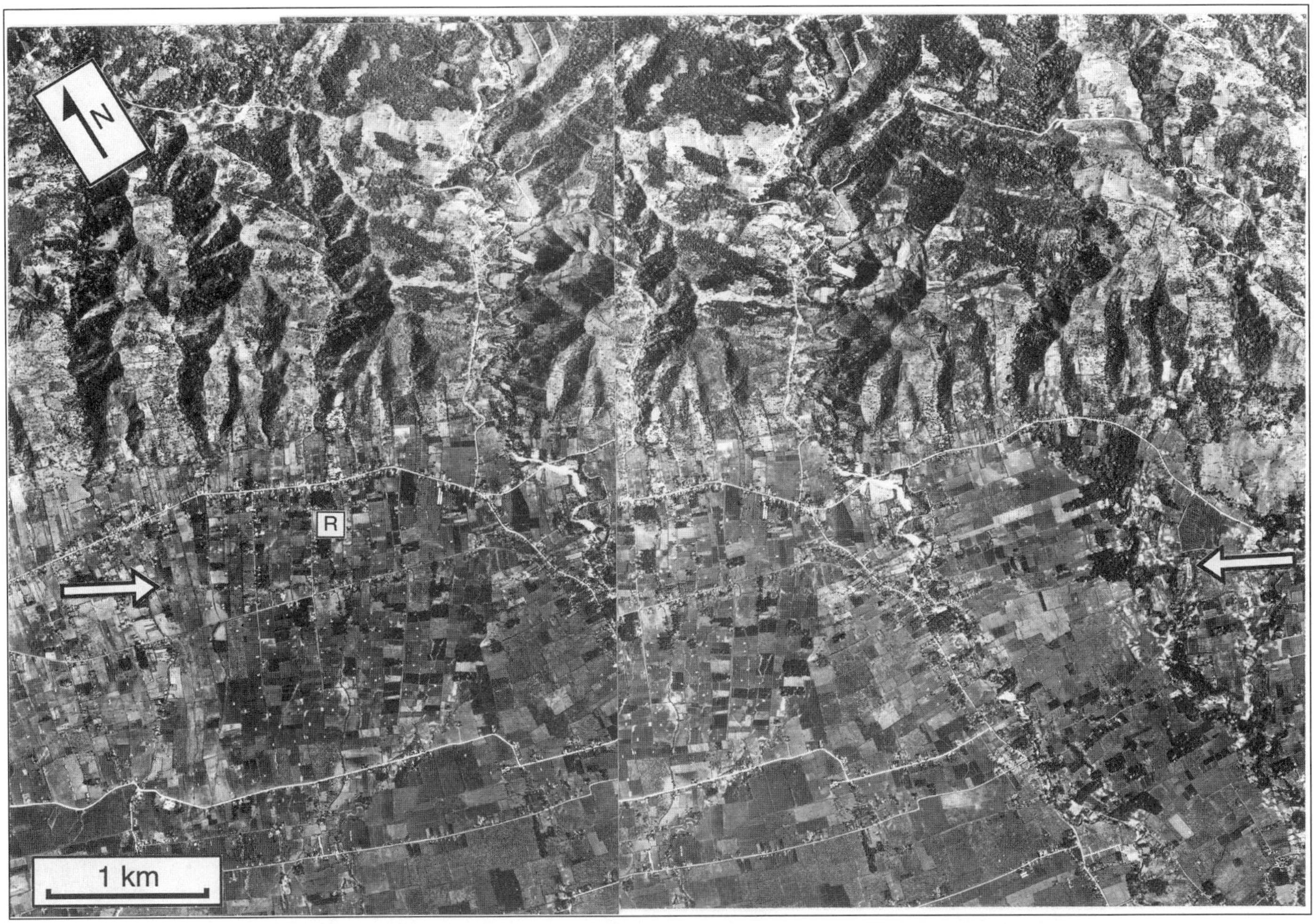

Figure 20. Stereopair of 1983 aerial photographs showing scarp (indicated by a pair of white arrows) in the area of Paso de Moca (cf. Fig. 11 inset for location). "R" indicates left-laterally active offset stream channel (see Fig. 30). Left-lateral offset of stream channel is 15 ± 3 m. Aerial photographs: roll 7, flight 9, February 16, 1983, nos. 623 and 624; original scale 1:40,000.

ured at profile 6 near Monte Adentro (Fig. 17B) is the highest we measured along the central Septentrional fault zone. This 20-km-long, north-facing segment of the fault scarp is significantly lower at the 4.1-m-high Rio Bacú site to the west and at the 4.7-m-high Arroyo Agua Fria site to the east (Fig. 17A). Stream diversion and blockage are common along this fault segment producing swampy ground on the north side of the fault scarp (Fig. 22). Figure 23 provides an oblique aerial view looking southward across a 50-m-wide area of swampy ground locally present on the north side of the scarp.

Scarp near Arroyo Agua Fria. Profile 7 near Arroyo Agua Fria at the western edge of the Rio Cenoví alluvial fan shows a 4.7-m-high, north-facing scarp in late Quaternary alluvium (Fig. 17B). A south-dipping surface that may have been tectonically tilted during one or more earthquakes is present on the southern, upthrown side of the fault. Road building and agriculture may have altered the natural slope of the scarp. Similarly tilted surfaces are present at the Tenares, Los Bejucos, and Los Arroyos sites to the east (Fig. 17B). Tilting of the southern block, northward facing of the scarp, and the similarity in scarp heights (4.6–6.4 m) at all four sites suggest that cumulative earthquake rupture effects may have been similar at the Tenares, Los Bejucos, and Los Arroyos sites described in the following paragraphs.

Scarp near Tenares. Profile 8 near Tenares shows a 4.6-m-high, north-facing scarp in late Quaternary alluvium (Fig. 17B). This down-to-the-north scarp forms a prominent, straight lineament in aerial photographs of the Tenares area (Figs. 24, 25). The fault scarp is overgrown with large trees and therefore may preserve a more natural slope than the previously described sites located in pastures and cultivated areas.

Scarp near Los Bejucos. Profile 9 near Los Bejucos shows a 4.6-m-high, north-facing scarp in late Quaternary alluvium (Fig. 17B). Erosion along stream channels at the base of the scarp has added several meters to the height of the fault

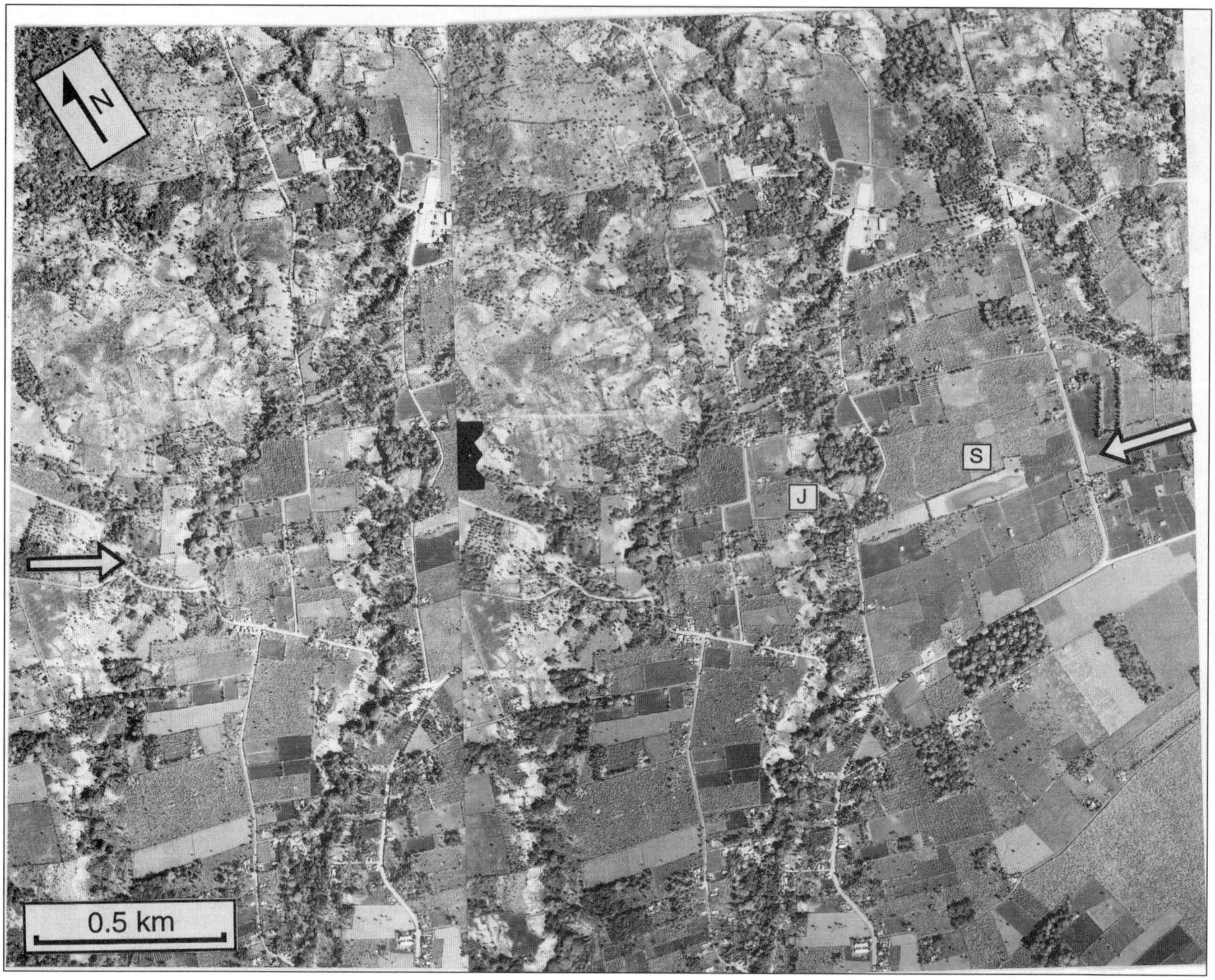

Figure 21. Stereopair of 1967 aerial photographs of scarp (indicated by arrows) near the Rio Juan Lopez west of the community of Los Turcos (see Fig. 11 inset for location). This rectangular sag pond at right of photographs indicated by "S" marks an inflection point between a north-facing scarp to the east and the south-facing scarp shown in the photographs. "J" indicates left-laterally offset fluvial terrace riser between terraces along the Rio Juan Lopez (see Fig. 26B). Left-lateral offset is 40 ± 10 m. Aerial photographs: DR-B67-V18-16, December 25, 1967, nos. 3432 and 3433; original scale 1:20,000.

scarp (pre-erosional level approximated on profile in Fig. 17B). Agriculture may have degraded the natural slope of the scarp.

Scarp near Los Arroyos. Profile 10 near Los Arroyos shows a 6.4-m-high, north-facing scarp in late Quaternary alluvium (Fig. 17B). Agriculture may have altered the natural slope of the scarp. The scarp is not continuous east of Los Arroyos and within the prominent fault valley separating the San Francisco ridge from the Cordillera Septentrional (Fig. 8). The absence of a scarp may reflect burial beneath alluvial fans shed from the Cordillera Septentrional, 1 km north of the projection of the fault. If the fault is buried by fans, this area might provide potential trenching sites.

Offset streams and fluvial terrace risers

We investigated terrace riser and stream offsets at four localities on the central Septentrional fault zone, within the region shown on Figure 11. Features of different ages are offset differing amounts, indicating repeated Quaternary motion across this fault. In addition, the offset terrace risers provide slip data that, coupled with minimum radiocarbon dates, pro-

Figure 22. Stereopair of 1967 aerial photographs showing two rivers northwest of Salcedo offset left-laterally by roughly the same amount (~400 m) (see Fig. 11 inset for location). Numbers 1 and 2 indicate offset rivers. The fault is uphill-facing (downthrown to the north) and tends to impound drainages in the depression between the fault and the mountain front. Aerial photographs: DR-B67-V18-16, December 25, 1967, nos. 3427 and 3428; original scale 1:20,000.

A.

B.

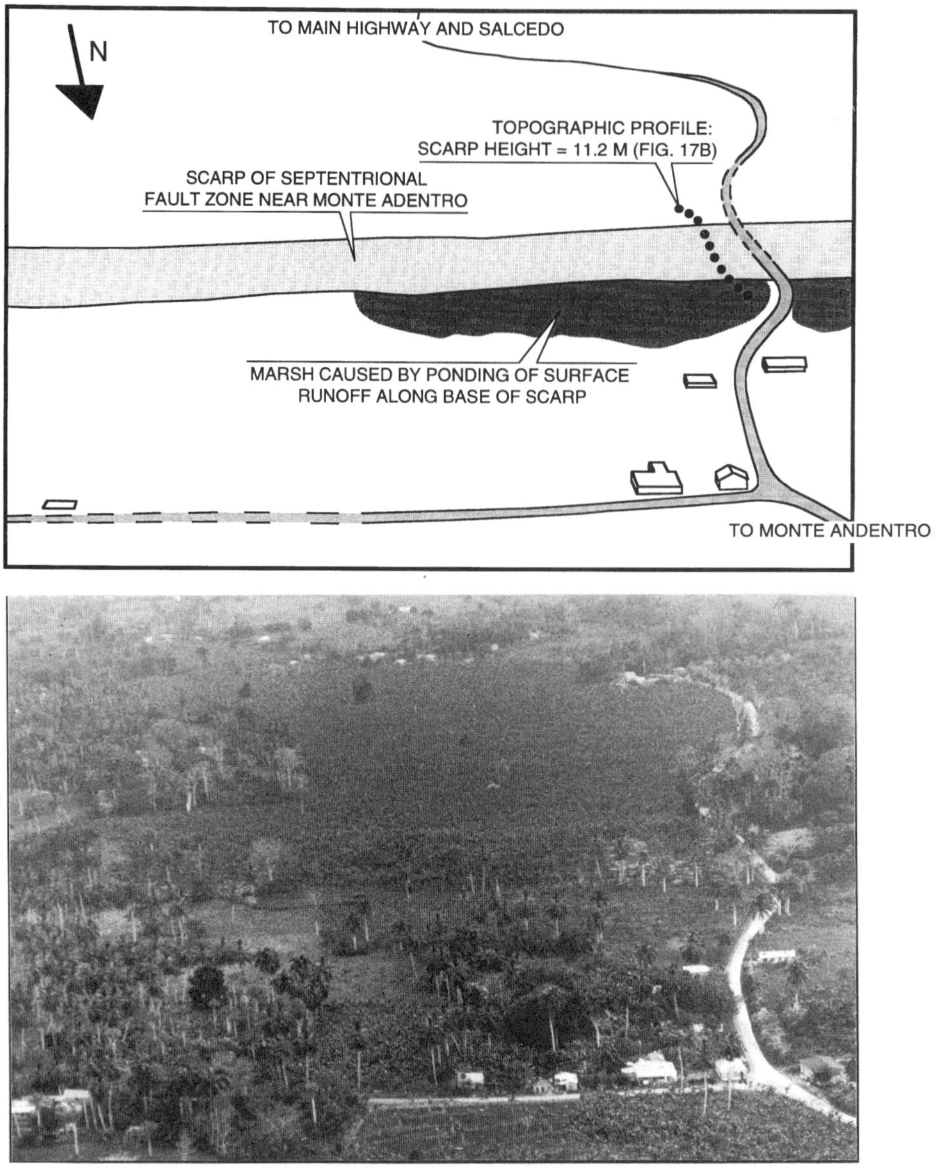

Figure 23. Oblique aerial view looking south across the scarp of the central Septentrional fault zone near Monte Adentro north of Salcedo. The scarp reaches its maximum elevation of 11.2 m in this area (cf. Fig. 17B).

vide maximum average Holocene rates of strike-slip motion on this ~22-km-long segment of the fault zone. The four offset localities and supporting maps and photos include:

(1) Rio Juan Lopez (general location on Figs. 11 and 26A; stereopair on Fig. 21; topographic map on Fig. 26B; ground-level photographs in Figs. 27 and 28);

(2) Rio Licey (general location on Figs. 11 and 26A; stereopair on Fig. 19; topographic map on Fig. 29);

(3) A small, unnamed stream near fault profile 4 (Fig. 17B) west of Rio Moca (general location on Fig. 11; stereopair on Fig. 20; topographic map on Fig. 30); and

(4) Two small, unnamed stream offsets east of fault profile 7 (Fig. 17B) and west of the Cenoví fan (general location on Fig. 11; stereopair on Fig. 22).

At all four sites, left-lateral offsets and fault scarps align with other features associated with the central Septentrional fault zone as summarized on the regional map of Figure 8. We made detailed plane-table maps at Rio Licey, Rio Juan Lopez, and the small stream west of Rio Moca to better quantify the amount of left-lateral offset at each site. We collected radiocarbon samples (gastropod shells) from shallow test pits dug into terraces at the Rio Juan Lopez (Fig. 26B) and Rio Licey (Fig. 29) that were analyzed by one of us (G. Burr) to constrain the ages of the offset terrace risers.

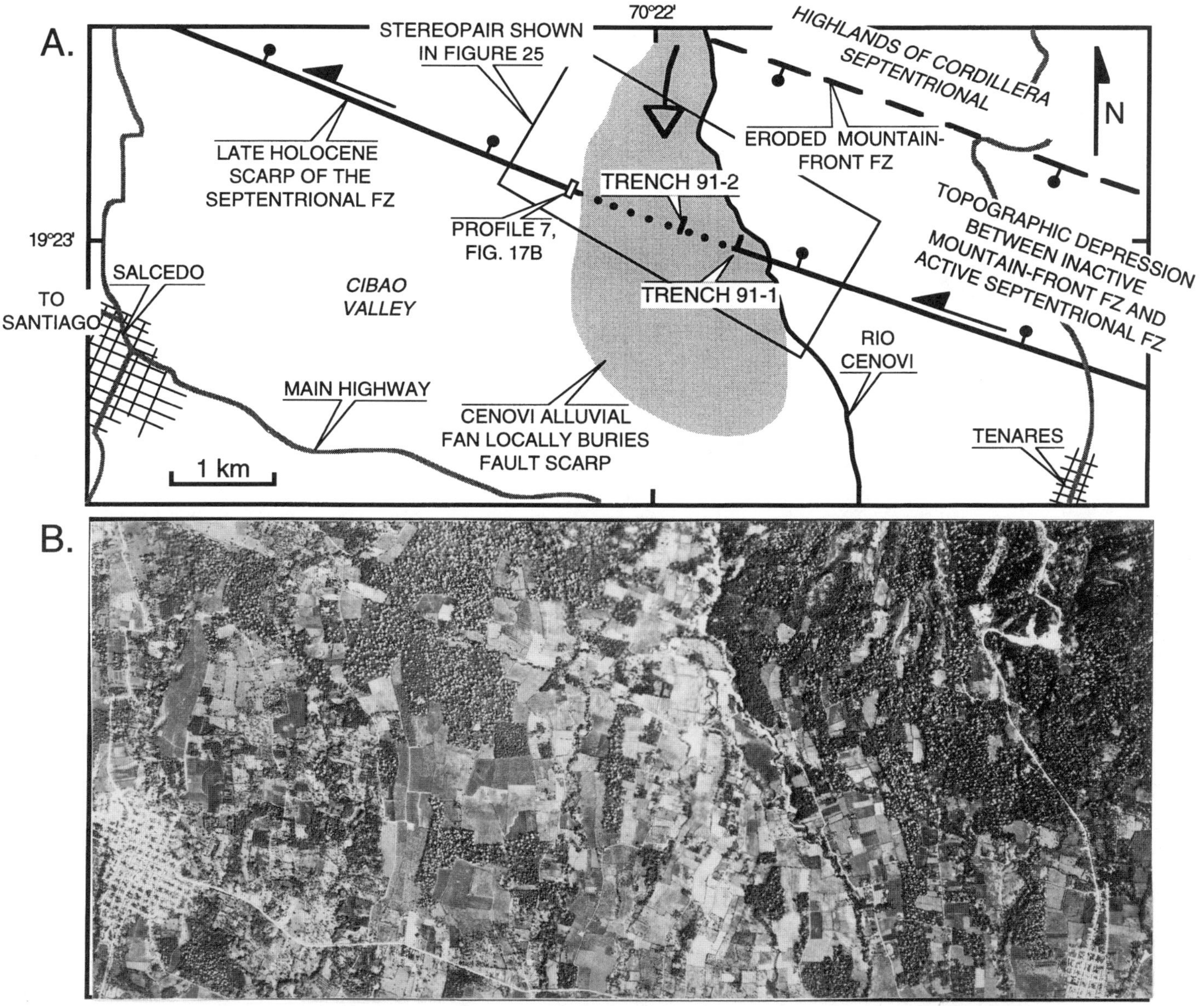

Figure 24. (A) Map showing the localized burial of the late Holocene scarp of the Septentrional fault zone by the Cenoví alluvial fan (buried fault is dotted). The head of the fan at the inactive mountain-front fault is about 2 km to the north of the scarp. The Rio Cenoví forms the present-day eastern edge of the alluvial fan. Trench sites 91-1 and 91-2 are indicated. (B) Aerial photograph of fan area. Aerial photograph: roll 7, flight 9, February 16, 1983, no. 626; original scale 1:40,000.

Rio Juan Lopez

Rio Juan Lopez is a perennial river that drains southward from the Cordillera Septentrional and crosses the Septentrional fault (Figs. 11, 21). Within this drainage, near the fault, are two terrace levels (Terrace 1 and Terrace 2, Fig. 26B) above the modern channel. The riser between these terraces is both left-laterally and vertically offset across the fault (Figs. 21, 26B).

The base of the terrace riser between Terraces 1 and 2 is left-laterally offset about 40 ± 10 m, as shown by the length of the bar that joins the offset ends of the riser on our detailed map (Fig. 26B). Both terraces are also vertically offset across the fault on the order of 4–6 m, with the northern side relatively upthrown, as shown by the south-facing scarp across Terrace 2, associated with the fault. Due to the left-lateral displacement, Terrace 1 north of the fault is juxtaposed against the higher Terrace 2 south of the fault, creating a north-facing scarp even though the north side is relatively uplifted (Fig. 26B). The photograph in Figure 27 shows a representative view looking north from the top of Terrace 2 across the south-facing fault scarp that breaks the surface of Terrace 2. The photograph in Figure 28

Figure 25. Stereopair of 1967 aerial photographs showing burial of the late Holocene scarp of the central Septentrional fault zone by the Cenoví alluvial fan east of the community of Ojo de Agua (see Fig. 24 for location). Rio Cenoví runs diagonally across eastern part of photograph and forms the eastern edge of the Cenoví fan. The scarp shown by arrows faces north on both sides of the fan and in excavation 91-2 through the fan. The locations of trench sites 91-1 and 91-2 are labeled. Aerial photographs: DR-B67-V17-17, December 26, 1967, nos. 3572 and 3573; original scale 1:20,000.

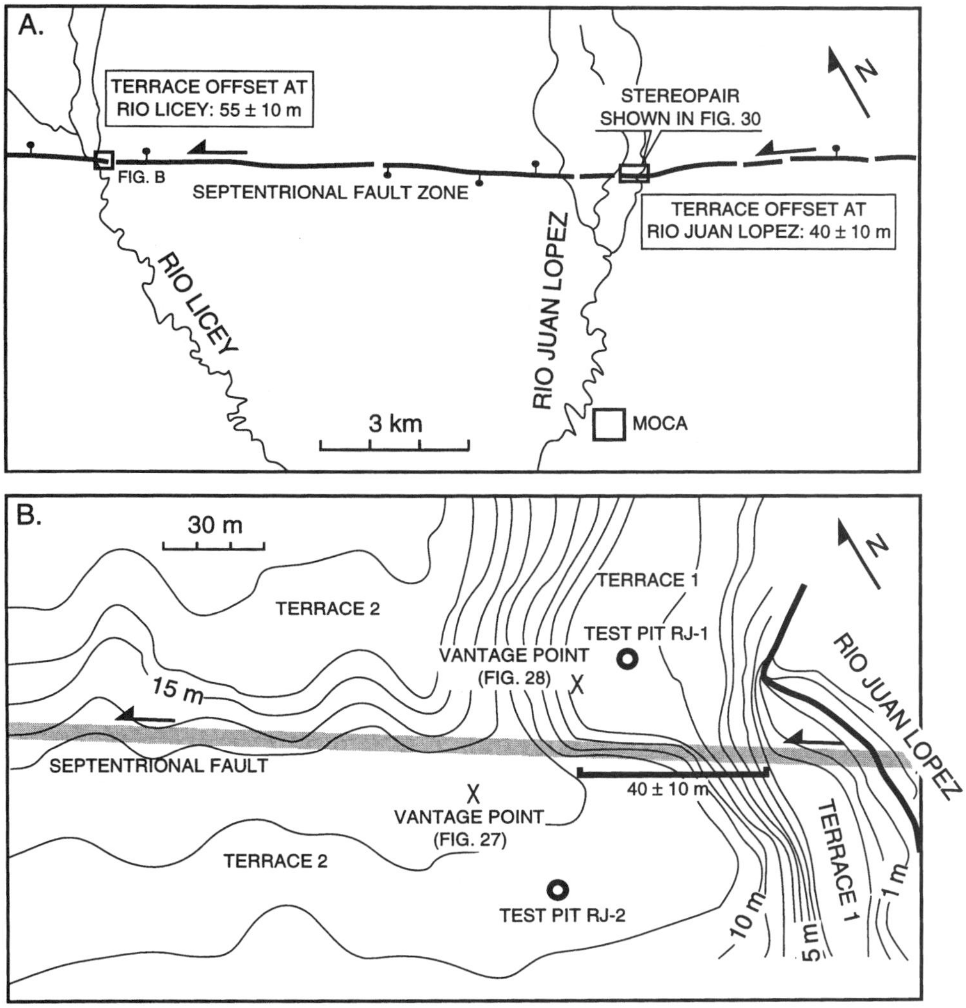

Figure 26. (A) Location map of offset terrace risers along the central Septentrional fault zone. (B) Hand-leveled topographic base map showing left-laterally offset of terrace riser along Rio Juan Lopez by 40 ± 10 m. A stereopair of this site is shown in Figure 21.

shows a view looking southeast from the riser between the two terraces across the north-facing fault scarp toward Terrace 2. The top of the riser between the two terraces is seen near the left edge of the photograph south of the fault. Vantage points for photographs in Figures 27 and 28 are shown in the map in Figure 26B.

Age estimate of terrace and rate of strike-slip offset. A shallow test pit (about 1 m^3) was hand excavated into the surface of each terrace (Fig. 26B). Although we found no charcoal, we collected a gastropod shell from the uppermost fluvial deposits below the modern soil exposed in test pit RJ-1, dug into the Terrace 1 surface north of the fault. The age of this sample provides an estimate of when the youngest fluvial unit associated with Terrace 1 was deposited, and, therefore, provides an estimate of when this terrace and the riser between Terraces 1 and 2 were abandoned and offset began to accrue. Analysis of this sample yielded a two-sigma, dendrochronologically calibrated radiocarbon age range of 2850–3190 B.P. (Table 1, sample 25). Because the shell was partially recrystallized, this must be viewed as a minimum age, and therefore provides a minimum estimate for the age of the offset. This minimum age estimate, combined with the measured offset of 40 ± 10 m provides a maximum average Holocene slip-rate estimate of 13 ± 4 mm/yr across the Septentrional fault at this location.

Rio Licey

Amount of horizontal offset. Rio Licey is another perennial river, located 8 km west of the Rio Juan Lopez, that flows southward across the Septentrional fault (Fig. 26A). Several terraces are present above the modern channel, as documented by our field observations and plane-table mapping (Fig. 29.) We suspect that the terraces labeled 2 and 3 on Figure 29 are probably part of a single terrace that has been vertically offset across the fault. The riser between Terrace 1 and the upper terrace(s) is left-laterally

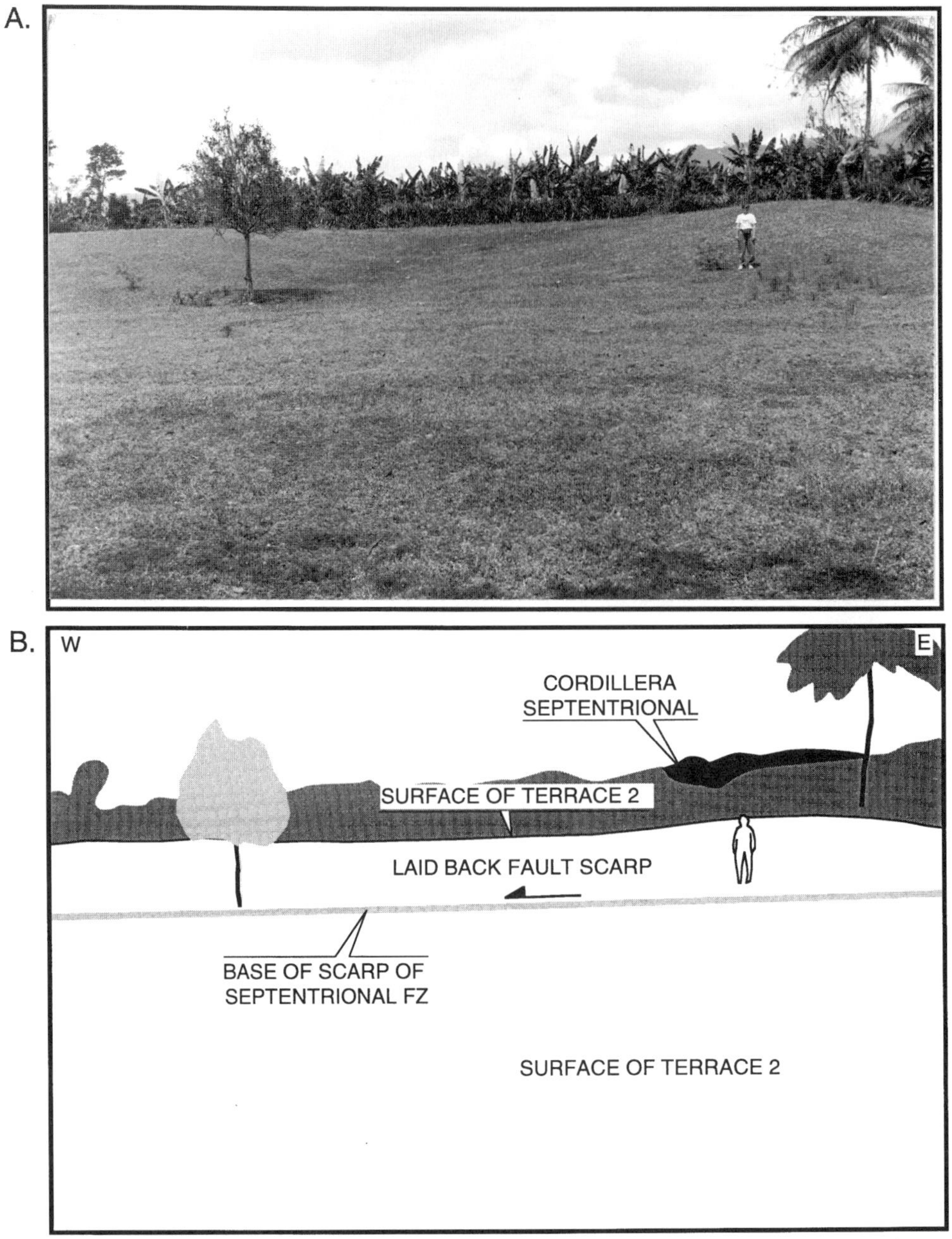

Figure 27. View to north across south-facing scarp in offset terrace surface 2 at the Rio Juan Lopez (see Fig. 26A for location and Fig. 26B for vantage point of photograph). The left-laterally offset terrace riser shown in Figure 28 and the Rio Juan Lopez is to the right of the field of view. The peaks of the Cordillera Septentrional at the inactive mountain front seen in the background of the photograph are about 2 km to the north of this locality.

offset as indicated by the thin arrows on the stereopair shown in Figure 19. The amount of offset is 55 ± 10 m based on our plane-table map (Fig. 29). The bar on this map represents the distance between the offset ends of the base of the riser. The south-facing fault scarp between the offset segments of the terrace riser is the result of strike-slip juxtaposition of higher Terrace 2 north of the fault against Terrace 1 south of the fault.

Age of terrace and rate of strike-slip offset. Three shallow test pits (each about 1 m^3) were hand excavated into the surfaces of the terraces (RL-1, RL-2, and RL-3 on Fig. 29). Although we found no charcoal in these test pits, we did collect another gastropod shell in pit RL-1, similar to that collected from RL-1 on the Rio Juan Lopez terrace. Radiocarbon analysis of this shell, collected from the youngest undisturbed fluvial deposits below the modern soil on Terrace 1, yielded a

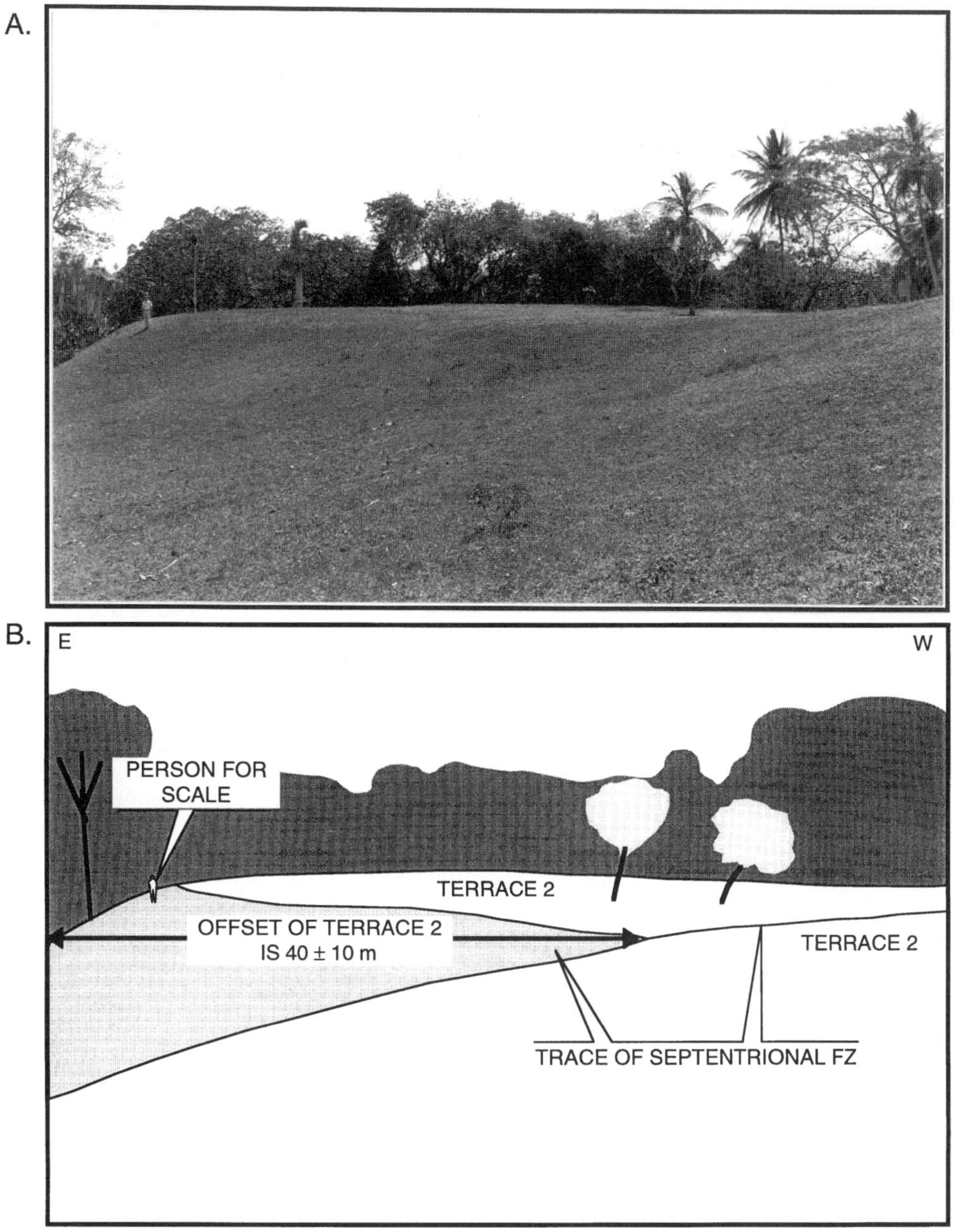

Figure 28. View to south from top of terrace level 2 on the Rio Juan Lopez to left-laterally offset riser of terrace 2 (gray area approximates plane of Septentrional fault zone; see Fig. 26A for location and Fig. 26B for vantage point of photograph). Terrace riser on southern block is to left of person; terrace riser on northern block is approximately at the vantage point of this photograph. Left-lateral offset of riser is 40 ± 10 m.

two-sigma, dendrochronologically calibrated radiocarbon age range of 2710–2180 B.P. (Table 1, sample 26). As for the Rio Juan Lopez terraces, the age of this sample provides an estimate of the time the terrace and riser were abandoned, and offset of the riser began to accumulate. This is a minimum age for the sample because the shell was partially recrystallized. This age combined with the measured offset allows calculation of a maximum average Holocene slip rate across the Septentrional fault of 23 ± 7 mm/yr.

That the two slip-rate estimates are not in close agreement (13 ± 4 and 23 ± 7 mm/yr) underscores the fact that they are both maximum rates. Both must be viewed only as maximum rates because each is based on a minimum age. Additional age control (preferably on charcoal samples) on the terraces is needed to provide better constraints on the slip rate of the Septentrional fault. These data suggest that the Holocene slip rate across the Septentrional fault is less than 13 ± 4 mm/yr.

TABLE 1. CALCULATED DATES FROM ^{14}C ANALYSIS OF CHARCOAL, ORGANIC-RICH SILT, AND CLAY HORIZONS, AND ONE GASTROPOD (91DRT2-20).

No.	Sample	^{14}C Age* (B.P.)	Calendar Date† (A.D. if Unlabeled)	Unit	Average Dates§
1	91DRT2-5A	890 ± 170	800–1330	10	
2	91DRT2-20	360 ± 70	1423–1657	30	
3	*91DRT2-14B&14B2*	*600 ± 60*	*1281–1420*	*30*	
4	91DRT2-106A West	1,050 ± 60	804–1154	30	
5	*91DRT2-17*	*7,100 ± 60*	*6091–5797 B.C.*	*30*	
6A	91DRT2-2B	240 ± 40	1519–1955	40	A.D. 1492–1795
6B	91DRT2-2B	370 ± 70	1422–1652	40	(Avg. of 6A–B)
7	91DRT2-17B West	690 ± 60	1219–1403	40	
8	91DRT2-16C East	570 ± 50	1294–1426	50	
9	91DRT2-7C	610 ± 50	1282–1410	50	A.D. 1277–1389
10A	91DRT2-105C	650 ± 50	1276–1407	50	(Avg. of 8, 9,
10B	91DRT2-105C	690 ± 60	1219–1403	50	and 10A–C)
10C	91DRT2-105C	770 ± 40	1164–1283	50	
11A	91DRT2-4D East	720 ± 50	1209–1390	60	A.D. 1207–1285
11B	91DRT2-4D East	780 ± 70	1038–1386	60	(Avg. of 11A–B
12	91DRT2-3C East	790 ± 70	1037–1300	60	and 12)
13	*91DRT2-1C West*	*890 ± 40*	*1036–1225*	*60*	
14	*91DRT2-17C West*	*920 ± 50*	*1002–1215*	*60*	
15	*91DRT2-17C West*	*1,000 ± 70*	*889–1194*	*60*	
16	*91DRT2-17,16*	*1,070 ± 50*	*781–1033*	*60*	
17	91DRT2-17D West	740 ± 50	1163–1387	65C	A.D. 1158–1259
18	91DRT2-18D East	780 ± 40	1163–1282	65C	(Avg. of 17, 18,
19A	91DRT2-op17	880 ± 40	1030–1255	65B	and 19A–C)
19B	91DRT2-op17	910 ± 50	1002–1221	65B	
19C	91DRT2-op17	930 ± 60	993–1221	65B	
20	91DRT2-17D West	1,310 ± 90	554–944	70	A.D. 644–775
21	91DRT2-1D West	1,310 ± 40	654–795	70	(Avg. of 20 and 21)
22A	91DRT2-15E West	2,190 ± 70	393–96 B.C.	80	
22B	91DRT2-15E West	2,280 ± 60	712–172 B.C.	80	397–210 B.C.
23A	91DRT2-7&8D	2,260 ± 50	405–202	80	(Avg. of 22A–B
23B	91DRT2-7&8D	2,270 ± 40	404–208 B.C.	80	and 23A–B)
24	*91DRT2-7D*	*3,160 ± 60*	*1601–1313 B.C.*	*80*	
25	93DR RJL P1-2	2,890 ± 50	2850–3190 B.P.		
26	93DR TA P1-1	2,380	2180–2710 B.P.		

Note: Dates of samples in italics are not in correct stratigraphic order, and we believe that these samples were contaminated with ancient carbon absorbed into the silt and clay that resisted pretreatment dissolution. Samples numbered 1–6B, 8–12, 14, 16–19C, and 22–26 were analyzed at the University of Arizona using accelerator mass spectrometry; samples numbered 7, 13, 15, 20, and 21, analyzed at the University of Texas, are bulk samples of organic-rich horizons dated by conventional radiocarbon analysis; numbers 6A, 8, 9, 10C, 11A–B, and 17–19A are best samples, hand-picked charcoal to eliminate contaminated clay and silt.

*Calculations assume a Libby half-life (5,568 yr). Sample and standard d^{13}C values are normalized to –25‰, except sample 2, a terrestrial gastropod with d^{13}C= –11.1‰, and samples 25 and 26, also terrestrial gastropods with d^{13}C= –10‰. Uncertainties are 1 standard deviation counting errors. Uncertainties include several sources of error including the blank, the random machine error, and counts.

†Dendrochronologically calibrated, calendar age ranges from program of Stuiver and Reimer (1993), Method B, 10 yr calibration curve, 2 standard deviation uncertainty.

§Weighted average using program of Stuiver and Reimer (1993), Method B. Calendar years, 2 standard deviation uncertainty.

Other examples of offset streams

West of the Rio Moca, we observed several offset streams associated with a well-expressed south-facing fault scarp (Fig. 20). These streams and their offsets are smaller than either the Rio Juan Lopez or the Rio Licey. We documented one of these small offset streams (shown by letter "R" in Fig. 20) by making a detailed plane-table map (Fig. 30). The western edge of the channel north of the fault is sharp and is indicated by one end of the black bar. South of

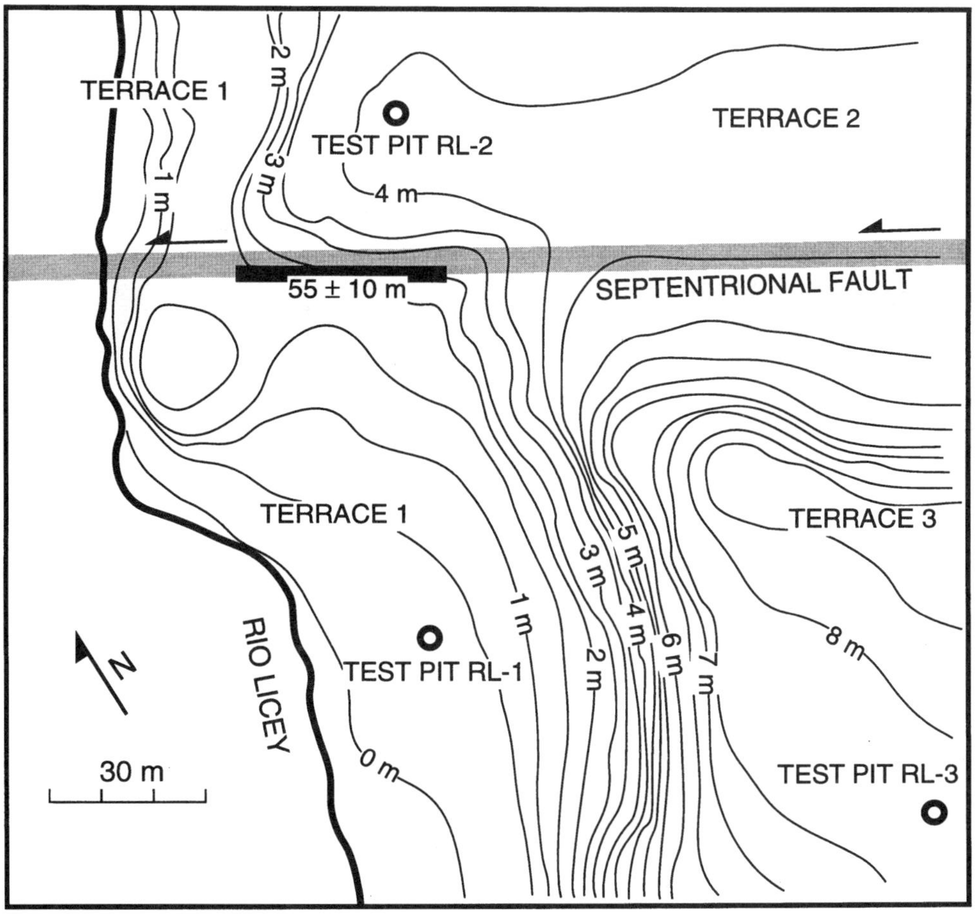

Figure 29. Contour interval (0.5 m) topographic map of left-laterally offset fluvial terrace riser of terrace along Rio Licey (cf. Fig. 26A for location; thin arrows indicate offset on stereopair in Fig. 19). Left-lateral offset is 55 ± 10 m.

the fault the offset western edge of the channel is less distinct. The eastern thin gray bar on Figure 30 shows the projection of the downstream segment of the thalweg back to the fault and represents the easternmost possible location of the offset, western channel edge. The western thin gray bar shows the westernmost possible location of the offset, western channel edge. The error in the position of the western channel edge south of the fault is the 3 meter distance between the two thin, gray bars (Fig. 30).These considerations imply the stream is left-laterally offset 15 ± 3 m.

Two perennial streams east of Rio Bacú each display a left-lateral deflection of about 400 m as they cross the Septentrional fault (Fig. 22). Fault profile 5 measured 500 m west of the western stream near Rio Bacú (Fig. 17B) shows that the scarp is north facing and has 4.1 m of relief. Marshy areas are formed along the north side of the fault by blockage of streams flowing southward from the Cordillera Septentrional into the Cibao Valley (Fig. 22).

Left-lateral stream and stream terrace offsets are common along the Septentrional fault zone. Those that we document here range in magnitude from 15 ± 3 m in Figure 30 to about 400 m in Figure 22. This range of offsets on streams of varying size and age indicates that slip along this fault has accrued throughout late Quaternary time. Maximum slip rates estimated by measuring and dating offset terrace risers at Rio Juan Lopez and Rio Licey are 13 ± 4 and 23 ± 7 mm/yr, respectively. These estimates provide maximum limits on the slip rate across the Septentrional fault, suggesting that the slip rate is less than 13 ± 4 mm/yr.

7. EVIDENCE FOR LACK OF LATE QUATERNARY MOVEMENT ON THE NORTHERN BRANCH OF THE WESTERN SEPTENTRIONAL FAULT ZONE

Scarps in bedrock

The northern branch of the western Septentrional fault system exhibits impressive examples of strike-slip fault geomorphology (Fig. 14). The most prominent scarps are found on the

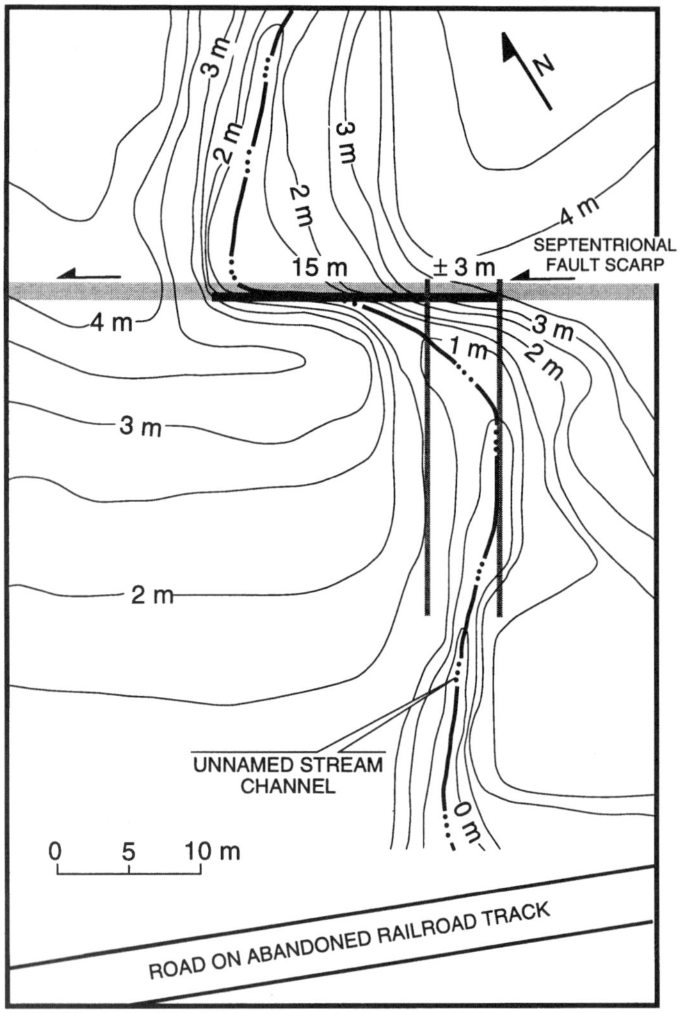

Figure 30. Contour interval (1 m) topographic map of left-laterally active offset of unnamed stream channel west of Paso de Moca near abandoned railroad track. Left-lateral offset of stream channel is 15 ± 3 m. A stereopair of this site is given in Figure 20.

Monte Cristi and Villa Vasquez fault zones in the western part of the area (Fig. 14). The mountain-front fault becomes less prominent or covered along much of its length in the westernmost area of the Cordillera Septentrional. In contrast, the eastern ends of the Monte Cristi and Villa Vasquez faults are covered and the western ends near the towns of Villa Vasquez and Monte Cristi become more prominent (Fig. 14). To document scarps along the western part of the Monte Cristi fault, we used a transit and rod to measure three fault scarp profiles at undisturbed or little-disturbed fault localities over a distance of 10 km (localities 1-3 on Fig. 14).

Four characteristics of the Villa Vasquez and Monte Cristi fault zones are remarkable and discussed in detail in the following paragraphs:

(1) At all localities both faults form a single, straight, and continuous scarp lacking oblique splays. This characteristic is best seen on the mosaic of 1:40,000 scale aerial photographs shown on Figure 16.

(2) At all localities visited, these single scarps of the Villa Vasquez and Monte Cristi faults occur in indurated, late Neogene sedimentary rocks that are gently folded and uplifted in low hills in this part of the Cibao Valley (Calais et al., 1992b) (Fig. 16).

(3) No evidence for Holocene offset or disturbance was found at localities where the bedrock fault trace is prominent on aerial photographs and where it intersects stream valleys or overlying, late Quaternary alluvium.

(4) The scarps of the Monte Cristi and Villa Vasquez fault zones are linked by the geomorphologically prominent, right-stepping, 12-km-long, 1.5-km-wide, Villa Vasquez push-up structure (Fig. 16).

Scarp of the Villa Vasquez fault zone east of the Villa Vasquez push-up. In this area the Villa Vasquez fault zone separates low hills composed of well-lithified sandstone and shale lithologies cropping out in the western Cordillera Septentrional from a pediment surface composed of older, coalesced fans and colluvium that fringes the northern edge of the Cibao Valley (Fig. 31). Field checking of the head of the alluvial fan deposited across the scarp at Arroyo Hondo (indicated by the northeast-pointing arrow in Fig. 31) showed no evidence for displacement. We attribute the youthful appearance of the Villa Vasquez fault zone in this area to the dry climate (Fig. 3) and the juxtaposition of resistant sandstone north of the scarp and less resistant shales south of the fault.

Scarps of the Villa Vasquez and Monte Cristi fault zones within the Villa Vasquez push-up. The overlap of the scarps of the Villa Vasquez and Monte Cristi fault zones bounding the push-up accounts for the slightly higher topographic elevation in the push-up and the area of badlands topography south of the Villa Vasquez fault zone and north of the town of Villa Vasquez (Figs. 16, 32). The distinctive badlands topography is produced by weathering of less resistant shale units observed south of the Villa Vasquez fault zone.

The southern edge of the badlands topography is marked by a linear edge that may be a strike-slip fault strand parallel to the Villa Vasquez and Monte Cristi fault zones (Figs. 16, 32). Both the Villa Vasquez and Monte Cristi fault zones form prominent fault valleys at the edges of the push-up that are locally occupied by streams. Arroyo El Charcazo obliquely crosses the push-up and is deflected at both faults in a left-lateral sense by 200–300 m as shown by the thin arrows on the stereopair in Figure 32. Field checking of both localities along Arroyo El Charcazo showed no evidence for displacement of young terraces overlying the lithified late Neogene sedimentary rocks forming the walls of the fault valleys. Landform variations in the push-up area appear closely related to fault-juxtaposed lithologies. The rugged area north of the fault is a sand and shale facies, the rugged area within the push-up is a sandier facies, and the low-lying badlands topography south of the fault is a shalier facies that is less resistant than the facies to the north of the Villa Vasquez and Monte Cristi fault zones.

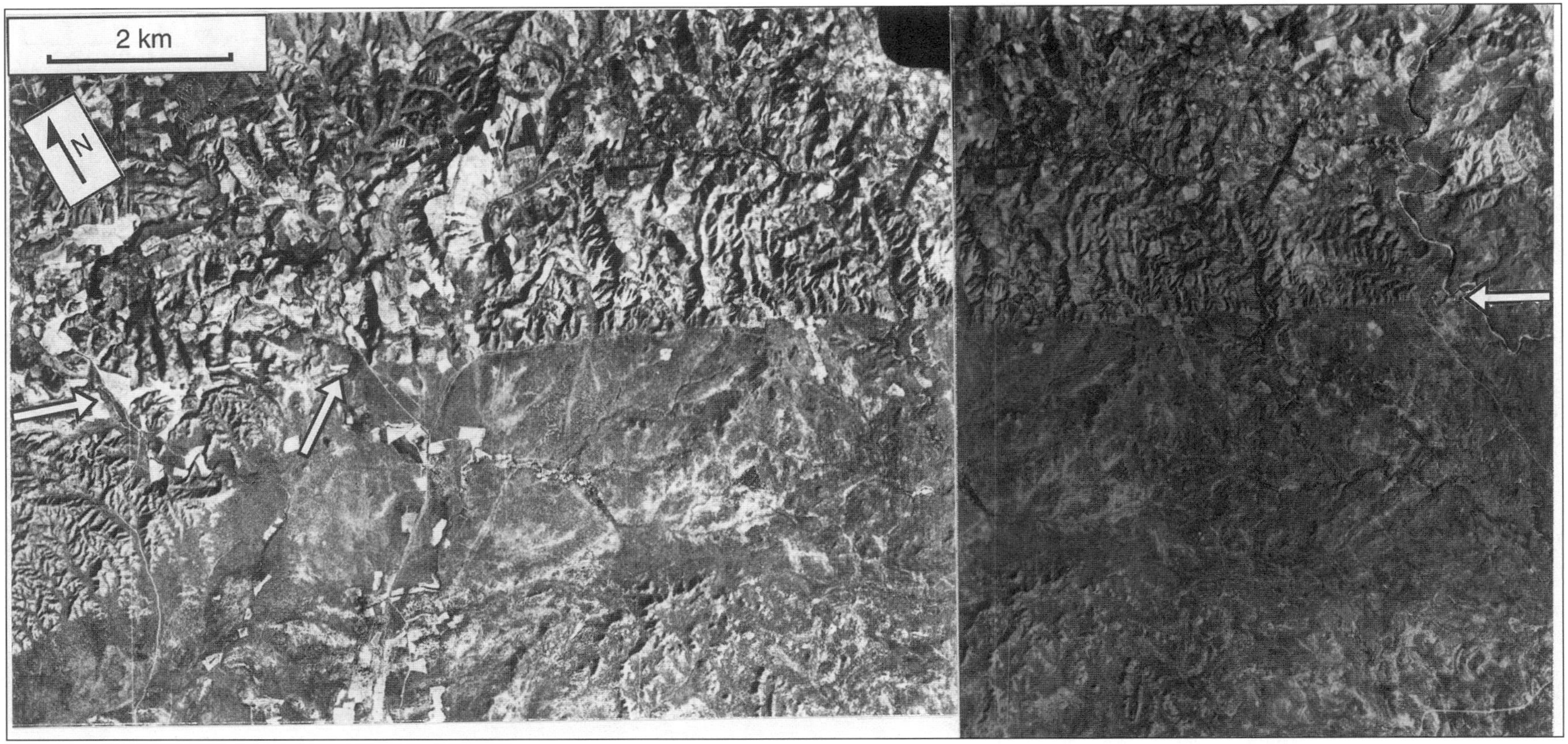

Figure 31. Stereopair of 1958 aerial photographs showing scarp of the Villa Vasquez fault zone east of the Villa Vasquez push-up (cf. Fig. 16 inset for location). Despite its fresh appearance, the scarp exhibits no evidence for late Pleistocene displacement. The fan deposited across the fault by the Arroyo Hondo at northeast-pointing arrow shows no evidence of displacement. The fresh appearance of this scarp is attributed to the lithologic contrast and differential erosion across the scarp and the lack of precipitation along this area of the Septentrional fault system (cf. Fig. 3). Aerial photographs: ICM-1(1958)-13-V28-12-29, December, 1958, nos. 1096 and 1097; original scale, 1:60,000.

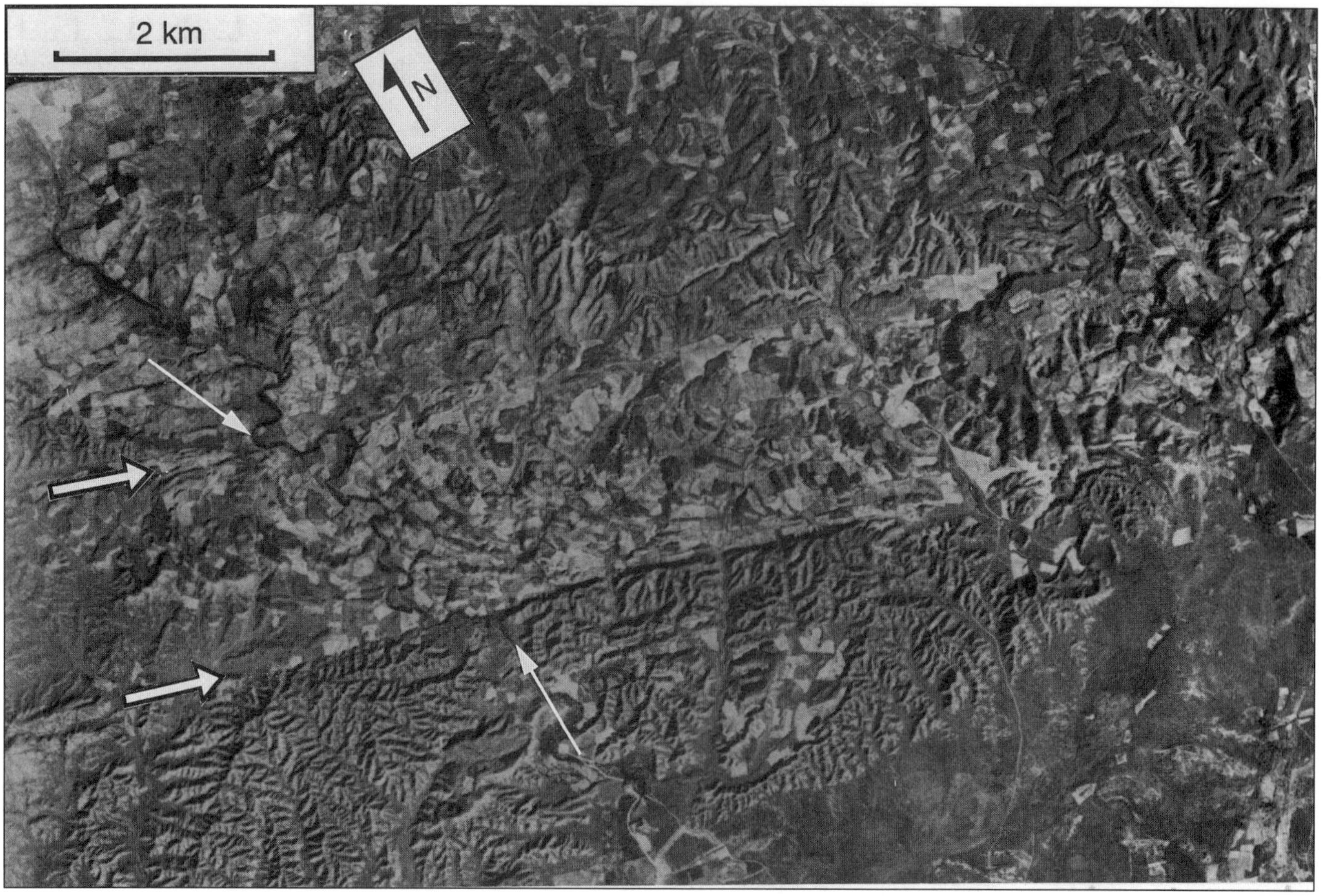

Figure 32 (this and opposite page). Stereopair of 1958 aerial photographs showing scarps (broad white arrows) of the Monte Cristi and Villa Vasquez fault zones bounding the Villa Vasquez push-up (cf. Fig. 16 inset for location). Despite their fresh appearance, neither scarp exhibits evidence for Holocene displacement. Stream (Arroyo el Charcazo) indicated by pair of thin white arrows exhibits left-lateral deflection of their channels but no evidence for displacement in young terrace deposits. The preservation of these scarps is attributed to the presence of resistant sedimentary lithologies along the scarp and the dry climate (cf. Fig. 3). Aerial photographs: ICM-1(1958)-13-V28-12-29, December, 1958, nos. 1096 and 1097; original scale, 1:60,000.

Scarp of the Monte Cristi fault zone west of the Villa Vasquez push-up and fault exposure at Arroyo Salado. The Monte Cristi fault zone 14 km west of the push-up is characterized by a prominent scarp in late Neogene sedimentary rocks (Fig. 33). We investigated an artificial exposure cut across the Monte Cristi fault zone during the construction of a dam across the Arroyo Salado (indicated by southwest-pointing arrow on Fig. 33). Three main lithologic units are present in the exposure and are shown in cross section in Figure 34A: Pliocene thinly bedded sandstone and shale south of the fault; resistant Pliocene sandstone north of the fault, and overlying late Pleistocene(?) fluvial gravel.

Horizontally bedded, late Pleistocene(?) fluvial gravel and minor sandstone sits unconformably upon moderately to shallowly dipping Pliocene sandstone and shale and appears to have formed a late Pleistocene(?) alluvial fan derived from the hills of the Cordillera Septentrional to the north and deposited at the mouth of the Arroyo Salado (Fig. 34A, B). Some of the fan gravels have been disturbed or completely removed for use in the landfill dam across the Arroyo Salado. Pliocene thinly bedded sandstone and shale is found south of the south-facing scarp of the Septentrional fault and is largely covered by its own talus. Dips are southward away from the fault scarp. High-angle faults are present that mainly dip northward and are unconformably overlain by the horizontal gravel unit. Pliocene thickly bedded sandstone is found north of the south-facing scarp of the Septentrional fault and is well exposed. Traverses showed that this unit underlies an extensive area of the low hills of the Cordillera Septentrional north of the fault. As observed in Villa Vasquez push-up area (Fig. 16), this unit appears to be much less erodible than the thinly bedded sandstone and shale south of the fault and therefore may contribute to the higher topography of the low, steep-sided hills north of the fault zone.

Shallow excavations and enlargement of the artificial fault

exposure at the Arroyo Salado site revealed that the Monte Cristi fault zone juxtaposes the thinly bedded sandstone and shale and thickly bedded sandstone units (Fig. 34B). The Monte Cristi fault zone is overlain by and therefore predates the late Pleistocene gravels capping the exposure. Faults within the Pliocene thickly bedded sandstone unit extend to the top of the exposure because the overlying alluvial fan gravels were removed during the dam excavation. However, we are confident that the gravel cap previously extended over this part of the outcrop and that these fault features are truncated by the fan gravels. Moreover, the lineament of the Monte Cristi fault zone shown on Figure 33 aligns well with the fault we indicate on the cross section in Figure 34A. Based on this observation along with the observations made in areas to the east, we conclude that Villa Vasquez and Monte Cristi fault zones of the western Septentrional fault system are inactive and that interplate motion is now accommodated along the largely buried Septentrional fault zone that extends through the southern Cibao Valley (Fig. 14). Although we did not examine exposures of the mountain-front fault in the field, we infer that this fault may also be inactive based on its embayed appearance in aerial photographs (Fig. 14).

Fault scarp profiles of the Monte Cristi fault zone. We surveyed three fault scarp profiles along a 3-km-long segment of the Monte Cristi fault zone using a rod and transit (Fig. 14). Profiles shown in Figure 35 reveal that all three localities show a scarp exposing late Neogene sedimentary rocks along both the upthrown and downthrown sides of the fault. Bedrock north of the fault at all three localities is similar to that seen at Arroyo Salado and consists of an indurated and thickly bedded sandstone. Bedrock south of the fault is a thinner bedded sandstone and shale sequence.

8. TIME CONSTRAINTS ON THE MOST RECENT EARTHQUAKE ALONG THE CENTRAL SEPTENTRIONAL FAULT ZONE

Geologic setting of trench sites on the Cenoví alluvial fan

We chose the Rio Cenoví fan as a site for detailed paleoseismic study because fine-grained alluvial fan sediments have buried the Holocene trace of the Septentrional fault across most of the width of the fan (Figs. 24, 25). We excavated trenches across the fault at two locations, 91-1 and 91-2 (Figs. 24, 25), exposing a sequence of alluvial and quiet-water sediments that record the most recent ground-rupturing earthquake. Radiocarbon analysis of charcoal contained within the fine-grained sedimentary layers allows time constraints to be placed on the age of this seismic event (Table 1).

Figure 33. Stereopair of 1958 aerial photographs showing scarps of the Monte Cristi fault zone west of the Villa Vasquez push-up (cf. Fig. 16 inset for location). Despite their fresh appearance, these scarps exhibit no evidence for displacement of late Pleistocene fan deposits or misaligned microtopography. An unfaulted late Pleistocene fan crosses the fault trace at the Arroyo Salado (indicated by southwest-pointing arrow, cf. Fig. 34 for cross section). Preservation of these scarps is attributed to contrasting resistance to erosion of juxtaposed lithologies along the scarp (indurated and thickly bedded sandstone to north, less resistant rocks to south) and the dry climate (cf. Fig. 3). Aerial photographs: ICM-1(1958)-13-V28-12-29, December, 1958, nos. 1093 and 1094; original scale, 1:60,000.

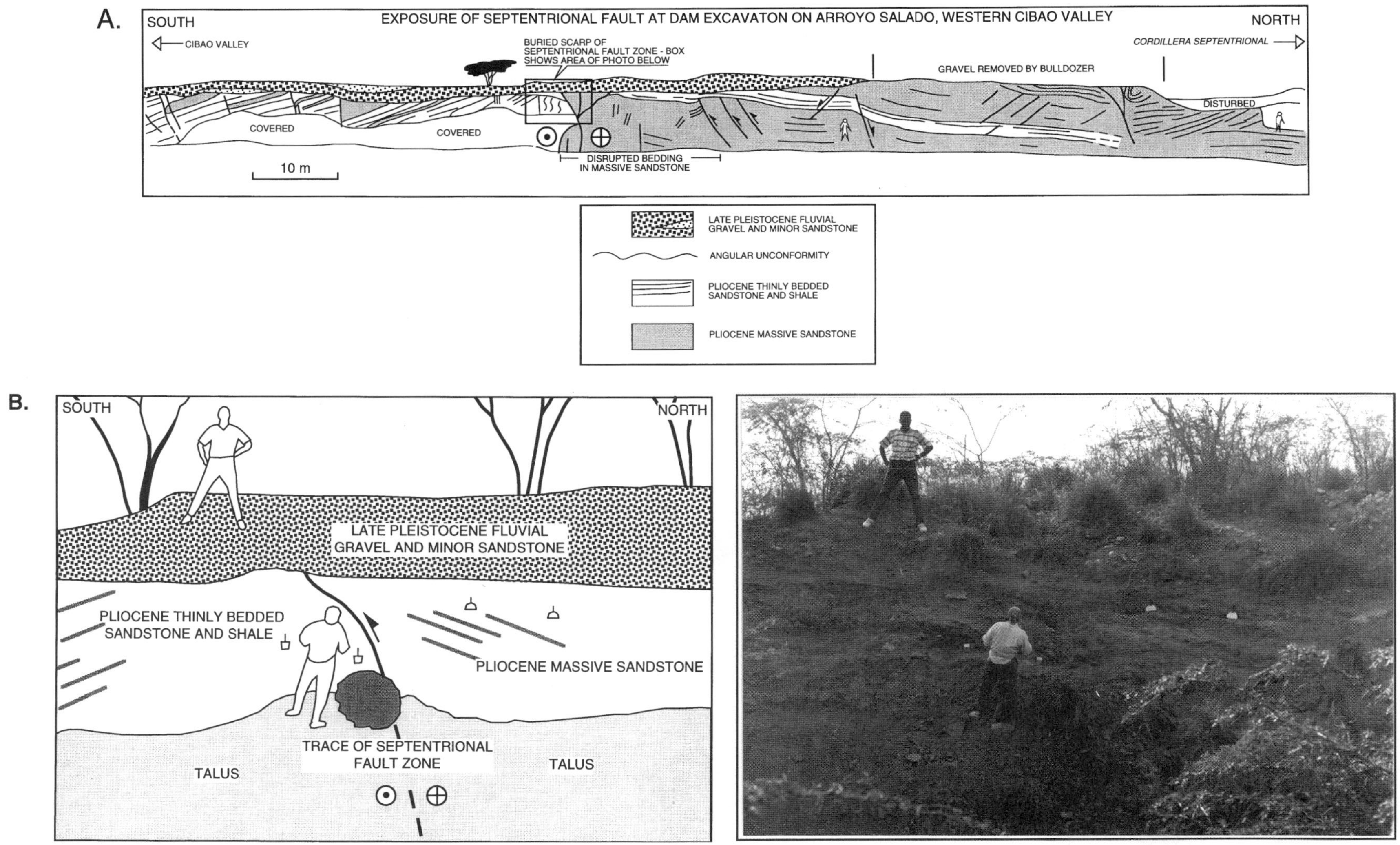

Figure 34. (A) Cross section along artificial exposure of the Monte Cristi fault zone at a dam excavation on the Arroyo Salado (section drawn from photographs; locality is numbered “2” on Fig. 14 and is shown by southwest-pointing arrow on Fig. 33). The fault juxtaposes more resistant sandstone to the north and less resistant sandstone and shale to the south and is buried by unfaulted late Pleistocene alluvium. The fresh appearance of the scarp on aerial photographs and in the field is attributed to the contrast in erodibility of the two rock types. (B) Interpretation and close-up photograph showing unfaulted late Pleistocene fan (fluvial gravel and minor sandstone) along the Monte Cristi fault zone.

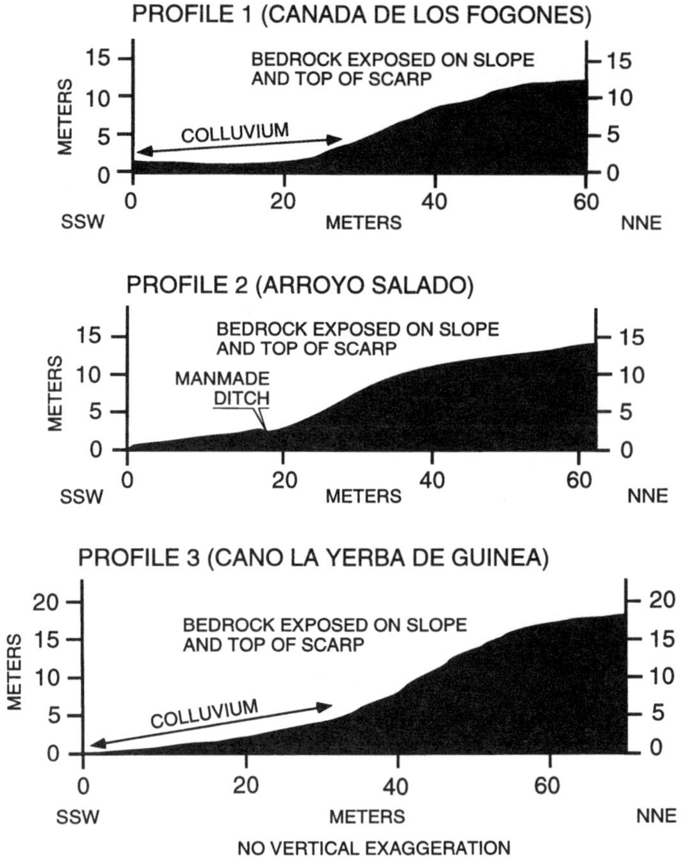

Figure 35. Leveled topographic profiles of the scarp of the Monte Cristi fault west of the Villa Vasquez push-up (see Fig. 14 for locations of profiles 1–3). In all three cases, indurated and thickly bedded sandstone exposed on the northeast side of the fault contributes to the remarkable preservation of the inactive scarp as seen in the stereopair in Figure 33.

The two trench sites, 91-1 and 91-2, are separated by about 750 m and were excavated in February, 1991. Preliminary results from trench 91-2 were published in a short article by Prentice et al. (1993). The more detailed observations presented in this chapter support the main conclusions of Prentice et al. (1993).

Paleoseismic results

Site 91-2. We excavated a trench across the fault scarp near the center of the Cenoví fan, about 900 m west of the Rio Cenoví (Fig. 24). This site is located in a privately owned plantain field north of the paved road that connects the community of Ojo de Agua to the main highway between Santiago and San Francisco de Macorís. Figure 36 is a 1993 aerial view of the site showing the location of the trench site relative to the paved Ojo de Agua road, its major cultural features, and the projected trace of the buried Septentrional fault zone on both sides of the trench site. Figure 37B provides a ground view from the south showing the excavated and shored trench.

The absence of the scarp at the 91-2 trench site is evidence that the most recent earthquake has been buried by subsequent alluvial fan sedimentation (Fig. 24). We sited the trench across the projected trend of the fault scarps at the eastern and western edges of the Rio Cenoví fan (Figs. 24, 25). Figure 37A is a surveyed map of this site that we made using a rod and transit to show the topography of the fan surface and to locate the trench relative to the reference points on the paved road. We logged the western wall of the trench at a scale of 1:20 with reference to a 1 m^2 grid as shown on the trench log in Figure 38.

This excavation exposed a 4-m-thick, well-bedded stratigraphic section of fluvial and calm-water sediment (Fig. 38). Three massive, organic-rich black clay horizons form prominent marker horizons that are shown in black on Figure 38 and partially on the photograph of the western trench wall in Figure 39. The lowermost stratigraphic units in the trench are faulted and folded and are overlain by undeformed units. The Septentrional fault zone, near the southwestern end of the trench, is about 6 m wide and consists of at least six fault strands over this distance. The faults bound a small graben with the southern strands downthrown to the north and the northern strands downthrown to the south. Offset units indicated on Figure 39 exhibit thickness changes across one of the fault planes, demonstrating strike-slip displacement. These high-angle faults displace units 80 and 85 and terminate in unit 70, and several fault fissures are filled with the black, organic-rich clay of unit 70 (Figs. 38, 39).

Units 70 and 80 are folded into a broad anticline over a lateral distance of about 20 m north of the fault zone (Fig. 38). Slightly sandy silt, clayey silt, and silty clay units 65A–D were deposited on the flanks of this anticline and pinch out over its crest (Fig. 40), making the contact between unit 60 and 70 indistinguishable over the crest of the anticline as well as on the upthrown block southwest of the fault zone. Unit 60/70, a clay unit, drapes the fault scarp. Horizon 50 thins over the crest of the anticline and on the southwest side of the fault zone, and both units 50 and 40 drape the fault scarp. Unit 30 thins over the fault zone to the southwest, while unit 20 continues unaffected over the fault zone (Fig. 38). A low-relief elongate rise, visible on the ground surface northeast of the trench, may correspond to this anticline where it has not yet been completely buried (Figs. 36, 37A).

These relationships indicate the occurrence of a ground-rupturing earthquake during the deposition of unit 70. The lower part of unit 70 was in place at the time of the earthquake; it was disrupted by faulting. Continued deposition of unit 70 filled the fissures with organic-rich, black clay. Cobbles and pebbles derived from unit 85, scattered in the clay of units 70 and 60/70 within a few meters northeast of the fault zone. These cobbles and pebbles probably rolled off the north-facing scarp formed in the loose gravel of unit 85 (unit 70 and older units were folded during this event), while subsequent sedimentary horizons (units 65A–D) were deposited over the folded and faulted ground surface (Fig. 38). Later sedimentation buried the deformed ground surface completely.

A.

B.

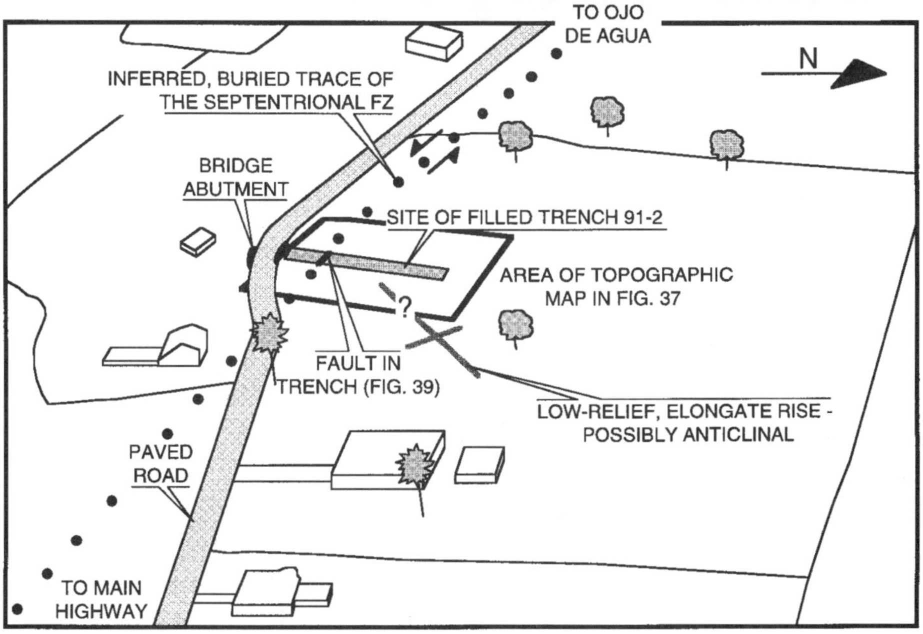

Figure 36. Aerial view taken in 1993 and interpretation of the area of trench site 91-2 in the middle of the Cenoví fan (see Fig. 24 for setting of Cenoví fan). In this area there is no topographic expression of the fault trace. Dots mark approximate position of buried fault trace and dash shows its mapped location in trench 91-2. Box shows location of topographic map in Figure 37A.

Radiocarbon analysis. The numbered triangles on Figure 38 show the locations of 24 samples collected for radiocarbon dating from trench 91-2. Table 1 shows the results of the 32 radiocarbon analyses of these samples and indicates that this stratigraphic section has been deposited over the last 2,400 yr (Prentice et al., 1993). We consider the 22 dates listed in normal print to be reliable because they are in stratigraphic order and multiple dates from individual units are indistinguishable. For units 80, 70, 65, 60, 50, and 40 we average multiple samples that yield indistinguishable ages to give the average age ranges that are summarized as a stratigraphic column in Figure 41.

Unreliable dates are indicated by italic print on Table 1. All of the bulk samples (samples numbered 7, 13, 15, 20, and 21) give ages older than charcoal collected from the same unit or picked out of the bulk samples; therefore we consider them to be unreliable. We believe that these samples were contami-

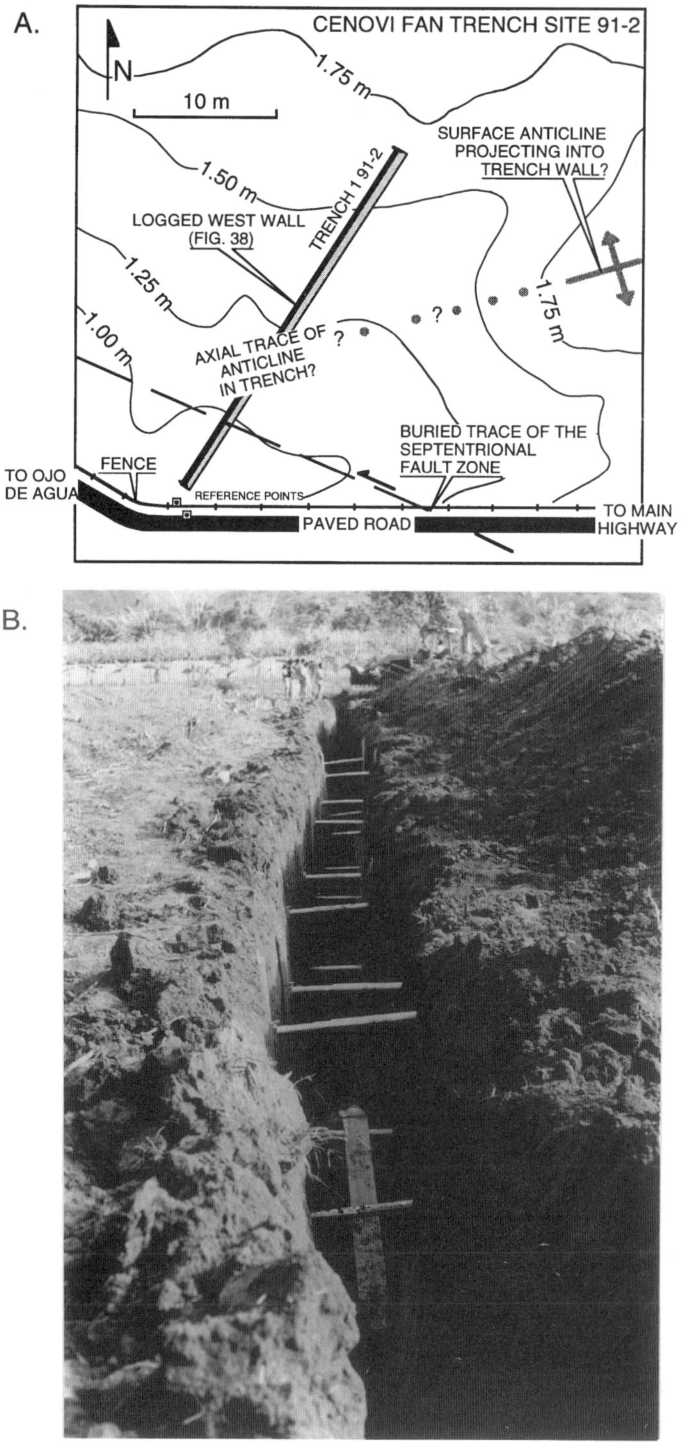

Figure 37. (A) Topographic map of site 91-2. Note the presence of the low ridge that may correspond to the buried anticline in late Holocene sediments deformed adjacent to the fault plane. (B) View looking north of trench site 91-2 on the Rio Cenoví fan. Seven-ft-long aluminum hydraulic shores expand from 28 to 46 in wide and prevent trench walls excavated by backhoe from collapsing. Shores also provide supports for planks used as catwalks during trench logging. Excavated active trace of the Septentrional fault zone is in foreground near area of plank.

nated with ancient carbon, derived from the late Neogene carbonate cap (Villa Trina Formation) that is being eroded in the drainage area of the Rio Cenoví (Fig. 37B). This carbonate material was absorbed and resisted pretreatment dissolution. We found that unconventionally rigorous pretreatment on charcoal samples that were hand picked to avoid contamination with ancient carbonate was needed to give reliable ages and that analyses of bulk samples of the organic-rich black clays did not yield useful results.

The radiocarbon ages of units 65B and C are indistinguishable, and therefore we averaged them to yield an age estimate of A.D. 1160–1260 (Table 1; Fig. 41). Because these units were deposited after the time of the earthquake, this represents a minimum amount of time since the most recent ground-rupturing earthquake occurred along this segment of the Septentrional fault. The age of unit 70, based on the weighted average of two indistinguishable sample ages, is A.D. 640–780 (Fig. 41). Although we suspect these samples were deposited before the time of the earthquake, the ambiguity resulting from the observation that deposition of unit 70 spanned the time of the event does not allow us to use this as a maximum limiting age with much confidence. The age of unit 80 (averaged from four indistinguishable sample ages) is 400–210 B.C. and provides a certain maximum limit for the date of this earthquake. Therefore, the most recent ground-rupturing earthquake at this site occurred before A.D. 1260, more than 730 yr ago (Fig. 41).

Site 91-1. We excavated a second trench across the fault scarp near the easternmost edge of the Cenoví fan, about 150 m west of the Rio Cenoví (Fig. 24). This site is located in a privately owned plantain field north of the paved road that connects the community of Ojo de Agua to the main highway between Santiago and San Francisco de Macorís. Figure 42 is an aerial view of the site showing the low, less than 1-m-high, north-facing scarp that trends about 110° across the gently sloping fan surface. We surveyed the site using a rod and transit to make a detailed site map, shown in Figure 43. This map shows the trench location relative to a reference point established on a concrete water well 30 m east-northeast of the trench site. We logged the eastern wall of the trench at a scale of 1:20 as shown on Figure 44.

The well-bedded section of fluvial sand and gravel, and calm-water clay and silt exposed in trench 91-1 is similar to the section seen at trench 91-2 (Fig. 38) and is offset across two fault traces near the southwestern end of the trench (Fig. 44). Facies changes across the southwestern fault demonstrate strike-slip displacement, and the large vertical separation suggests a component of normal fault slip. Units between the two fault traces dip about 20° toward the northeast. These sand and gravel units are absent on the northeast side of the fault located near meter-mark 8; only massive clay and silty clay units are exposed in the northeastern part of the trench. Another fault trace that is less distinct is shown between meter-marks 11 and 12.

A massive, organic-rich, black clay horizon overlies the alluvial stratigraphy and forms a prominent marker horizon and the ground surface near the southwestern end of the trench. This horizon, shown in black on Figure 44, is not certainly faulted; however, its massive nature would make the detection of faults within the unit difficult. For this reason, we are uncertain how high in the section faulting extends in this exposure. The geometry of the unit suggests that it may have been folded as a result of an earthquake. Alternatively, it may, at least in part, have been deposited after a faulting event because it thickens toward the northeast on the down-dropped side of the fault zone. Because it is a clay horizon, it may not have been deposited horizontally and could have draped the fault scarp.

Surprisingly, the scarp is not directly associated with any of the faults exposed in the trench, but occurs 4–5 m southwest of the nearest fault (Fig. 44). This suggests that the scarp has retreated due to fluvial erosion subsequent to the most recent earthquake. If so, then at least the lower part of the black clay horizon was in place at the time of the most recent earthquake, and the upper part of the unit has been planed off by subsequent erosion.

We collected bulk samples of the organic-rich black clay horizon, but found no charcoal or other material suitable for radiocarbon analysis in any other units (Fig. 44). Because our experience with bulk samples from trench 91–2 showed that they provide erroneous ages (Table 1), and because of our uncertainty about the relationship of the black clay to the most recent faulting event, none of these samples was analyzed.

9. DISCUSSION

Seismic hazard of the Septentrional fault zone

The most recent ground-rupturing earthquake along the central Septentrional fault at site 91-2 occurred after deposition of unit 70 and before deposition of unit 65 (Fig. 38). Radiocarbon ages constrain the date of this earthquake to be after the interval A.D. 645–775 and before the interval A.D. 1160–1260 (Fig. 41). Thus, the most recent ground-rupturing earthquake occurred more than 730 yr ago. The lack of deformation on horizons younger than those deposited 730 yr ago means that this segment of the central Septentrional fault zone was probably not responsible for historical earthquakes that produced strong shaking in the Cibao Valley in 1564, 1783, 1842, 1887, and 1897 (Scherer, 1912) (Fig. 1C).

Estimation of slip rate along the Septentrional fault zone

Global data. Slip rates can be estimated for the Septentrional fault in several ways. One method is to use a total plate boundary slip rate based on the DeMets et al. (1990) Nuvel-1A global compilation of plate-rate data. We estimate that the Septentrional fault rate is at least 5 mm/yr, based on the lowest estimate of total plate-boundary slip rate of 9 mm/yr by DeMets et al. (1990) and the assumption that about half of the interplate motion could be accommodated by other subparallel strike-slip faults in Hispaniola like the Enriquillo-Plantain Garden fault zone (Mann et al., 1995) and faults within the north Hispaniola deformed belt (Dolan et al., this volume) (Fig. 2A). Based on these two assumptions, a slip deficit of more than 3.6 m of potential slip has accumulated at the 5 mm/yr rate in the more than 730 yr since the last ground-rupturing earthquake. Because these estimates are based on a minimum estimated slip rate, the amount of accumulated slip and the implied seismic hazard may be significantly higher.

Data from terrace riser and stream offsets. Left-lateral stream channel offsets occur along the trace of the Septentrional fault zone and range in magnitude from 15 to 400 m. Maximum rates of stream offset determined at Rio Juan Lopez (Fig. 26B) and Rio Licey (Fig. 29) range from 13 ± 4 to 23 ± 7 mm/yr, respectively, and provide maximum rates of slip over a period of several thousand years. Because these are maximum rates, the true slip rate is not necessarily within this range but is most likely lower than this range. For example, at a rate of 23 mm/yr, the central segment of the Septentrional fault zone would have accumulated an unrealistically high amount of strain (16.8 m) in the last 730 yr of its quiescence.

Data from GPS surveys. The southernmost GPS site in Hispaniola at Cabo Rojo has moved at a rate of about 20.6 ± 1.6 mm/yr over the past decade eastward relative to a site on the North America plate at Grand Turk Island (Fig. 2B). This rate is inferred to represent the entire North America–Caribbean relative plate motion as the two sites span much of the plate boundary seismogenic zone (Fig. 1A), both of the major east-west strike-slip faults in Hispaniola (Fig. 2A), and the offshore north Hispaniola deformed belt (Dixon et al., 1998) (Fig. 2B). If the Septentrional fault zone is assumed to accommodate all 21 mm/yr of interplate motion, there would be an unreasonably high amount of strain (15.3 m) accumulated on the Septentrional fault zone in the last 730 yr. Its possible that several meters of this strain was released during the historical earthquake of 1842 that affected an elongate zone parallel to the southern branch of the western Septentrional fault zone and its westward extension (Fig. 1C). This inferred onland rupture of the Septentrional terminated somewhere west of our trench site on the central Septentrional fault zone and will be the target of future trenching along the western part of the Septentrional fault zone (Fig. 14).

Simple elastic strain models by Dixon et al. (1998) using all available GPS data predict a slip rate of 8 ± 3 mm/yr for the Septentrional fault zone, 4 ± 3 mm/yr for the North Hispaniola deformed belt, and 8 ± 3 mm/yr for the Enriquillo-Plantain Garden fault zone in Southern Hispaniola.

Regional-scale pattern of earthquakes in the northeastern Caribbean

Sykes et al. (1982) pointed out that Caribbean–North America interplate hypocenters less than 50 km in depth are more common west of longitude 71°W than east of 71°W (Fig. 45). One explanation for this observation is that the area of shallow earthquakes west of 71°W in central and western His-

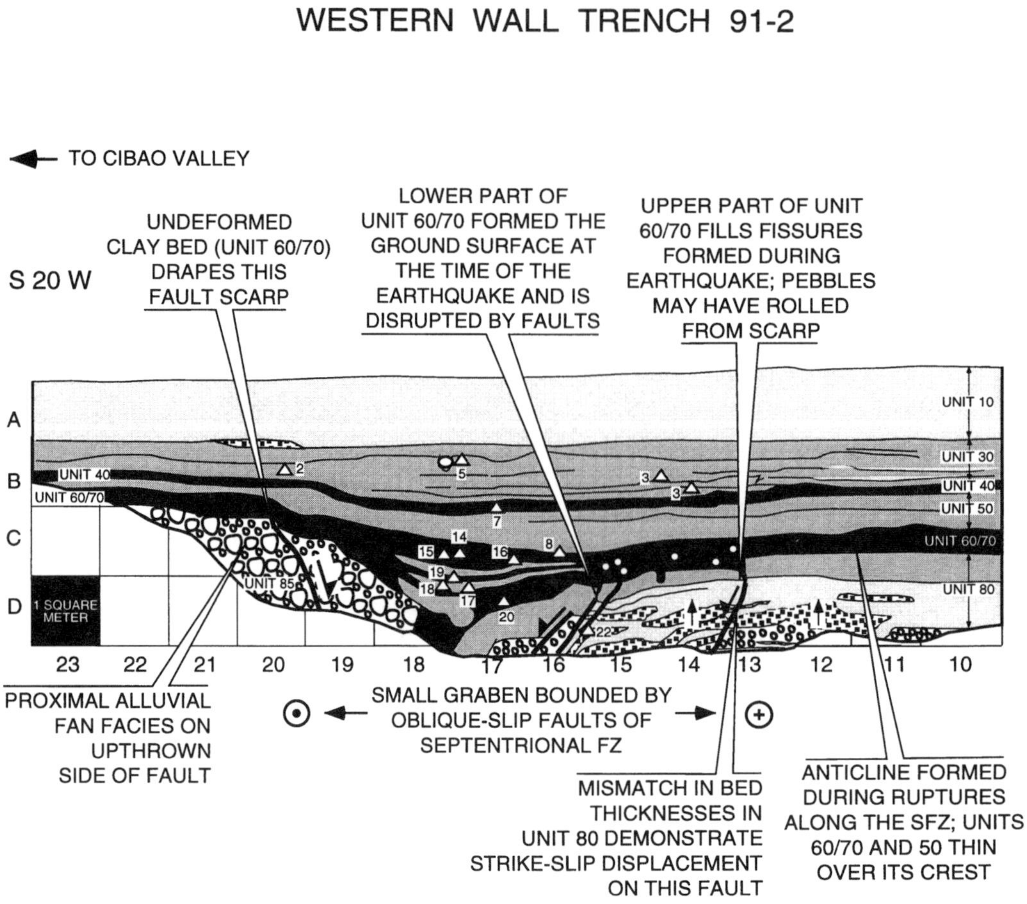

Figure 38 (this and opposite page). Log of western wall of trench 91-2, Septentrional fault zone, at the Rio Cenoví trench site. See text for discussion and Figure 41 and Table 1 for a summary of radiocarbon-dated stratigraphic units.

paniola reflects crustal strike-slip faulting and thrusting within the collision zone between the southeastern Bahamas platform and the eastward-moving Caribbean plate in central and western Hispaniola (Dolan et al., this volume). Crustal convergence in western Hispaniola is manifested by late Quaternary uplift of coral reefs and the high topographic elevations of central Hispaniola (Fig. 45). Mann et al. (1995) have proposed that this late Neogene oblique collisional event between the Bahamas platform and Hispaniola activated motion along the Enriquillo-Plantain Garden fault zone across southern Hispaniola, tectonically isolated the elongate Gonave microplate, and allowed the Caribbean plate south of Hispaniola to continue unimpeded around the Bahamas collisional area (Fig. 45). The east-southeastward direction of GPS velocities at Capotillo in north-central Hispaniola (Fig. 2B) may reflect the local deflection of Hispaniola arc rocks around the southeastern promontory of the Bahamas platform and is consistent with the especially broad seismogenic zone in Hispaniola (Fig. 1A).

East of the Bahamas-Hispaniola collision area and Gonave microplate, underthrusting of oceanic crust of the Atlantic Ocean is still occuring in a west-southwestward direction with deep earthquakes demarcating an irregular Benioff zone to depths of 250 km (McCann and Sykes, 1984; Calais et al., 1992a, this volume; Dolan and Wald, this volume). Because crustal convergence is accommodated by underthrusting of oceanic material into the mantle, there is little crustal convergence and accompanying late Quaternary crustal uplift.

Origin of instrumentally recorded earthquakes in the area of the Septentrional fault zone

The Septentrional fault zone straddles longitude 70°W and therefore overlies zones of both crustal convergence and underthrusting at the Puerto Rico trench and its western extension, the North Hispaniola deformed belt (Dolan and Wald, this volume; Fig. 2B). The transitional nature of this strike-slip–subduction setting makes the distinction between strike-slip and subduction-related earthquakes difficult to determine with either instrumentally recorded events (Fig. 46A, B) or historical earthquake records (Fig. 46C).

Based on relocation and analysis of 61 aftershocks and

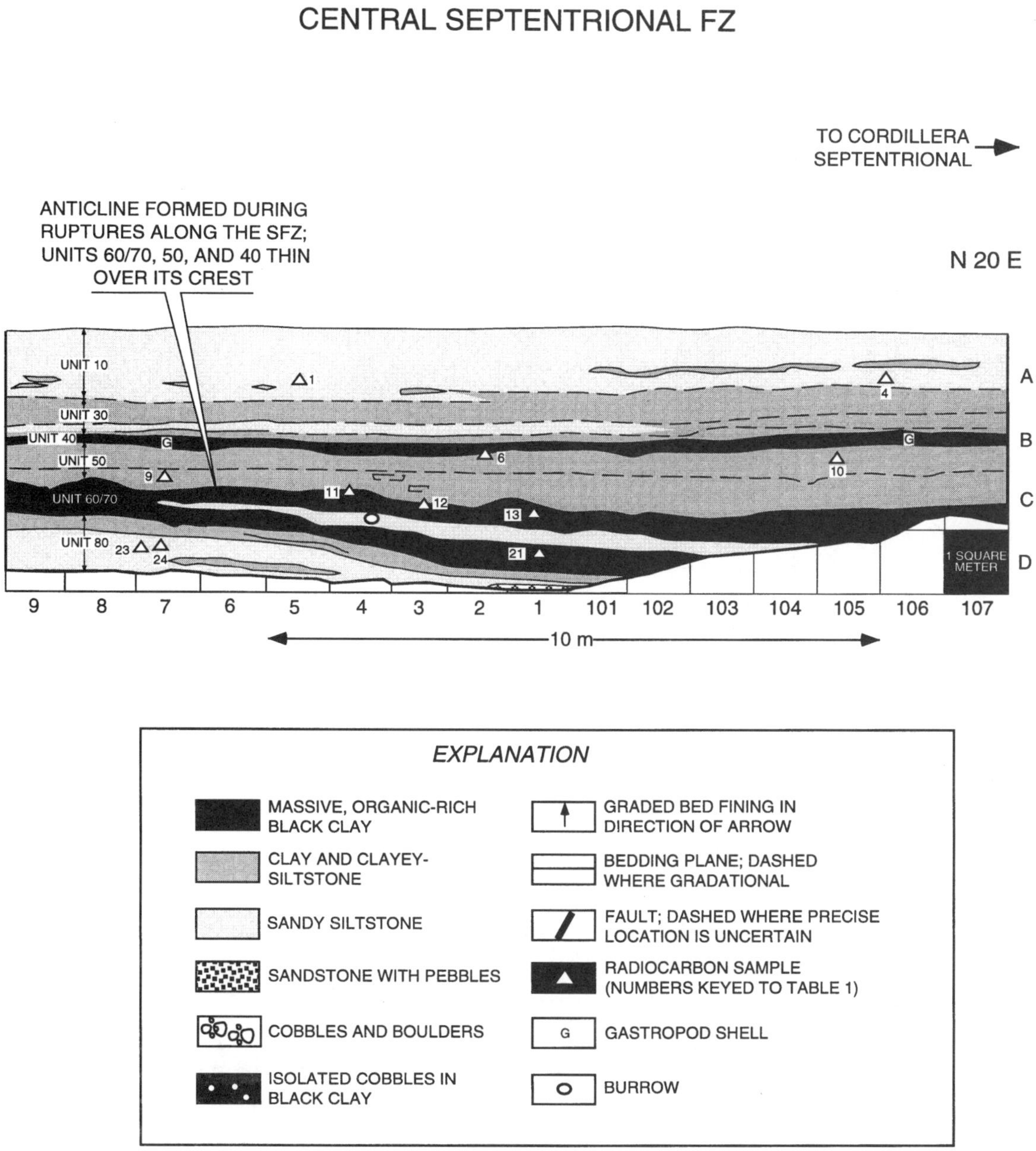

one foreshock, Russo and Villaseñor (1995) argue that the main shock of the great 1946 earthquake ($M_s = 7.8$) was located 15 km south of the south Samana Bay fault zone along the southern margin of Samana Bay (Fig. 4). Their interpretation of a strike-slip mechanism supports the idea of near east-west relative plate motion and a South Samana Bay strike-slip fault source (Fig. 4). In contrast, Dolan and Wald (1997, this volume) proposed that the tectonics of this area are dominated by strain partitioning above a south- to southwest-dipping thrust interface along the top of the underthrust North America slab rather than by a steep, east-west strike-slip fault as proposed by Russo and Villaseñor (1995). They interpreted records from the 1946 main shock as oblique-slip motion on a low-dipping, but obliquely slipping, thrust surface off the northeast coast of Hispaniola. This low-dipping thrust surface corresponds to the offshore surface trace of the frontal thrust of the North Hispaniola deformed belt mapped by Dolan et al. (this volume). We agree with this interpretation of the 1946 event and their inference that increased coupling related to attempted subduction of the southeastern Bahamas carbonate platform beneath Hispaniola is the tectonic mechanism for producing the 1946 and other large magnitude events in northeastern Hispaniola.

The description of the fault trace and the paleoseismicity of the Septentrional fault zone presented in this chapter is consistent with the model proposed by Dolan et al. (this volume)

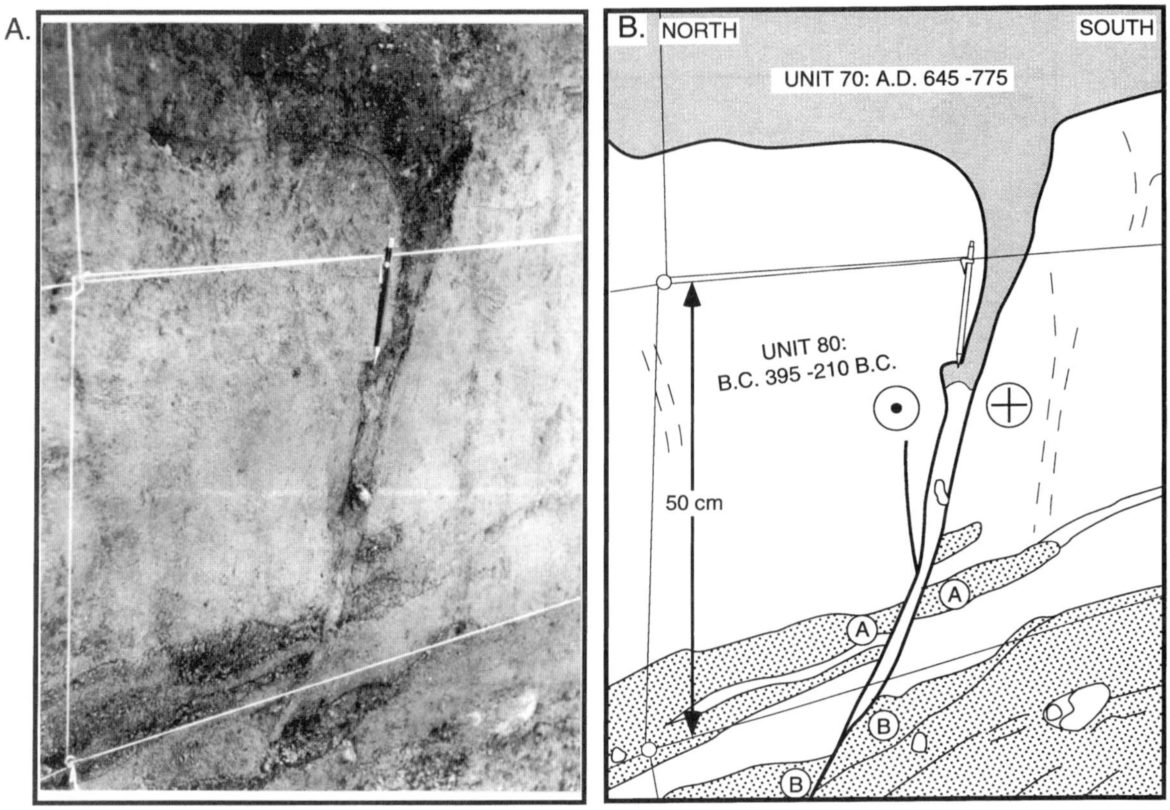

Figure 39. (A) Photograph of active trace of the Septentrional fault zone in the eastern wall of trench 91-2. (B) Line drawing of photograph in A. The north-dipping fault terminates as a fissure filled with organic-rich black clay of unit 70 (see Fig. 38 for entire log of trench). Distance between horizontal nails is 1 m; distance between vertical nails is 50 cm.

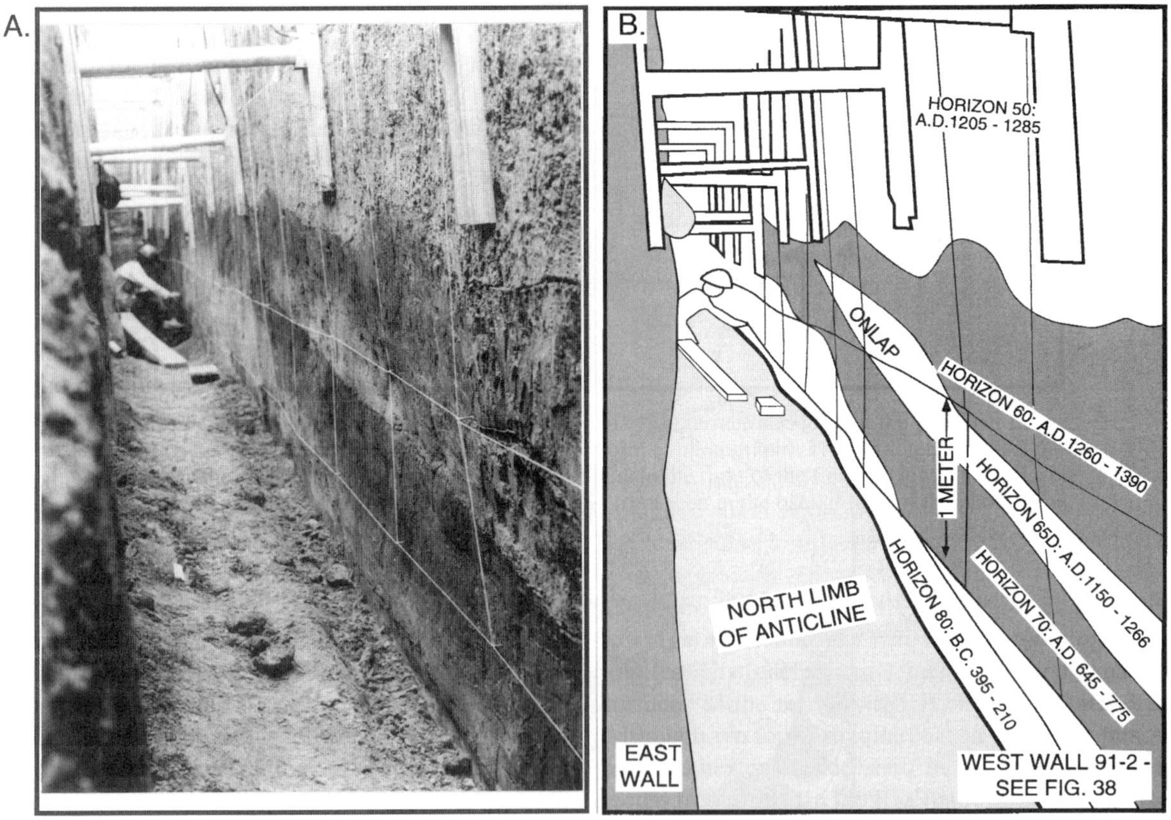

Figure 40. View looking south at stratigraphic relationships in western wall of trench 91-2 between horizons 70 (lower dark horizon), 65D (middle light horizon), and 60 (upper dark horizon) (compare to log of trench 91-1 in Fig. 38). Upper row of string grid consists of 1 m^2 panels; lower row is 1 m × 0.5 m. The earthquake that occurred when horizon 70 formed the ground surface resulted in folding of horizon 70 north of the fault. Horizons 65D and 60 onlap the topographic high formed by the folded ground surface during the time of onlap between A.D. 1150 and 1390.

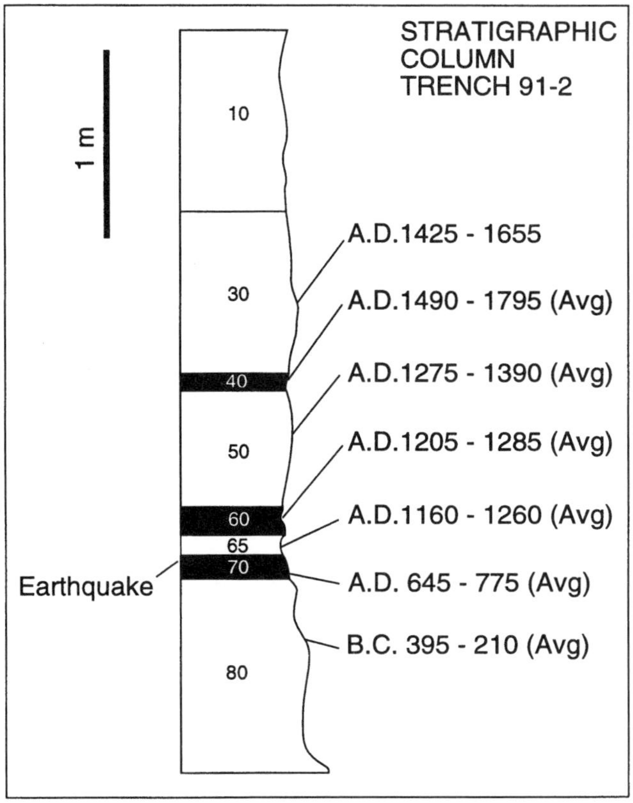

Figure 41. Stratigraphic columns showing units in trench site 91-2 on the Rio Cenoví fan and summarizing radiocarbon dates along with the relationship of dated units to the last ground-rupturing earthquake. Reliable radiocarbon dates used to make tectonic inferences are summarized in Table 1.

for strain partitioning between the offshore collisional zone (North Hispaniola deformed belt) and the onshore Septentrional fault zone. These two features form the northern and southern edges of the Septentrional microplate, an elongate sliver that we infer to be moving eastward relative to stable North America at a relatively slower rate than the areas of the Gonave microplate to the south (Fig. 46D). GPS results show a similar pattern of a slower relative motion along sites in northern Hispaniola (5.2 ± 1.3 to 11.7 ± 1.3 mm/yr) than along one in southern Hispaniola (20.6 ± 1.6 mm/yr), but discrepancies in rates may reflect a more complex pattern of elastic strain effects adjacent to locked faults like the Septentrional and Enriquillo-Plantain Garden fault zones (Fig. 2B) (Dixon et al., 1998).

Given the long quiescence of the Septentrional fault zone established by Prentice et al. (1993) and in this chapter, it seems reasonable to assume that the majority of the instrumentally recorded earthquakes shown in Figure 46A and B are derived from the proposed shallowly dipping, Benioff zone formed between the overriding Septentrional microplate of northern Hispaniola and the underriding southeastern Bahamas platform (Dolan et al., this volume).

Origin of historical earthquakes in the area of the Septentrional fault zone

The shapes of felt areas of historical earthquakes compiled by Scherer (1912) and Kelleher et al. (1973) define two provinces in northern Hispaniola (Fig. 44C). Large historical earthquakes, east of about longitude 70°W including the 1946 and 1953 shocks, form subcircular felt areas, whereas the 1842 and 1887 earthquakes form more elongate felt areas of earthquakes to the west of about 70°W. In the strain partitioning model (Dolan et al., this volume), historical events east of longitude 70°W occur on a Benioff zone in a tectonic province of oceanic plate subduction accompanied by little late Quaternary tectonic uplift (Fig. 45). Historical events west of longitude 70°W occur mainly on transpressional strike-slip zones that include the Septentrional fault and its western extension off the north coast of Haiti (Calais and Mercier de Lépinay, 1995). The 1842 earthquake, which killed thousands of people and damaged all major towns in the western Cibao Valley and along the north coast of Haiti (Scherer, 1912), may have been caused by left-lateral slip on the proposed active southern branch of the Septentrional fault zone (Fig. 14).

It is interesting to note that few shallow instrumentally recorded earthquakes occur west of 70°W (Fig. 45A). This observation is consistent with our finding that the central Septentrional fault zone at longitude 70°20'W has been locked for 730 yr and may be accumulating strain. Most shallow earthquakes occur east of 70°W, overlap with a zone of earthquakes deeper than 50 km, and are probably associated with subduction along the North Hispaniola deformed belt and Puerto Rico trench (Dolan et al., this volume) (Fig. 45B).

Period of earthquake clustering in Hispaniola and relevance for future earthquakes on the Septentrional fault zone

A compilation of the historical seismicity from 1500 to 1900 by Chalas-Jiménez (1989) (Fig. 46) supports the interpretation of Kelleher et al. (1973) for northern and southern seismic zones in Hispaniola that strike roughly east-west (Fig. 1C). Mann et al. (1984, 1995) proposed that both zones may correspond to historical strike-slip faulting along the Septentrional and Enriquillo-Plantain Garden fault zones that lie within the zone of crustal convergence west and strike-slip faulting associated with the motion of the Gonave microplate west of 70°W (Fig. 45).

Chalas-Jiménez (1989) shows distinct clusters of felt areas with modified Mercalli intensity > VIII for both the northern and southern zones (Fig. 47). For the southern zone, the "clusters" consist of both single events and up to three events spanning periods up to 25 yr. A longer term cluster of about 50 yr is suggested for the latter half of the 19th century along the northern seismic zone that corresponds to the Septentrional fault system. However, trenching and historical studies are needed to verify that any of these events produced shallow rupture along

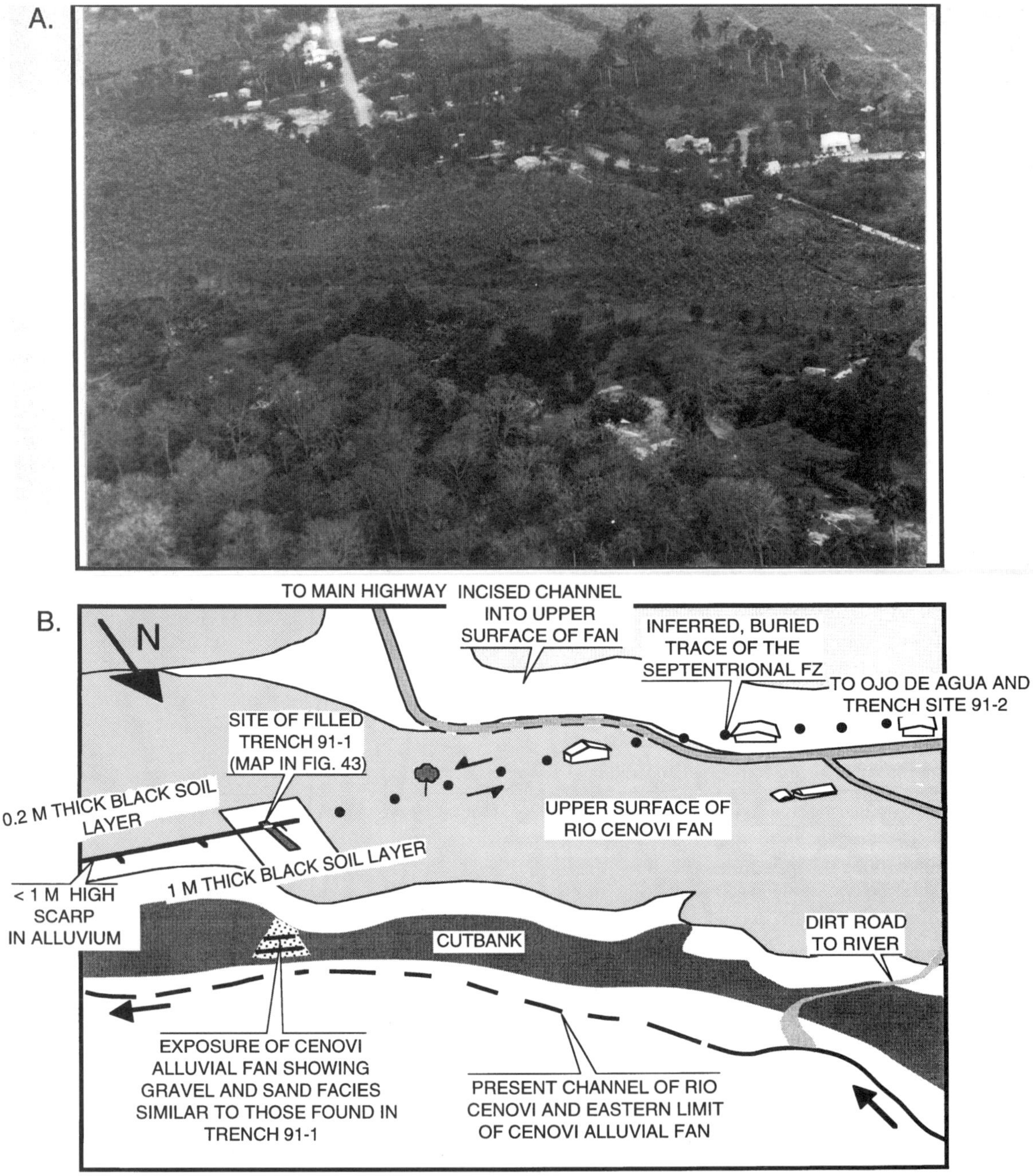

Figure 42. (A) Oblique aerial view taken in 1993 of the area of trench site 91-1 at the eastern edge of the Cenoví fan. (B) Line drawing of the photograph in A. Note the exposure of the alluvial fan sand and gravel facies similar to those found in trench 91-1. The fault forms a low scarp near the Rio Cenoví but loses expression as the fan sediments thicken to the west. Dots mark approximate position of buried fault trace. Box indicates topographic map of the site in Figure 43.

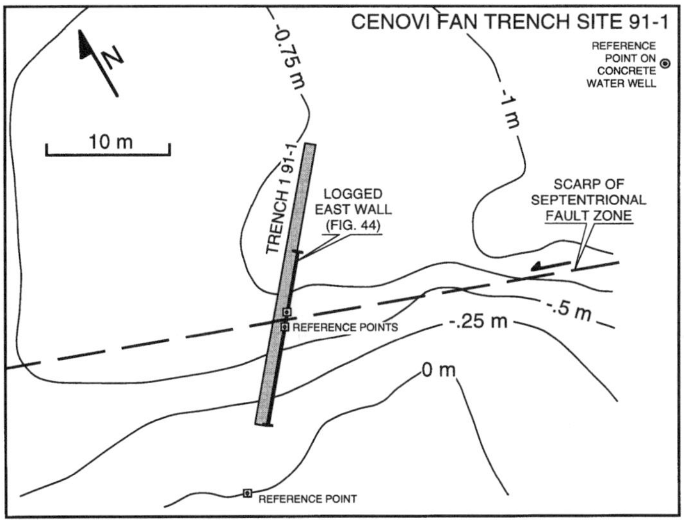

Figure 43. Topographic map of site 91-1. The trench was sited on the low scarp at the eastern edge of the Cenoví fan.

the Septentrional or Enriquillo-Plantain Garden fault zones. Some of the events may have occurred on subcrustal Benioff zones related to underthrusting of Atlantic lithosphere beneath northern Hispaniola (Dolan et al., this volume) or underthrusting of Caribbean lithosphere beneath southern Hispaniola (Byrne et al., 1985; Fig. 45B).

10. CONCLUSIONS

The main conclusions of this study of the Septentrional fault system in the northern Dominican Republic are as follows:

(1) The Septentrional fault system consists of several distinct active and inactive, left-lateral strike-slip fault strands. The active left-lateral trace of the fault system, which we have named the Septentrional fault zone, can be traced from Samana Bay at the eastern end of the Cibao Valley to near the town of Guayabín in the western Cibao Valley. Inactive traces of the Septentrional fault system include the mountain-front fault, which forms the northern edge of the Cibao Valley, and two faults in the western Cibao Valley that we have named the Villa Vasquez and Monte Cristi fault zones. The Septentrional fault system and the Motagua fault zone of Central America are the largest subaerial exposures of the 3,200-km-long North America–Caribbean strike-slip plate boundary.

(2) Offset streams and terraces and mismatches between stratigraphic horizons in trench sites 91-1 and 91-2 confirm late Holocene left-lateral offset along the central Septentrional fault zone. Offset streams and terraces at two localities indicate max-

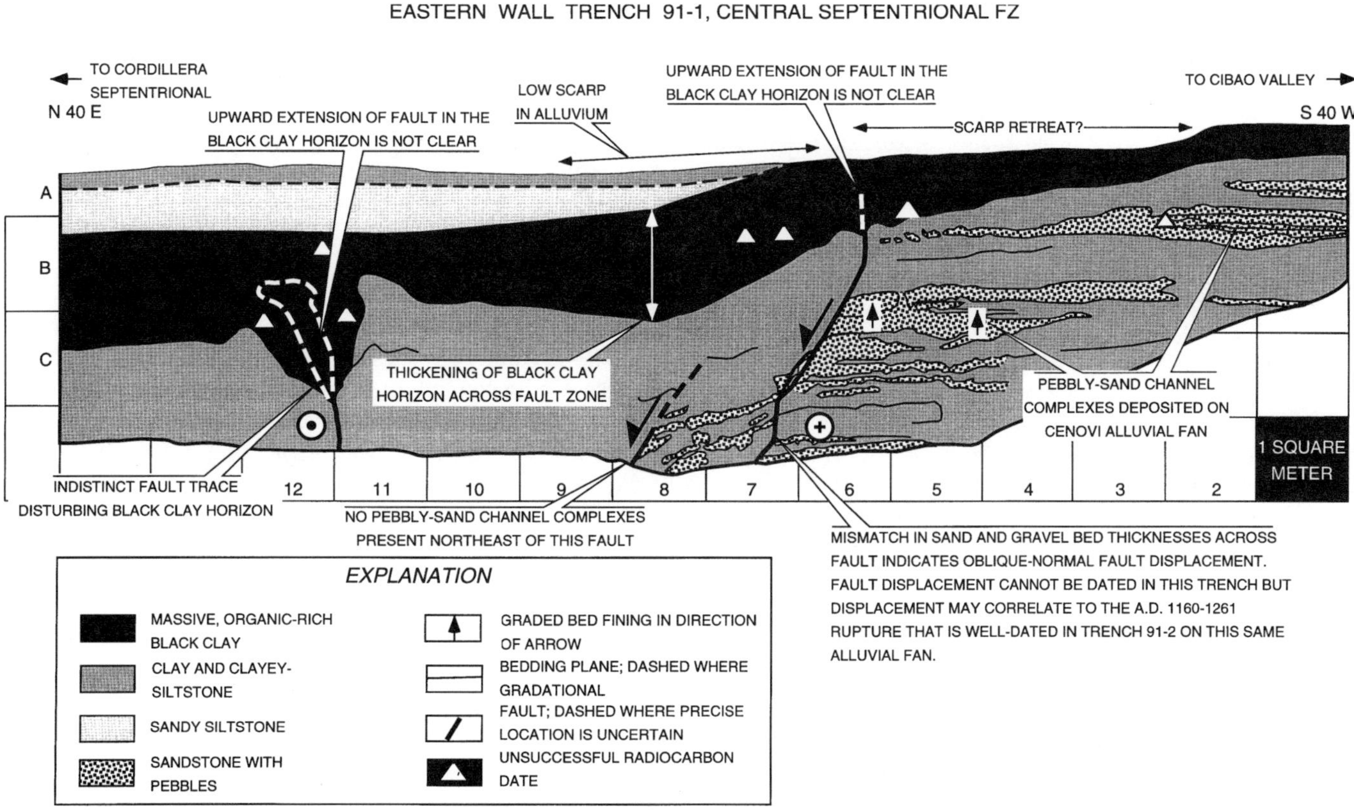

Figure 44. Log of eastern wall of trench 91-1, Septentrional fault zone, at the Rio Cenoví trench site. See text for discussion.

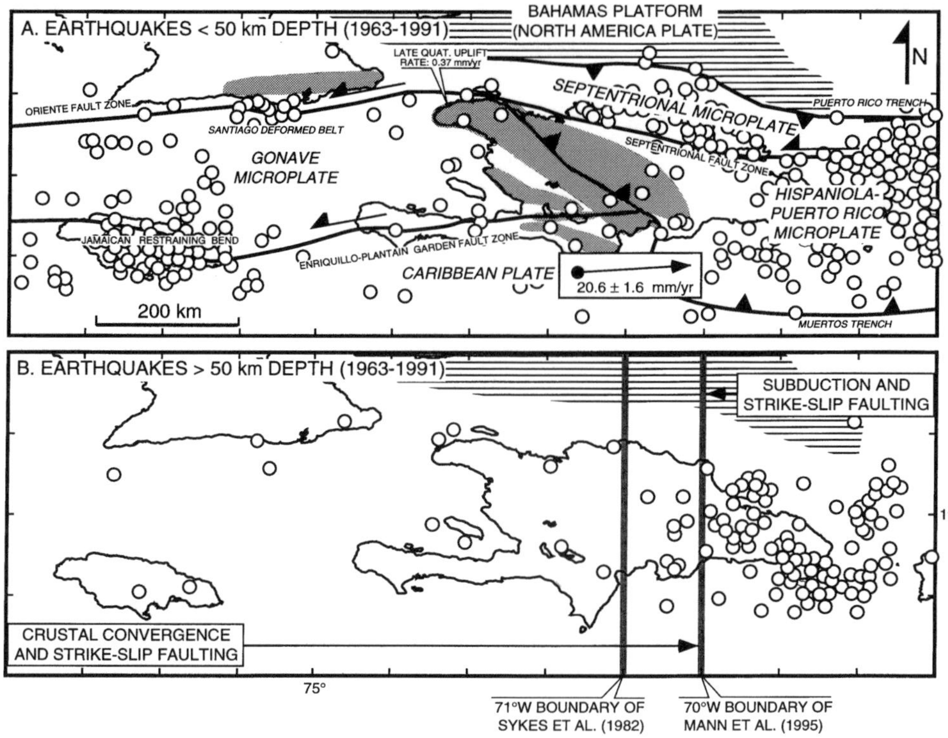

Figure 45. (A) Earthquake epicenters < 50 km depth and major plate boundary structures in the northeastern area of the North America–Caribbean plate boundary (earthquakes have magnitudes > 4.5 and are from the International Seismic Commission database). (B) Earthquake epicenters > 50 km depth and major plate boundary structures. The 71°W longitude was proposed by Sykes et al. (1982) as the boundary separating deep focus from shallow earthquakes in the northeastern Caribbean. Mann et al. (1995) shifted this boundary about 100 km eastward to 70°W longitude to include all of the topographic uplift of the Cordillera Central of the Dominican Republic. This thrust-bound, 3-km-high feature exhibits the highest topography in the northeastern Caribbean, known late Quaternary uplift at its eastern termination in Haiti, and suspected late Quaternary uplift at its southeastern termination in the Dominican Republic. Although some intermediate to deep earthquakes occur west of 70°W, most are confined to the area of subduction in eastern Hispaniola and Puerto Rico. The area of greatest Quaternary uplift coincides with crustal convergence and strike-slip faulting at the edges of the Gonave microplate to the west of 70°W.

imum slip rates averaged over thousands of years range from 13 ± 4 mm/yr at Rio Juan Lopez and 23 ± 7 mm/yr at Rio Licey. Because these are maximum rates, the true slip rate is not necessarily within this range but is probably lower than this range. The stream and terrace offset data support the earlier finding of Prentice et al. (1993) that the Septentrional fault zone is a source of high seismic potential in this densely populated region.

(3) The central Septentrional fault zone forms a straight and continuous fault scarp deforming late Quaternary sediments. Regional seismic reflection profiles across the Cibao Valley indicate that the active trace of the fault is one of many subvertical faults deforming late Neogene sedimentary rocks of the basin (Edgar, 1991). It is possible that other fault traces are present but have not been well preserved because of rapid tropical weathering and/or erosion related to major river systems. We favor a single active strand in the Cibao Valley because parallel strands would be expected to offset and deflect the subperpendicular drainage in a way that we do not observe.

(4) Although interplate slip appears concentrated on a single fault strand through the Cibao Valley, it is likely that North America–Caribbean interplate motion is distributed on at least two other major structures: submarine faults in the North Hispaniola deformed belt (Dolan et al., this volume) and the Enriquillo-Plantain Garden fault zone (Fig. 2B).

(5) GPS surveys in a study by Dixon et al. (1998) indicate that 20.6 ± 1.6 mm/yr of dominately left-lateral motion is occurring across the island of Hispaniola and the North Hispaniola deformed belt (Fig. 2B). Stations closer to the Septentrional fault reflect slower rates ranging from 11.7 ± 1.3 to 5.2 ± 1.3 mm/yr (Fig. 2B). Interpretation of the north-to-south differences in GPS-based rates and directions in Hispaniola and their discrepancy with previous Caribbean–North America plate model predictions by DeMets et al. (1990) (Fig. 2B) is not straightforward because of the following factors: (a) geodetically measured velocities at locations near locked, or nonslipping, strike-slip faults like the Septentrional and possibly

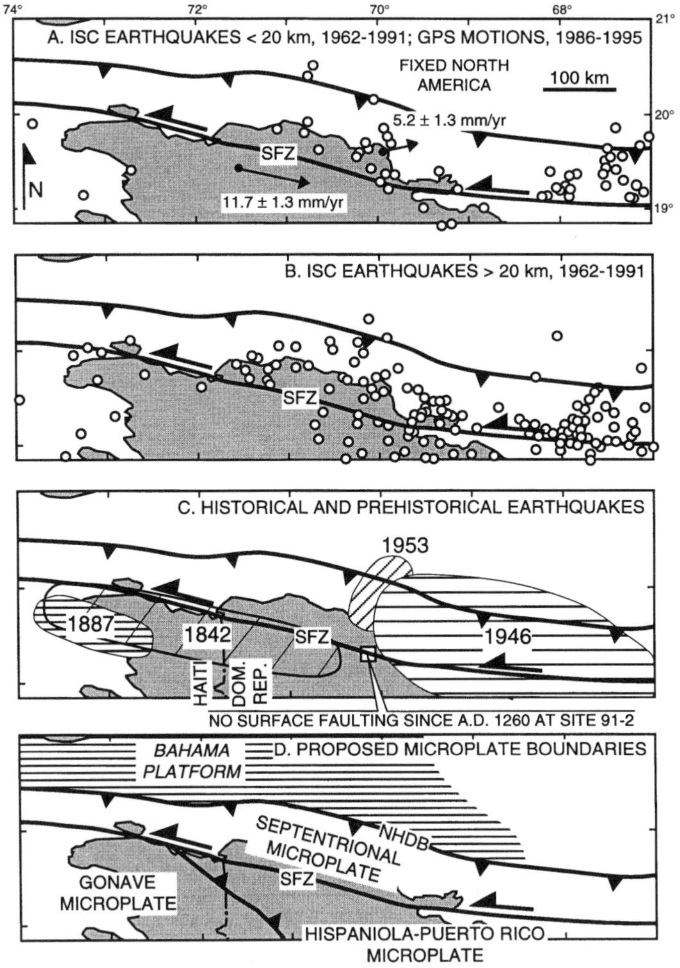

Figure 46. (A) ISC earthquakes < 20 km depth from the ISC catalog (1963–1992). (B) Earthquakes > 20 km depth from the ISC catalog (1963–1992). Deeper events in the vicinity of the Septentrional fault system may be related to southwestward underthrusting of the Bahama Platform beneath Hispaniola. (C) Felt areas of historical earthquakes from Kelleher et al. (1973) compared to the location of trench sites 91-1 and 91-2. The large earthquakes in 1946 and 1953 were probably related to southwestward underthrusting of the Bahama platform beneath Hispaniola (Dolan and Wald, this volume). The more elongate felt area of the 1842 earthquake parallels the Septentrional fault system and suggests a strike-slip origin. (D) Possible microplate boundaries in northern Hispaniola. Total North America–Caribbean plate motion may be partitioned between thrusts and possibly strike-slip faults that form the leading edge of the north Hispaniola deformed belt (NHDB) and the Septentrional fault system. Rates and directions of plate motion are from GPS studies by Dixon et al. (1998).

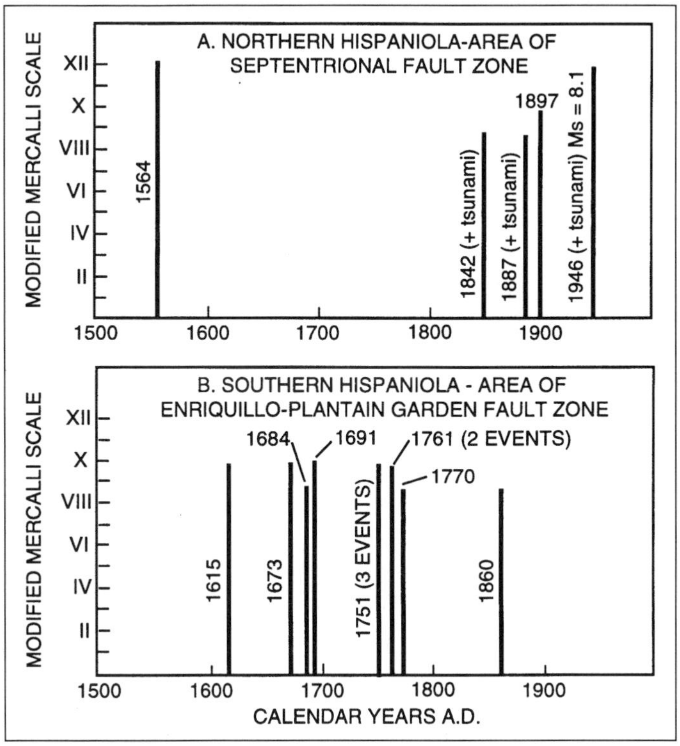

Figure 47. (A) Plot of calendar year versus modified Mercalli intensity scale of major historical earthquakes that have affected northern Hispaniola (Dominican Republic and Haiti) in the area of the Septentrional fault system (from Chalas-Jiménez, 1989). There is a 300-yr quiescence following the 1564 event and a clustering of events in the late 19th and 20th centuries. Although all of these felt areas are located in the vicinity of the Septentrional fault system, some of them, such as the 1946 event, probably occurred on faults of the North Hispaniola deformed belt that lie subparallel to the Septentrional fault zone. (B) Plot of calendar year versus modified Mercalli intensity scale of major historical earthquakes that have shaken the area of the Enriquillo-Plantain Garden fault zone in southern Hispaniola (Dominican Republic and Haiti). Events cluster into four groups with most activity in the 17th and 18th centuries. As in northern Hispaniola, some of these events may be generated by plate motion partitioned into thrust belts that lie parallel to the Enriquillo-Plaintain Garden fault zone.

the Enriquillo-Plaintain Garden fault zone to the south will be slower than locations outside these locked zones (Dixon et al., 1998); (b) North America–Caribbean motion is distributed on other faults parallel to the Septentrional such as the offshore frontal thrust of the North Hispaniola deformed belt and the Enriquillo-Plantain Garden fault zone; and (c) the neotectonic structure of Hispaniola is characterized by the presence of active microplates formed in response to the ongoing oblique collision of arc-type basement rocks in Hispaniola with the Bahamas carbonate platform; these microplates are predicted to have rates and directions of relative motion that differ from the North America and Caribbean plates (Mann et al., 1995) (Fig. 46D). These three conditions help explain the apparent complexity of the GPS results when compared with previous North America–Caribbean plate models by DeMets et al. (1990) and the type of long-term, geologic fault slip data that is presented here.

(6) The western and central segments of the active Septentrional fault zone are defined by a major change in the strike and character of the fault zone near the town of Navarette (Fig 13). The central trace strikes to the west-northwest (110°), is relatively straight, and exhibits several small sag ponds (Fig. 8). The western trace strikes more westerly (100°), is a more dis-

continuous fault, and is interrupted by larger and more substantial pull-apart basins (Fig. 14). It is interesting to note that the eastern limit of the historical felt zone of the 1842 earthquake as compiled by Kelleher et al. (1973) terminates near this major fault strike change in the central Cibao Valley (Fig. 46C). More detailed analysis of historical damage reports as well as fault trenching of the southern branch of the western Septentrional fault zone is needed to establish whether the western fault segment ruptured at this time up to its divergent junction with the central Septentrional fault zone.

(7) An important question is whether historical earthquakes of A.D. 1564, 1783, 1842, 1887, and 1897 that produced strong shaking in the Cibao Valley (Scherer, 1912) are related to strike-slip movement on the Septentrional fault system or related to the thrust displacement on the offshore North Hispaniola deformed belt (Dolan et al., this volume). The lack of a ground-rupturing earthquake for more than 730 yr on the central Septentrional fault zone as determined by Prentice et al. (1993) and in this chapter indicates that many of the instrumental and historical earthquakes as summarized on Figure 46C are produced by the offshore thrust belt. West of longitude 70°W, shallow instrumental shocks are infrequent, which may indicate the accumulation of seismic energy that will be released abruptly during the next major strike-slip earthquake.

ACKNOWLEDGMENTS

Funding from the National Science Foundation for this study included EAR-9004779 to Mann, Taylor, and Prentice for trenching and EAR-9316181 to Mann for collaborative GPS work with C. DeMets, T. Dixon, P. Jansma and E. Calais. Funding from the National Earthquake Hazards Reduction Program of the U.S. Geological Survey included contract 1434-93-G-2341 to Mann for collaborative trenching work with Prentice. We greatly appreciate logistic and scientific support provided by the following individuals and their institutions: Ings. Dionisio Sanchez and Federico Sirí (Universidad Catolica Madre y Maestra, Santiago, Dominican Republic); Ings. Gerald Ellis and Ivan Tavares (Dirección General de Minería, Santo Domingo, Dominican Republic); Ings. Salvador Brouwer and Jorge Jiménez (Falconbridge Dominicana, S. A., Santo Domingo and Bonao, Dominican Republic), and David Gilbrecht (The Plank Company, Austin, Texas). We thank the landowners and citizens of the Cibao Valley for their patience and cooperation during our fault trenching operations. This chapter was significantly improved by reviews by James Dolan, Grenville Draper, Patricia McCrory, and Martitia Tuttle. UTIG contribution number 1384.

REFERENCES CITED

Barka, A., and Kadinsky-Cade, K., 1988, Strike-slip fault geometry in Turkey and its influence on earthquake activity: Tectonics, v. 7, p. 663–684.

Bucknam, R., Plafker, G. and Sharp, R., 1978, Fault movement (afterslip) following the Guatemala earthquake of February 4, 1976: Geology, v. 6, p. 170–173.

Byrne, D., Suarez, G., and McCann, W., 1985, Muertos trench subduction: Microplate tectonics in the northeastern Caribbean? Nature, v. 317, p. 420–421.

Calais, E., and Mercier de Lépinay, B., 1995, Strike-slip tectonic processes in the northern Caribbean between Cuba and Hispaniola (Windward Passage): Marine Geophysical Researches, v. 17, p. 63–95.

Calais, E., Béthoux, N., and Mercier de Lépinay, B., 1992a, From transcurrent faulting to frontal subduction: A seismotectonic study of the northern Caribbean plate boundary from Cuba to Puerto Rico: Tectonics, v. 11, p. 114–123.

Calais, E., Mercier de Lépinay, B., Saint-Marc, P., Butterlin, J., and Schaaf, A., 1992b, La limite de plaques décrochante nord caraïbe en Hispaniola: Evolution paléogéographique et structurale cénozoïque: Bulletin of the Geological Society of France, v. 163, p. 309–324.

Chalas-Jiménez, J. A., 1989, Probabilidad de un sismo catastrófico en la República Dominicana: Historia sísmica, predicción y magnitudes esperadas: Revista Geofísica (Instituto Panamericano de Geografía e Historía), v. 31, p. 185–203.

DeMets, C., Gordon, R. G., Argus, D. F., and Stein, S., 1990, Current plate motions: Geophysical Journal International, v. 101, p. 425–478.

de Zoeten, R., and Mann, P., 1991, Structural geology and Cenozoic tectonic history of the central Cordillera Septentrional, Dominican Republic, *in* Mann, P., Draper, G., and Lewis, J., eds., Geologic and tectonic development of the North America–Caribbean plate boundary in Hispaniola: Geological Society of America Special Paper 262, p. 265–279.

Dillon, W. P., Austin, J. A., Jr., Scanlon, K. M., Edgar, N. T., and Parson, L. M., 1992, Accretionary margin of northwestern Hispaniola: Morphology, structure, and development of part of the northern Caribbean plate boundary: Marine and Petroleum Geology, v. 9, p. 70–88.

Dixon, T., and Daily, M., 1981, Analysis of SEASAT-SAR image of the northeastern Dominican Republic: Photo-Interpretation, v. 5, p. 4.3–4.7.

Dixon, T., Gonzalez, G., Lichten, S., and Katsigris, E., 1991, First epoch measurements with the Global Positioning System across the northern Caribbean plate boundary zone: Journal of Geophysical Research, v. 96, p. 2397–2415.

Dixon, T., Farina, F., DeMets, C., Jansma, P., Mann, P., and Calais, E., 1998, Relative motion between the Caribbean and North American plates and related boundary zone deformation from a decade of GPS observations: Journal of Geophysical Research, vol. 103, p. 15, 157-15, 182.

Dolan, J. F., and Wald, D. J., 1997, The 1946 Hispaniola earthquake and the tectonics of the North America–Caribbean plate boundary zone, northeastern Hispaniola: Comment: Journal of Geophysical Research, v. 102, p. 785–792.

Donnelly, T., Crane, D., and Burkart, B., 1968, Geologic history of the landward extension of the Bartlett trough—Preliminary notes *in* Saunders, J. B., ed., Transactions of the 4th Caribbean Geological Conference (Trinidad), p. 225–228.

Edgar, N. T., 1991, Structure and geologic development of the Cibao Valley, northern Hispaniola, *in* Mann, P., Draper, G., and Lewis, J., eds., Geologic and tectonic development of the North America–Caribbean plate boundary in Hispaniola: Geological Society of America Special Paper 262, p. 281–299.

Evans, C., 1986, Facies evolution in a Neogene transpressional basin: Cibao Valley, Dominican Republic [Ph.D. thesis]: Coral Gables, Florida, University of Miami, 103 p.

García, S., 1976, Geografía Dominicana, Amigo del Hogar: Santo Domingo, Dominican Republic, Amigo del Hogar, 272 p.

Grindlay, N., Mann, P., and Dolan, J., 1997, Researchers investigate submarine faults north of Puerto Rico: Eos (Transactions of the American Geophysical Union), v. 78, p. 404.

Jennings, C. W., 1975, Fault map of California with locations of volcanoes, thermal springs, and thermal wells, California Geologic Data Map Series, Map no. 1, scale 1:750,000 (fourth printing, 1988).

Joyce, J., 1991, Blueschist metamorphism and deformation on the Samana Peninsula: A record of subduction and collision in the Greater Antilles: *in* Mann, P., Draper, G., and Lewis, J., eds., Geologic and tectonic development of the North America–Caribbean plate boundary in Hispaniola: Geological Society of America Special Paper 262, p. 47–76.

Kelleher, J., Sykes, L., and Oliver, J., 1973, Possible criteria for predicting locations and their application to major tectonic boundaries of the Pacific and the Caribbean: Journal of Geophysical Research, v. 78, p. 2547–2585.

Larue, D. K., Joyce, J., and Ryan, H., F., 1990, Neotectonics of the Puerto Rico trench: Extensional tectonism and forearc subsidence: *in* Larue, D. K., and Draper, G., eds., Transactions of the 12th Caribbean Geological Conference, St. Croix, U.S. Virgin Islands: Miami, Miami Geological Society, p. 231–247.

Lynch, J. J., and Bodle, R. R., 1948, The Dominican earthquakes of August 1946: Bulletin of the Seismological Society of America, v. 38, p. 1–17.

Mann, P., Burke, K., and Matumoto, T., 1984, Neotectonics of Hispaniola: Plate motion, sedimentation, and seismicity at a restraining bend: Earth and Planetary Science Letters, v. 70, p. 311–324.

Mann, P., Schubert, C., and Burke, K., 1990, Review of Caribbean neotectonics, *in* Dengo, G., and Case, J. E., eds., The Caribbean region: Boulder, Colorado, Geological Society of America, The Geology of North America, v. H., p. 307–338.

Mann, P., Draper, G., and Lewis, J., 1991, An overview of the geologic and tectonic development of Hispaniola, *in* Mann, P., Draper, G., and Lewis, J., eds., Geologic and tectonic development of the North America–Caribbean plate boundary in Hispaniola: Geological Society of America Special Paper 262, p. 1–28.

Mann, P., Taylor, F., Edwards. R. L., and Ku, T., 1995, Actively evolving microplate formation by oblique collision and sideways motion along strike-slip faults: An example from the northeastern margin of the Caribbean plate: Tectonophysics, v. 246, p. 1–69.

Masson, D. G., and Scanlon, K. M., 1991, The neotectonic setting of Puerto Rico: Geological Society of America Bulletin, v. 103, p. 144–154.

McCann, W. R., 1985, On the earthquake hazards of Puerto Rico and the Virgin Islands: Bulletin of the Seismological Society of America, v. 75, p. 251–262.

McCann, W. R., and Sykes, L. R., 1984, Subduction of aseismic ridges beneath the Caribbean plate: Implications for the tectonics and seismic potential of the northeastern Caribbean: Journal of Geophysical Research, v. 89, p. 4493–4519.

McCann, W. R., Nishenko, S. P., Sykes, L. R., and Krause, J., 1979, Seismic gaps and plate tectonics: Seismic potential for major boundaries: Pure and Applied Geophysics, v. 117, p. 1082–1146.

Mullins, H. T., Breen, N., Dolan, J., Wellner, R. W., Petruccione, J. L., Gaylord, M., Andersen, B., Melillo, A. J., Jurgens, A. D., and Orange, D., 1992, Carbonate platforms along the southeast Bahamas-Hispaniola collision zone: Marine Geology, v. 105, p. 169–209.

Osiecki, P., 1981, Estimated intensities and probable tectonic sources of historic (pre-1898) Honduran earthquakes: Seismological Society of America Bulletin, v. 71, p. 865–881.

Plafker, G., 1976, Tectonic aspects of the Guatemala earthquake of 4 February, 1976: Science, v. 193, p. 1201–1208.

Prentice, C., Mann, P., Taylor, F. W., Burr, G., and Valastro, S., Jr., 1993, Paleoseismicity of the North American-Caribbean plate boundary (Septentrional fault), Dominican Republic: Geology, v. 21, p. 49–52.

Robinson, E., Versey, H. R., and Williams, J. B., 1958, The Jamaica earthquake of March 1, 1957 *in* Weaver, J., ed., Transactions of the 2nd Caribbean Geological Conference: Mayaguez, Puerto Rico, University of Puerto Rico, p. 50–57.

Rodríguez-Demorizi, E., 1979, Mapas y Planas de Santo Domingo: Estudio preliminar y adiciones de Lic. Pedro J. Santiago: Fundación Rodríguez Demorizi, vol. XIII, Editora Taller, Santo Domingo, 371 p.

Rosencrantz, E., and Mann, P., 1991, SeaMARC II mapping of transform faults in the Cayman trough, Caribbean Sea: Geology, v. 19, p. 690–693.

Russo, R., and Villaseñor, A., 1995, The 1946 Hispaniola earthquake and the tectonics of the North America–Caribbean plate boundary zone, northeastern Hispaniola: Journal of Geophysical Research, v. 100, p. 6265–6280.

Saunders, J. B., Jung, P., and Biju-Duval, B., 1986, Neogene paleontology in the northern Dominican Republic: Field surveys, lithology, environment, and age: Bulletins of American Paleontology, v. 89, no. 323, 79 p.

Scherer, J., 1912, Great earthquakes in the island of Haiti: Seismological Society of America Bulletin, v. 2, p. 161–180.

Schwartz, D. P., and Coppersmith, K. J., 1984, Fault behavior and characteristic earthquakes: Examples from the Wasatch and San Andreas faults: Journal of Geophysical Research, v. 89, p. 5681–5698.

Schwartz, D., Cluff, L, and Donnelly, T., 1979, Quaternary faulting along the Caribbean–North American plate boundary in Central America: Tectonophysics, v. 52, p. 431–445.

Sieh, K. E., 1981, A review of geological evidence for recurrence times of large earthquakes: *in* Simpson, D. W. and Richards, P. G., eds., Earthquake Prediction—An international review, Maurice Ewing Series volume 4, American Geophysical Union, Washington, D.C., p. 181–207.

Sieh, K. E., and Jahns, R. H., 1984, Holocene activity of the San Andreas fault at Wallace Creek, California: Geological Society of America Bulletin, v. 95, p. 883–896.

Stuiver, M., and Reimer, P. J., 1993, Extended ^{14}C database and revised CALIB 3.0 ^{14}C age calibration program: Radiocarbon, v. 35, no. 1, p. 215–230.

Sykes, L. R., McCann, W. R., and Kafka, A. L., 1982, Motion of the Caribbean plate during the past 7 million years and implications for earlier Cenozoic plate motions: Journal of Geophysical Research, v. 87, p. 10656–10676.

White, R., 1985, The Guatemala earthquake of 1816 on the Chixoy-Polochic fault: Seismological Society of America Bulletin, v. 75, p. 455–473.

Winslow, M., Guglielmo, G., Jr., Nadai, A., Vega, L., and McCann, W., 1991, Tectonic evolution of the San Francisco Ridge of the eastern Cibao basin, northeastern Hispaniola: *in* Mann, P., Draper, G., and Lewis, J., eds., Geologic and tectonic development of the North America–Caribbean plate boundary in Hispaniola: Geological Society of America Special Paper 262, p. 301–313.

Manuscript Accepted by the Society January 9, 1998

Geological Society of America
Special Paper 326
1998

Strike-slip tectonics and seismicity along the northern Caribbean plate boundary from Cuba to Hispaniola

E. Calais, J. Perrot, and B. Mercier de Lépinay
CNRS, UMR Géosciences Azur, 250 rue Albert Einstein, Valbonne 06560, France

ABSTRACT

We present a compilation of marine geophysical data (Seabeam bathymetry, side-scan sonar, seismic-reflection profiles) collected along the northern Caribbean plate boundary from the southern margin of Cuba to the northern margin of Hispaniola. We compare the structural observations with the seismicity to quantify the kinematic regime along the plate boundary. The marine data show a narrow zone of active deformation extending along a major left-lateral strike-slip fault, the Oriente fault. The most prominent structural feature is the Santiago deformed belt, a transpressive deformation area that extends along the southern Cuban margin and displays active folding and reverse faulting. Dextral offsets in the Oriente fault trace create tensional strain and local pull-apart basins. The earthquake distribution and source parameters are well correlated with the active tectonic structures. They show a combination of north-south compression and left-lateral shear starting at about 78°W long. and continuing all the way to the east along the southern Cuban margin. Our results indicate the existence of a small convergence component added to the mostly strike-slip displacement between the Caribbean and North America plates. This convergence component is responsible for large-scale transpressive structures such as the Santiago deformed belt and is compatible with local tensional structures due to the geometry of the main fault trace. On the basis of instrumental and historical seismicity records, we compute a seismic slip rate of 13–15 mm/yr along the northern Caribbean plate boundary. This result agrees well with the 12 ± 3 mm/yr NUVEL-1 estimate and could indicate that most of the stress along the plate boundary is released seismically.

INTRODUCTION

The Caribbean domain and Central America form a small lithospheric plate between North and South America (Fig. 1), which was first identified by Molnar and Sykes (1969) on the basis of seismicity. Although the North and South America plates show little relative motion, the Caribbean plate is moving eastward relative to them at a rate of 12 ± 3 mm/yr (DeMets et al., 1990). This displacement occurs along two major strike-slip fault zones: the Bocono and El Pilar faults along its southern boundary, and the Polochic-Motagua, Swan, and Oriente faults along its northern boundary. The Caribbean plate is bounded to the east by the Lesser Antilles subduction zone, and to the west by the Central America subduction zone. Despite this apparently well-constrained kinematic framework, the current motion of the Caribbean plate is one of the more poorly known among all major plates. The Caribbean plate does not have any active spreading centers, except a short segment of spreading ridge in the Cayman trough. Moreover, most of its northern transform boundary is located at sea, and detailed mapping of its trace has been undertaken only recently.

We present in this chapter a synthesis of the active tectonic structures along the northern Caribbean plate boundary from Cuba to Hispaniola, identified from marine geophysical data

Calais, E., Perrot, J., and Mercier de Lépinay, B., 1998, Strike-slip tectonics and seismicity along the northern Caribbean plate boundary from Cuba to Hispaniola, *in* Dolan, J. F., and Mann, P., eds., Active Strike-Slip and Collisional Tectonics of the Northern Caribbean Plate Boundary Zone: Boulder, Colorado, Geological Society of America Special Paper 326.

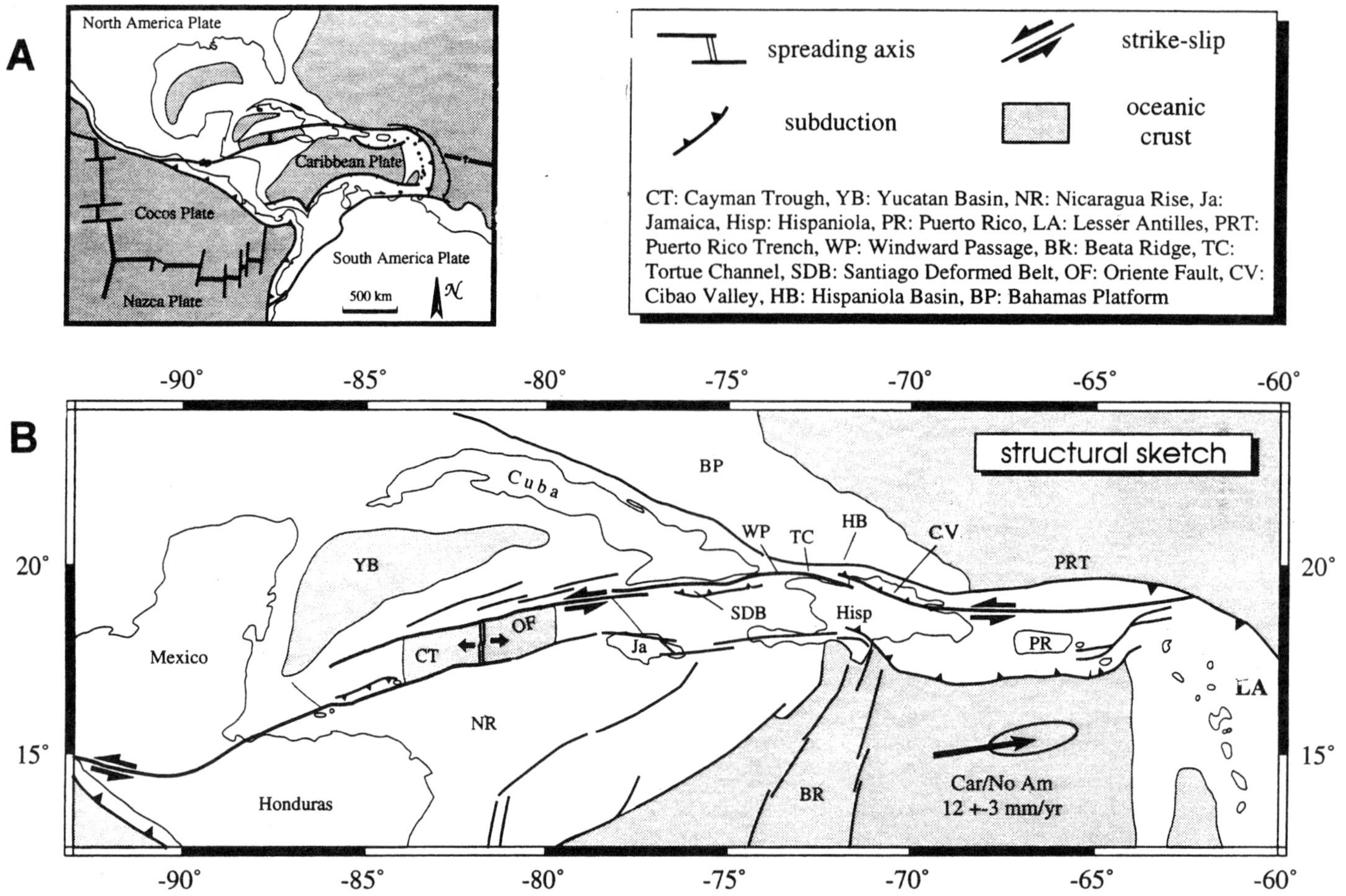

Figure 1. (A) Kinematic framework of the Caribbean plate. (B) Major active faults along the northern Caribbean plate boundary.

(seabeam bathymetry, seismic reflection, side-scan sonar) and teleseismic data. This comprehensive data set allows us to image in great detail the present-day trace of the plate boundary and its associated structures and to characterize the deformation regime along it.

THE NORTHERN CARIBBEAN PLATE BOUNDARY, GEOLOGICAL SETTING

The northern boundary of the Caribbean plate consists of two major strike-slip systems, the Swan and the Oriente fault zones (Fig. 1). The Swan fault is the marine extension of the Polochic-Motagua fault zone of Central America. It is connected to the Oriente fault by the mid-Cayman spreading center, a short oceanic spreading ridge that has been generating the oceanic crust of the Cayman trough since late Eocene (Rosencrantz et al., 1988). The Oriente fault follows the southern Cuban margin and the northern coast of Hispaniola. It continues onland into northern Dominican Republic (Mann et al., this volume) and merges farther east with the Puerto Rico subduction trench (Dolan et al., this volume).

From the mid-Cayman spreading center to the western end of the Cuban margin, the geometry of the Oriente fault is poorly known. It has been thought to follow the foot of the southern escarpment of the Cayman ridge, as shown by some seismic reflection profiles (Goreau, 1983) and by recent seismicity (Fig. 2). Further east, the plate boundary trace continues along the southern margin of Cuba where it corresponds to a prominent topographic scarp, the Oriente Wall. This scarp extends from the summits of the Sierra Maestra of Cuba (2,000 m) into a deep (6,500 m) and narrow east-west–trending marine basin, the Oriente Deep (Case and Holcombe, 1980; Mercier de Lépinay et al., 1989). The Oriente Deep was first interpreted as an active pull-apart basin (Mann and Burke, 1984) on the basis of its topography and its close spatial association with the Oriente transform zone. However, seismic reflection, magnetism, and gravity profiles led Goreau (1983) to the conclusion that the southern Cuban margin is structurally complex and shows mainly compressive tectonics. He described folds and north-dipping reverse faults affecting the sediments of the Oriente Deep, and he noticed a high negative gravimetric anomaly parallel to the Deep, slightly shifted to the north. On the basis of

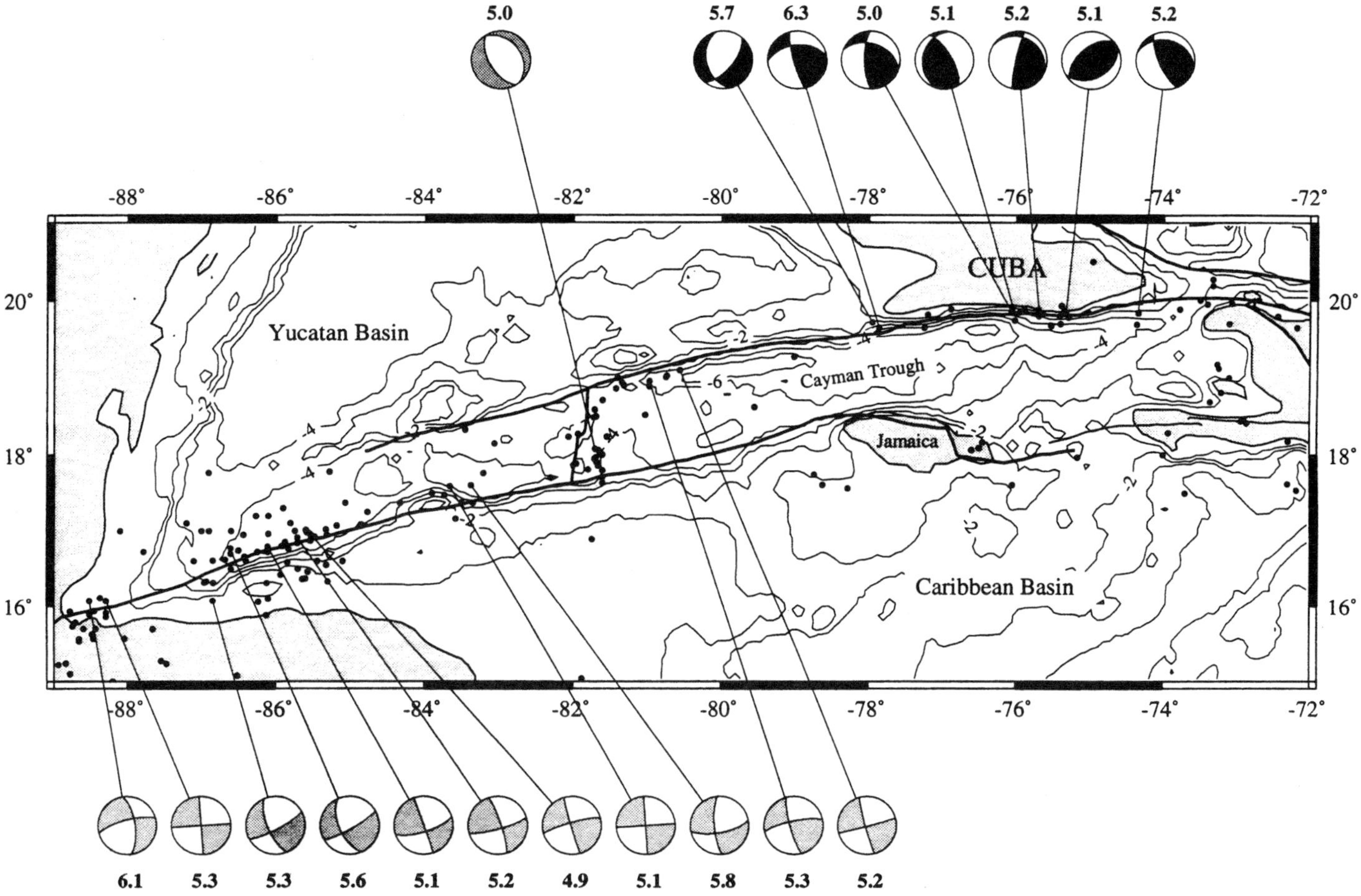

Figure 2. Seismicity (black dots, ISC catalog, 1960–1993, Ms > 3.5) and earthquake focal mechanisms (CMT catalog, 1977–1993) in the Cayman trough and along the southern Cuban margin. Bathymetry labeled in thousand meters.

these observations, he compared this area to a subduction zone and proposed the existence of a major south-verging thrust parallel to the southern Cuban margin.

East of the Cuban margin, the plate boundary trace crosses the Windward Passage, a 90-km-wide strait between the southeastern Cuban coast and the northwestern peninsula of Haiti. Prior to our investigations, few data were available in the Windward Passage. On the basis of two single channel seismic lines, Goreau (1983) defined the main morphostructural units of the Windward Passage, namely, the Windward Passage Sill and the Windward Passage Deep. Two additional multichannel seismic lines (written communication, R/V *F. H. Moore*, University of Texas at Austin, cruise CT2, 1989) led Mann and Burke (1984) to interpret the Windward Passage Deep as an active pull-apart basin. They also postulated the Tortue channel (Fig. 1) to be the eastward prolongation of the Oriente fault, although other authors claimed that it was connected with the western end of the North America plate subduction under Puerto Rico and Hispaniola (Mascle and Letouzey, 1990).

THE PLATE BOUNDARY TRACE AND ITS ASSOCIATED STRUCTURES

Figure 3 displays a tectonic map of the active structures in the study area, established from a morphostructural analysis of Seabeam bathymetry, 50 single channel seismic lines, and a full high-resolution side-scan sonar coverage in the Windward Passage area. Figure 4 shows the location of some of the figures referenced in this chapter. We identify four major structural domains and describe them from west to east in the following sections.

The Cabo Cruz basin

In the western part of the study area, the plate boundary trace is divided into two parallel faults that trend N80°E. These faults bound the northern and southern margin of a rectilinear depression, the Cabo Cruz basin (Fig. 5). This basin is internally divided into a series of oblique horsts and grabens delimited by normal faults trending N45° to 50°E. The southern border of the basin is the eastward extension of the Oriente fault that follows

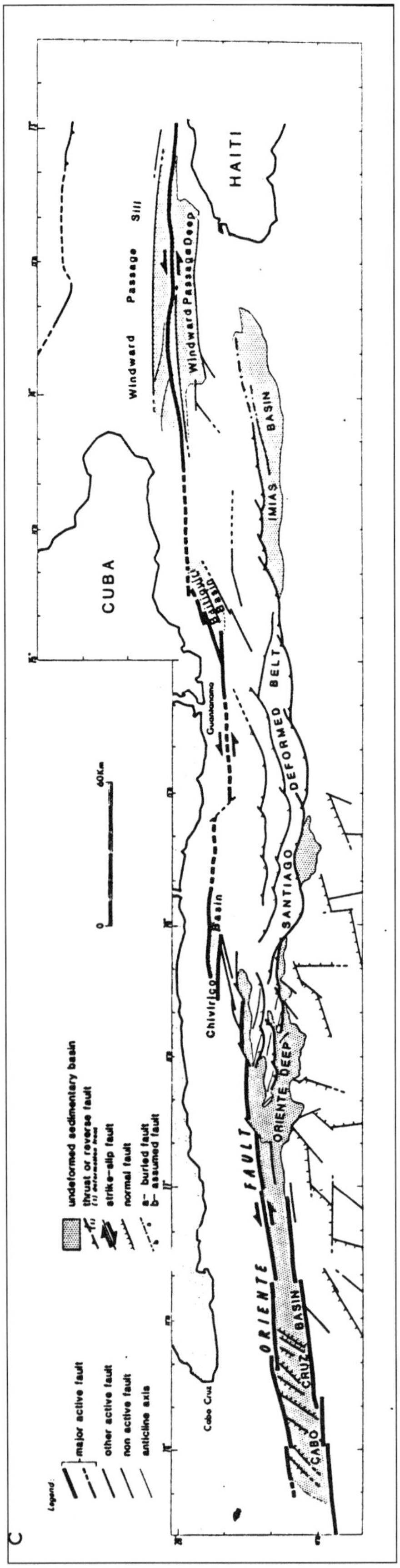

Figure 3. Tectonic map of the northern Caribbean plate boundary along the southern Cuban margin and through the Windward Passage.

the foot of the southern flank of the Cayman ridge (Fig. 1). This fault disappears to the east around 77°05W but is relayed by the fault that bounds the basin to the north. The northern fault does not extend westward of 78°20W but from this point on continues to the east along the Cuban margin. We interpret the Cabo Cruz depression as a pull-apart basin created by left-lateral shear along the two dextrally offset faults that bound it to the north and to the south. Left-lateral displacement along these faults imposes tensional strain in the relay area and causes extension and subsidence. In that context, the N45°–50°E–trending normal faults mapped within the basin can be interpreted as conjugate Riedel fractures.

Oriente Deep

East of the Cabo Cruz basin and morphologically connected to it, the Oriente Deep is an east-west–trending depression, located between 6,000 and 6,500 m deep (Fig. 3). To the north, the Oriente Deep is bounded by a prominent scarp, the Oriente Wall (Figs. 5, 6, and 7). The irregular slope and segmented trace of the Oriente Wall on map is due to two fault sets: (1) N90°E–trending faults showing significant vertical offsets, and (2) N60°E–trending faults relaying the previous ones in a sinistral offset pattern. These faults represent the trace of the plate boundary and have a significant left-lateral slip component in addition to the obvious vertical displacement along them (Calais and Mercier de Lépinay, 1991). Two small pull-apart basins (Chivirico and Baitiquiri basins, Fig. 3) have developed on tensional relays along the main left-lateral strike-slip fault. The Oriente Deep is filled with a sedimentary sequence thicker than 2,000 m in the northern part of the basin, along the active fault trace. The sediments thin to the south where they onlap the north-dipping basin substratum (Fig. 8). The sedimentary infilling of the Oriente Deep is most probably terrigenous, as suggested by the well-defined and continuous reflections on the seismic lines. The sediment supply mainly comes from the adjacent Cuban mountains (Sierra Maestra), but also from Jamaica by the way of submarine canyons (Mercier de Lépinay et al., 1989).

Santiago deformed belt

The flat bottom of the Oriente Deep is interrupted at 76°40′W by a narrow elliptic ridge, trending N130°E (Figs. 6 and 7). Seismic-reflection profiles show that this ridge is an anticline with its axial plane slightly inclined toward the south (Fig. 8). Further east, the compressional deformation progressively affects a greater width of the basin, defining a narrow east-west–trending band of hills (Fig. 8). The seismic-reflection profiles show that each of these hills corresponds to an anticline fold, 300–600 m high above the bottom of the basin, about 10 km long and 3 km wide. Continuing to the east, the deformation involves the whole width of the Oriente Deep and the basement of the Cuban margin, forming the Santiago Promontory (Fig. 7). These compressional tectonic structures extend as far east as the

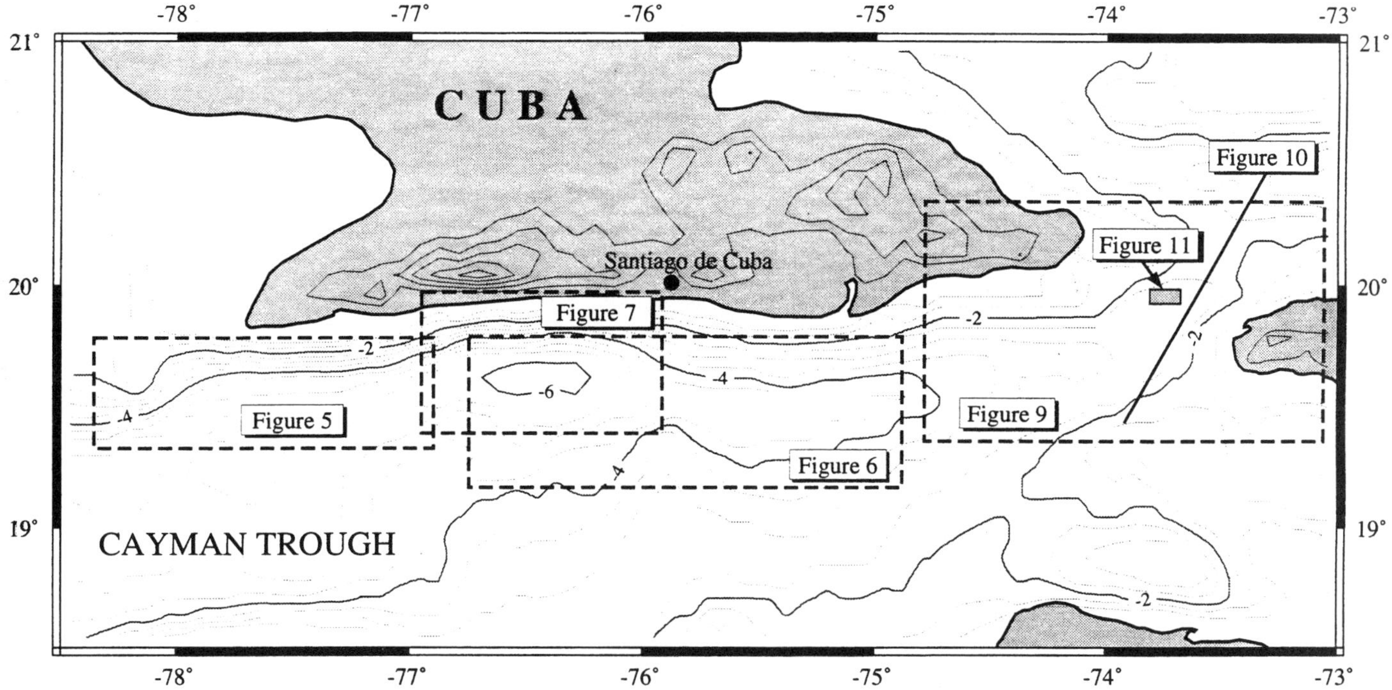

Figure 4. Location of Figures 5–7 and 9–11.

Windward Passage and form the Santiago deformed belt (Fig. 3). This 300-km-long and 10–30-km-wide belt extends parallel to the Oriente fault, that marks its northern boundary. Its southern boundary is a deformation front that usually corresponds to the surface trace of a north-dipping thrust or reverse fault. The deformation of the Santiago deformed belt is active and caused by transpression along the left-lateral Oriente strike-slip fault (Calais and Mercier de Lépinay, 1990, 1991).

Windward Passage

Between Cuba and Hispaniola, the Oriente fault runs straight through the Windward Passage into the Tortue channel (Figs. 3 and 9). Seismic reflection profiles and side-scan sonar imaging show that the two scarps that bound the Windward Passage Deep to the north and to the south are not associated with deformation of the recent basin sediments. We therefore believe that these scarps are not active faults (Fig. 10), which discards the interpretation of this basin as an active pull-apart (Mann and Burke, 1984). The Oriente fault actually crosses the Windward Passage Deep in its middle, affecting its youngest sediments. This structural arrangement of a single strike-slip fault cutting through a sedimentary basin has been classically reproduced in analogous clay and sand models (e.g., Odonne and Vialon, 1983) and is here naturally offered to direct observation. The main fault trace strikes N80°E in the western part of the basin, N90°E in its central part, and N95°E in its eastern part. Seismic profiles show that this fault is a south-verging reverse fault in the western part, which becomes vertical in the middle part and continues as a north-verging reverse fault in the eastern part as it enters the Tortue channel (Calais and Mercier de Lépinay, 1995). It is associated with minor folding and reverse faulting displaying a positive flower structure pattern (Fig. 10). The high-resolution side-scan sonar images show that the main fault is made of successive rectilinear segments. A slight offset and change of direction occur at each relay between two successive segments, giving the appearance of one continuous fault trace (Fig. 11). These relays are marked by a more intense deformation over a 500 m–1 km wide area, contrasting with the low deformation level along the rectilinear segments themselves. The deformation is mostly characterized by en echelon sigmoidal tensional fractures. Their S-shape and N60°E general trend indicate left-lateral displacement along the main east-west–trending strike-slip fault.

TRANSPRESSION IN THE ORIENTE DEEP

In the Oriente Deep, the lack of submarine erosion has clearly preserved the tectonic structures in the morphology. Our full Seabeam coverage of this area together with a dense seismic reflection coverage allowed a detailed analysis of the deformation and of its evolution in time and space.

Morphostructures

The western part of the Santiago deformed belt forms a narrow band of three N90°E–trending hills rising about 300 m above the bottom of the Oriente Deep. Each of them corresponds to a set of right-handed en echelon folds, which are clearly visible on the detailed bathymetric map (Fig. 7). This

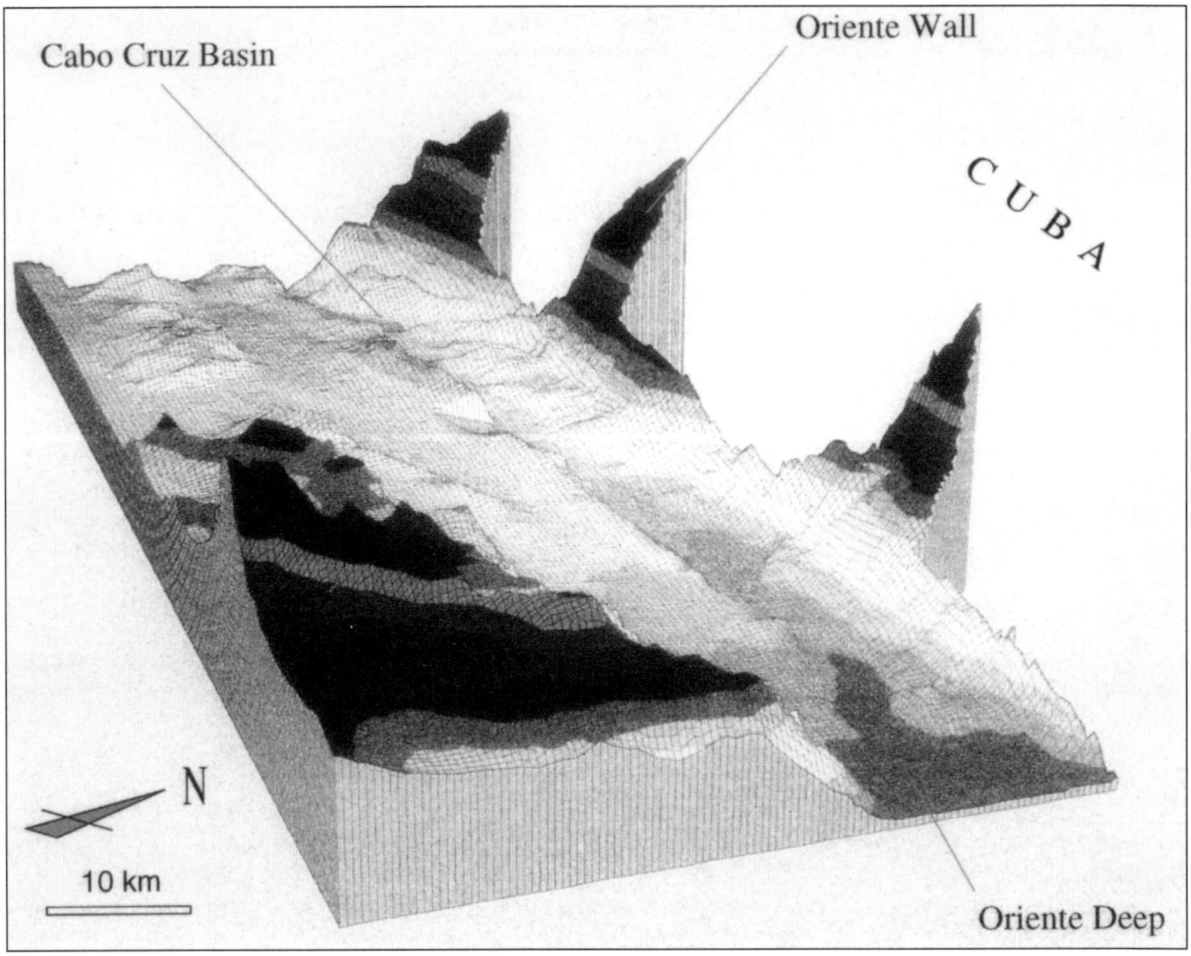

Figure 5. Three-dimensional bathymetry of the Cabo Cruz basin area.

arrangement is typical for folding in a left-lateral strike-slip tectonic regime. The axial trend of the folds, however, significantly varies within each set. On both the central and southern fold sets, the westernmost anticline trends N125°E–N130°E, making an angle of 35°–40° with the main strike-slip fault direction. Both are of small size and have a rectilinear axis. The other anticlines are bigger and have a curved axis striking 5°–20° from the main strike-slip fault direction. The stretched sigmoidal shape of these curved fold axes indicates left-lateral shear. The anticlines' wavelength is 9 km in the northern fold set, 7 km in the central fold set, and 5 km in the southern fold set, but remains constant within each fold set. The size of the anticlines shows that same gradient: they are 12 km long × 3 km wide in the northern fold set but only 5 km long × 2 km wide in the southern one. We interpret this geometry as the result of progressive northward deepening of the sedimentary infilling of the Oriente Deep, as suggested by analog models (Wilcox et al., 1973; Cobbold, 1975; Odonne and Vialon, 1983; Richard and Cobbold, 1989). The trend of the central set of folds occurs on an extension of the fault bounding the Cabo Cruz basin to the south. This suggests that this fault may extend farther east in the subsurface and act as a preexisting discontinuity that would passively initiate the folding.

SPACE AND TIME EVOLUTION OF FOLDING AND REVERSE FAULTING

An upper sequence of diverging reflections onlaps southward onto the northern flank of the anticlines (B′ on profile 59 shown in Fig. 8), with a thickness that progressively increases from the folds' crest down along their flank. This geometry characterizes a syntectonic sequence, deposited and progressively tilted during the folding of the underlying sequence. A sequence showing the same seismic characteristics is visible on profile 105 (A shown in Fig. 8), but is covered on that profile by a sequence of parallel reflections that uncomformably overlies the fold. A third sequence can be identified on profiles 59 and 105 (C or C′ shown in Fig. 8), with continuous and parallel reflections clearly affected by folding on both profiles. The most recent sediments on profile 59 are therefore syntectonic, whereas they are post-tectonic on profile 105. This result indicates that the fold on profile 59 is currently growing, whereas the fold on profile 105 is not. It suggests that

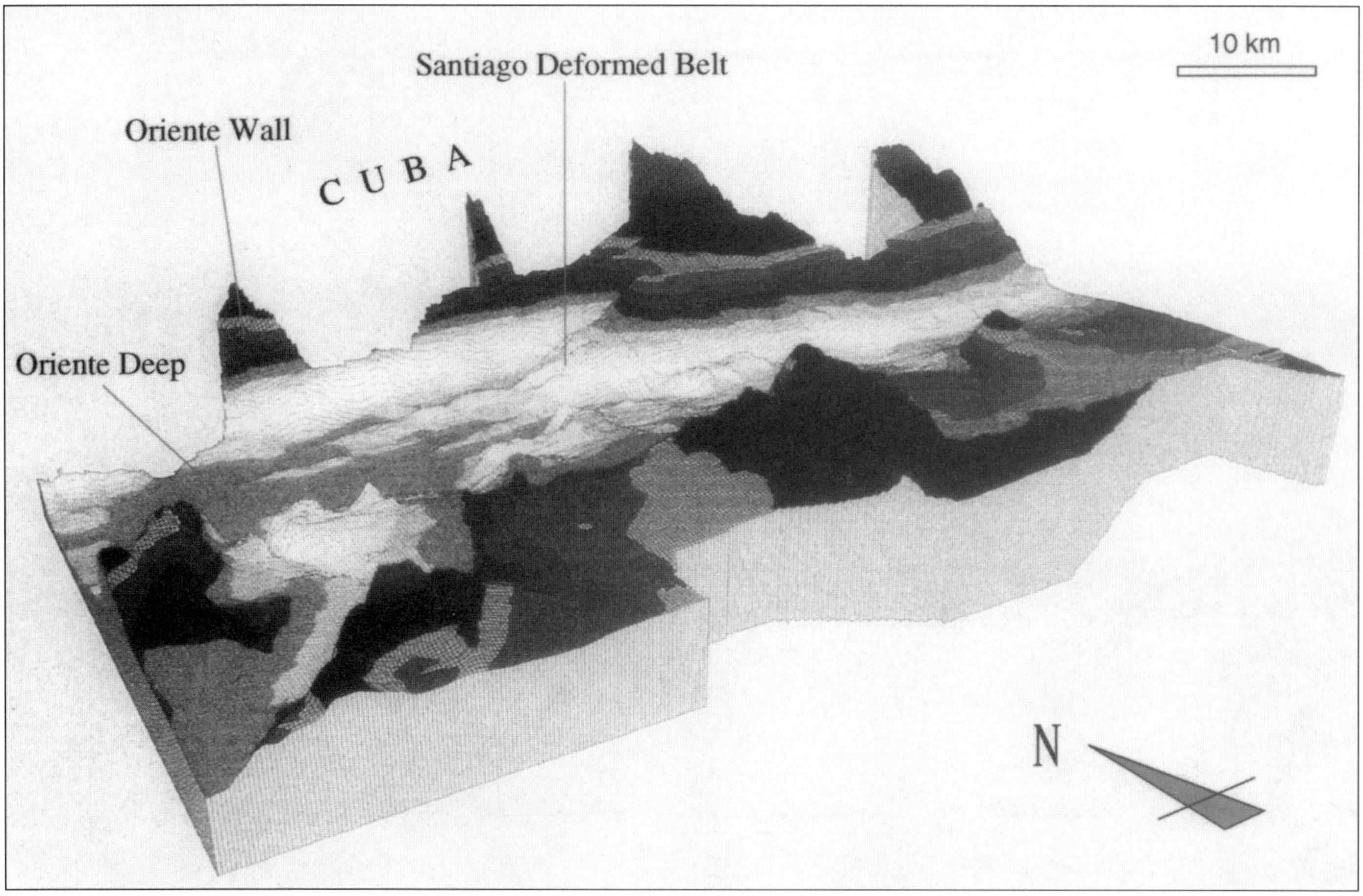

Figure 6. Three-dimensional bathymetry of the Oriente Deep area.

folding is diachronous and propagates westward. Profile 105 shows an active reverse fault at the base of its southern flank that affects the youngest sediments of the Oriente Deep (Fig. 8). Such a fault is not visible on profile 59, where folding is still active. These observations suggest that folding in the Oriente Deep is a short-lived process and that the compressive strain is taken up by reverse faulting as soon as the folds reach a sufficient size.

EARTHQUAKE DISTRIBUTION AND FOCAL MECHANISMS

Instrumental seismicity

We used the instrumental seismicity data published in the International Seismological Center (ISC) catalogs that provide location, time, and magnitude of global earthquakes for the past 35 yr (1960–1995). The focal mechanisms were taken from the Centroid Moment Tensor (CMT) solutions published by Dziewonsky et al. (e.g., 1981) and cover from 1977 to the present.

Earthquake distribution reveals rather little seismicity along the Swan and Oriente faults in the Cayman trough (Fig. 2). Shallow earthquakes associated with both faults show pure left-lateral strike-slip displacement along east-west–trending vertical fault planes. One focal mechanism in the vicinity of the mid-Cayman spreading center shows east-west extension related to oceanic accretion at the ridge axis. The seismicity pattern along the Oriente fault changes as it reaches the southern Cuban margin. The focal mechanisms of all but one of the events of the southern Cuban margin show a strike-slip component, as along the Cayman trough. It is, however, always associated with either a thrust component (in all but one case) or an extension component (in one case, the Cabo Cruz basin). One earthquake shows pure thrusting. All these events share a focal plane that trends between N30ºE and N130°E and dips between 21° and 69° to the north. Given the active north-dipping south-verging thrusts that we mapped in that area (cf. supra), we correlate this nodal plane with the fault plane and interpret these events as oblique thrusting on north-dipping rupture planes.

Two large earthquakes have recently been recorded in the Cabo Cruz basin area. The first one (mb = 5.7, August 26, 1990) has a focal mechanism that essentially shows east-west extension, with a little strike-slip component. It is probably caused by normal faulting related to extension in the Cabo Cruz basin. We propose to correlate the N30°E–trending nodal plane with faults of the same direction that we observed within the basin. In that hypothesis, these faults would mostly have a normal component,

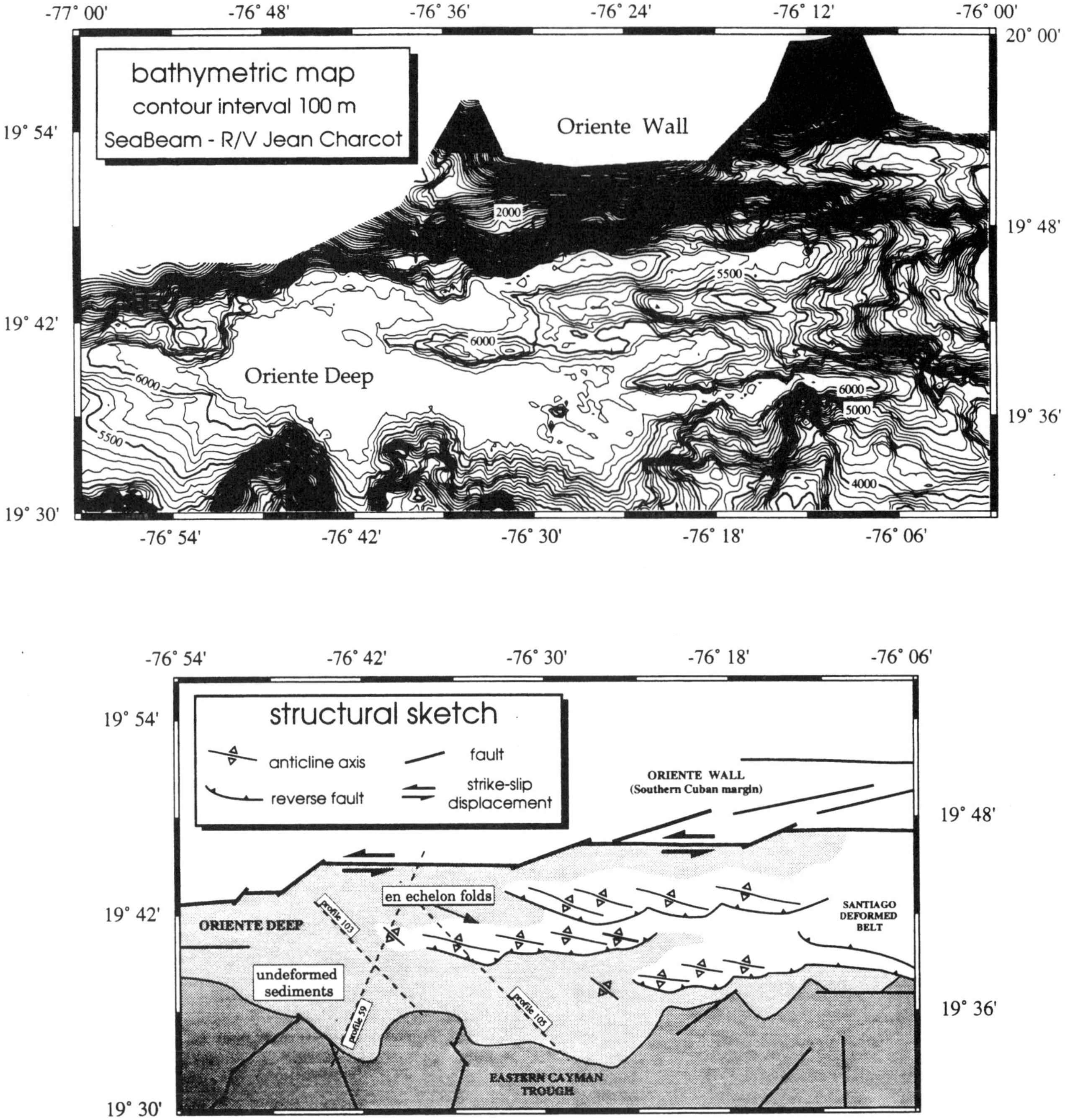

Figure 7. Detailed bathymetric map of the Oriente Deep and morphostructural interpretation. The dashed lines give the location of the seismic profiles of Figure 8.

combined with little left-lateral strike-slip. This result is in accordance with an interpretation of these faults as Riedel fractures associated with left-lateral shear along major east-west–trending faults. The most recent and biggest earthquake along the southern Cuban margin is a mb = 6.3 that struck the Cabo Cruz basin area on May 25, 1992 (Perrot et al., 1997). The main shock probably occurred along the northern edge of the basin, although the accuracy of teleseismic determinations does not give a precise enough location to undoubtedly point out the associated rupture. We modeled that earthquake using a hybrid ray-tracing method (Perrot et al., 1994) that enables us to use a two-dimensional crustal model around the seismic source. We used the crustal model obtained by Edgar et al. (1971) from seismic-refraction data. The best fit between the observed and synthetic seismo-

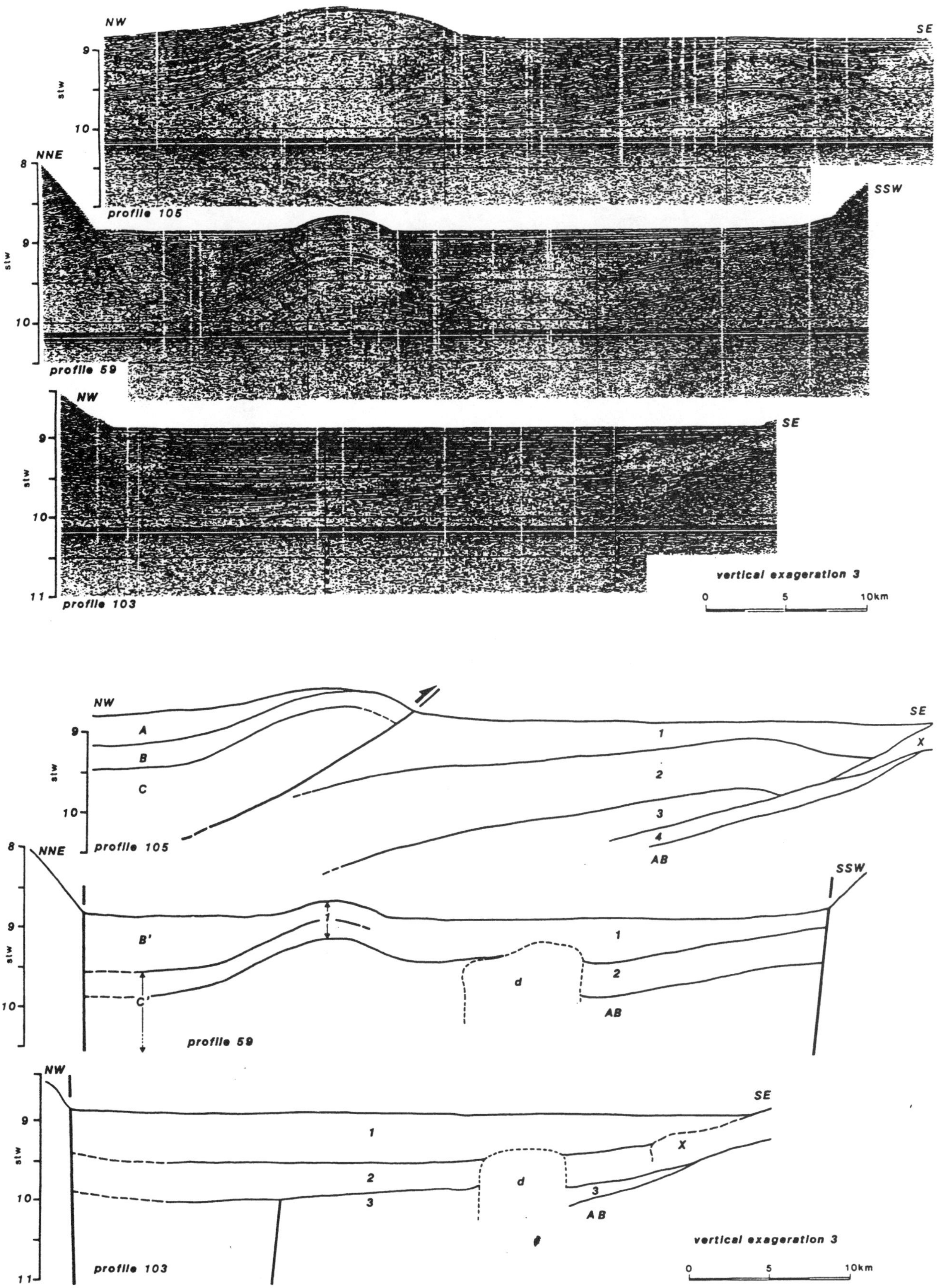

Figure 8. Single-channel seismic-reflection profiles 59, 103, and 105 across the Oriente Deep and corresponding interpretation.

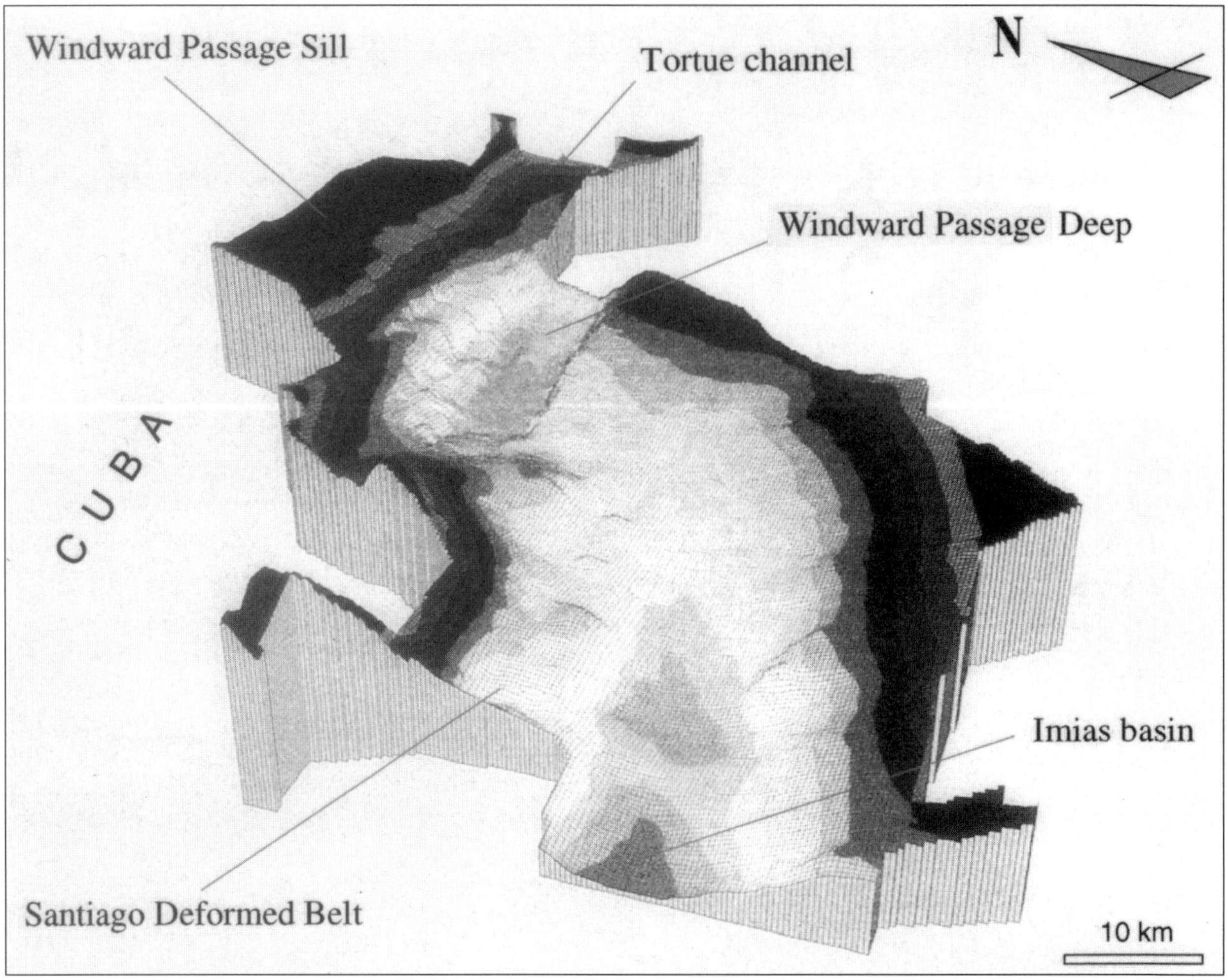

Figure 9. Three-dimensional bathymetry of the Windward Passage area.

grams is obtained for a 19-km-deep hypocenter with a focal plane trending N80°E and dipping 55° to the north (Fig. 12). Because of its coincidence with the Oriente fault direction and the consistency of this focal mechanism with others farther east, we interpret this focal plane as the rupture plane.

Historical seismicity

Reports of earthquakes in the Greater Antilles date as far back as the 16th century when the region was being colonized by the Spanish. On the basis of these reports, Salterain (1883) already noticed the significant seismic activity in southern Cuba. Before the 20th century, the only populated area in that region was the city of Santiago de Cuba (Fig. 4), which was founded in 1514. The history of Santiago therefore provides a precise calendar for major earthquakes in that region. Taber (1922, p. 211) noted that while strong shocks were usually felt over a large region, small events have been reported only from Santiago. The location of these old events is not accurate, but because the Oriente fault is the only major active fault in the area, it seems reasonable to assume that they occurred along it, as Taber already proposed in 1922: "Most of the earthquakes that have damaged property in eastern Cuba have probably their origins along the precipitous scarp that forms the north side of the Bartlett [Cayman] Trough." The first earthquake reported in Salterain's listing that caused serious damage to Santiago occurred in 1578. Two other strong and damaging shocks followed in 1675 and 1677, two years apart. One year later, on February 11, 1678, a strong shock caused extensive destruction in the city. In 1755, a severe earthquake was followed by a tsunami that flooded Santiago. The strongest earthquake ever felt in this region occurred on June 11, 1766, and almost completely destroyed the city of Santiago. Salterain lists 18 strong earthquakes between 1777 and 1852 and quotes a witness of the August 20, 1852, earthquake who reported that "the first shock consisted of a strong up and down motion, causing many of the inhabitants to flee from the city." This last information, and especially the 1755 tsunami, suggests that some of the historical

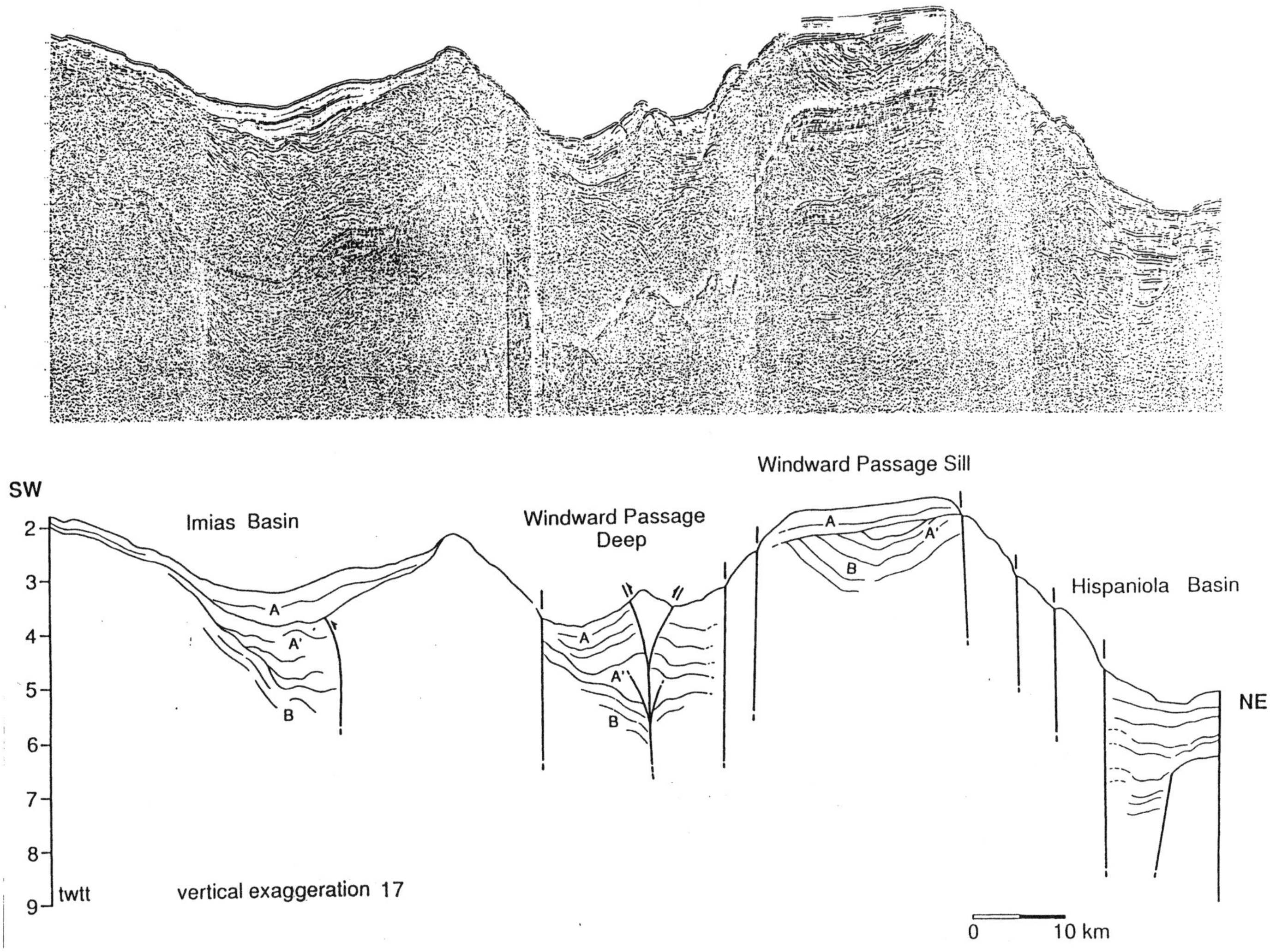

Figure 10. Multichannel seismic-reflection profile crossing the Windward Passage Deep and corresponding interpretation.

earthquakes of southern Cuba may have had a thrust component, as the instrumental seismicity has shown for most of the recent earthquakes in that region.

The earliest earthquake in the southern Cuban region for which an estimate of source parameters is available is dated 1887 (Mocquet, 1984). In about a century, five earthquakes with a magnitude Ms greater than 6.5 occurred along the southern Cuban margin (Table 1, from Mocquet, 1984). This provides us with a century of quantitative information on major earthquakes in this region.

SEISMIC SLIP, STRESS AND STRAIN

Seismic slip rates

The seismic energy released during earthquakes provides an estimate of the displacement due to earthquakes on a strike-slip fault (e.g., Brune, 1968; Thatcher et al., 1975; Anderson, 1979). In the following paragraphs, we estimate the slip rate along the southern Cuban margin from 78°W to 74°W (Fig. 2).

The average seismic slip is given by

$$\bar{u} = \frac{1}{\mu A \Delta T} \sum_{\Delta T} M_o \qquad (1)$$

where A is the total fault area, μ the shear modulus, ΔT the data time span, and M_o the seismic moment (Brune, 1968). We used $\mu = rV_s^2 = 3.3 \times 10^{11}$ dyn/cm^2, a 440-km-long × 15-km-deep fault plane (from 78°E to 74°E). We applied the above equation to the data listed in Table 1 and in the ISC catalog (1960–present). From 1887 to the present, we obtain an average seismic slip rate of 15 mm/yr, fairly close to the 12 ± 3 mm/yr NUVEL-1 slip estimate along the Oriente fault (Table 2, p. 211). However, using only the ISC data we obtain an average seismic slip rate of only 6.3 mm/yr. This probably indicates that the 35 yr span of the ISC catalog does not provide a record of the seis-

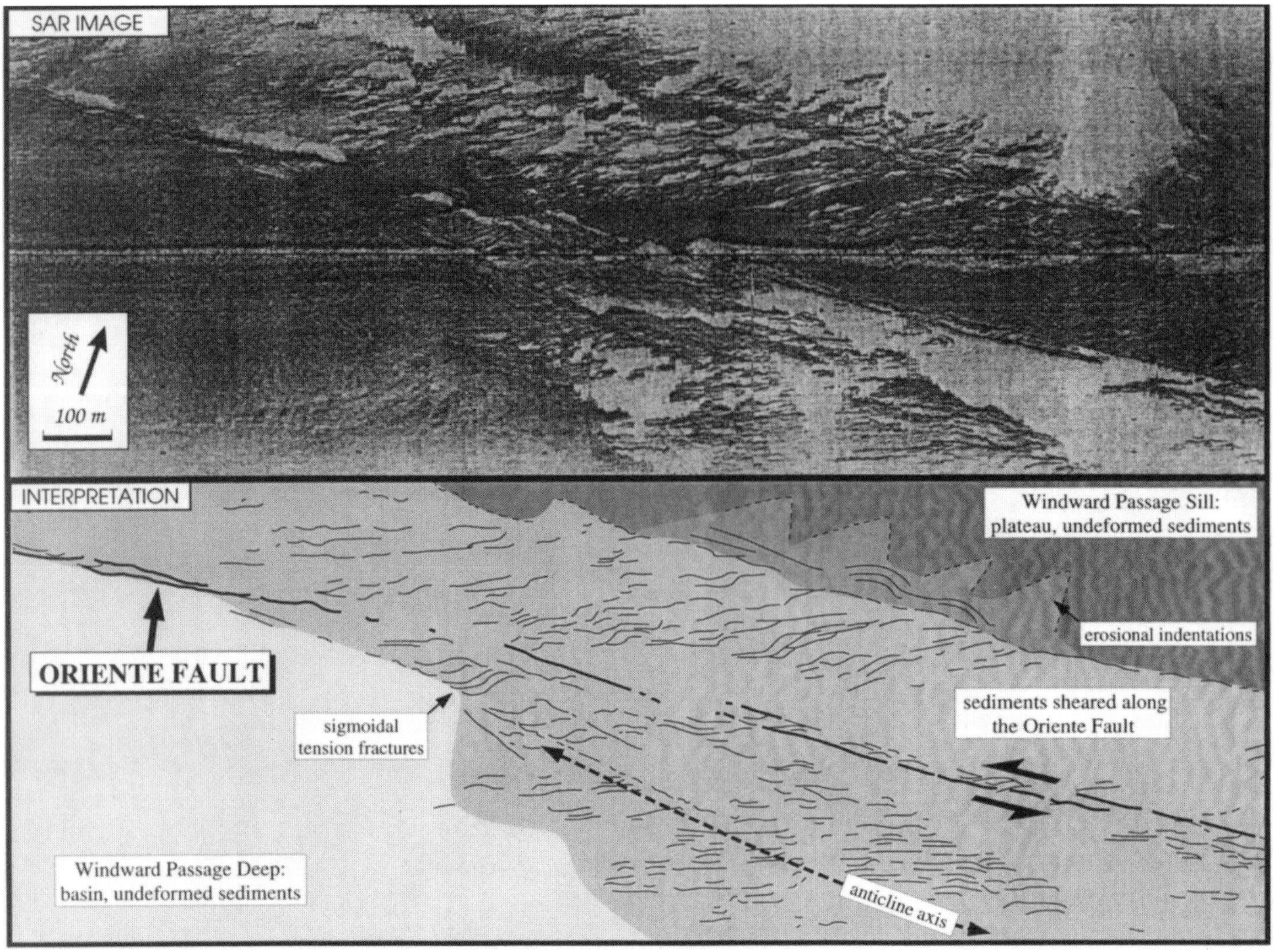

Figure 11. Side-scan sonar image of a segment of the Oriente fault in the Windward Passage Deep (SAR = side aperture radar).

micity typical of longer time spans. However, it could also mean that the magnitudes of the historical events have been overestimated.

To test these hypotheses, we used an alternative method based on the frequency-magnitude and moment-magnitude relations that characterize the seismic area (Anderson, 1979). In a given seismic area where $\log(N) = a - b\, m_b$ and $\log(M_o) = c\, m_b + d$, the average seismic slip is given by:

$$\bar{u} = \frac{1}{\mu A}\frac{10^{\alpha}10^{(1-\beta)\gamma_{max}}}{(1-\beta)\ln 10}, \quad (2)$$

with $\beta = b/c$, $\alpha = a + db/c - \log(c)$, and $\gamma_{max} = \log(M_o^{max})$. We used the ISC catalog and obtained for the southern Cuban margin $\log(N) = 3.6 - 0.6m_b$, $\log(M_o) = 0.9m_b + 19.4$ (Fig. 13), and an average seismic slip rate of 13.3 mm/yr (Table 3).

This value is about twice as great as the 6.3 mm/yr obtained by summing the seismic moments provided by the ISC catalog only, but it agrees well with the 15 mm/yr obtained from using the historical seismicity and with the NUVEL-1 slip estimate of 12 ± 3 mm/yr. We therefore believe that the 6.3 mm/yr value was biased by a peculiar short-term seismicity pattern over the past 35 yr and that the magnitudes of the historical events were probably correct.

Seismic stress tensor

We computed a stress tensor by inverting the source parameters of 6 $m_b > 5$ thrust events of the southern Cuban margin (Angelier and Mechler, 1979). The source parameters are listed in Table 3 and the principal components of the resulting stress tensor are shown on Figure 14. We obtain a principal compressive stress s_1 trending N240°E and dipping 13° to the southwest and a principal tensional stress s_3 trending N134°E and dipping 50° to the southeast. This stress tensor characterizes a combination of thrust and strike-slip and is consistent with the tectonic observations. The principal compressive

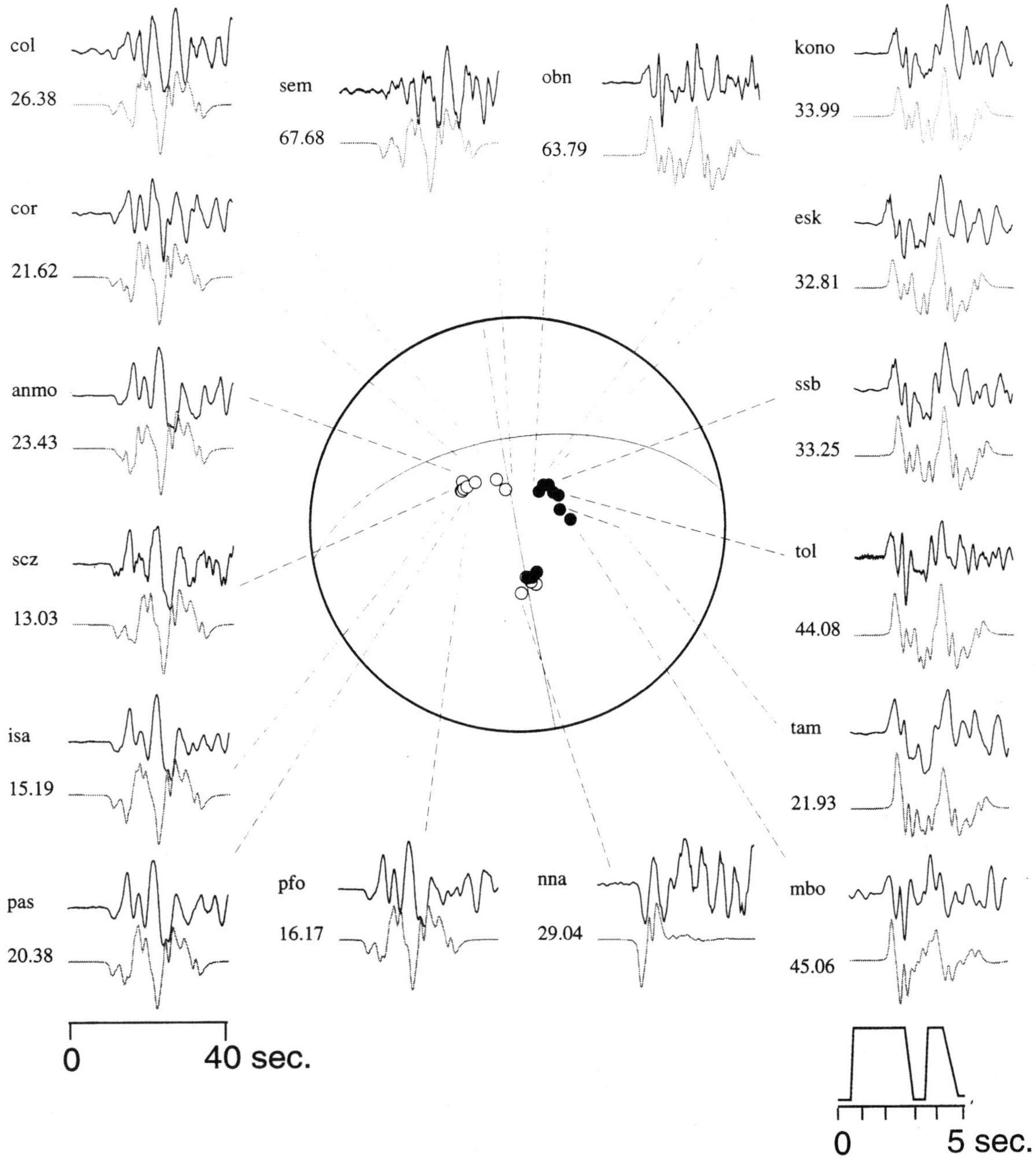

Figure 12. Focal mechanism of the Ms = 7.0 "Cabo Cruz" earthquake (May 25, 1992) from P-waves modeling. The observed (top) and synthetic (bottom) seismograms are displayed for each station with the station name on the left. Seismic moments are in 10^{-18} N.m. The source time function is at the bottom right of the figure.

stress is subperpendicular to the main strike-slip fault and perpendicular to the axis of the folds that we imaged in the Oriente Deep (Fig. 7).

This same stress pattern has been described along the San Andreas fault system in California (Mount and Suppe, 1987), where a small component of convergence is added to the general strike-slip displacement between the Pacific and North America plates. We tentatively apply the same interpretation to the northern Caribbean plate boundary along the southern Cuban margin and thus infer a small convergence component to be added to the general strike-slip displacement between the Caribbean and North America plates in that region. This convergence component could explain the transpressive tectonic regime revealed by the morphostructural analysis as well as the teleseismic data along the southern Cuban margin.

TABLE 1. HISTORICAL EARTHQUAKES REPORTED ALONG THE SOUTHERN CUBAN MARGIN*

Date	Location	Moment (10^{26} dyn.cm)	Magnitude M_s
Sept. 23, 1887	South of Cuba	25	7.5
Feb. 20, 1917	19.5°N/78.5°W	14	7.4
Feb. 3, 1932	19.8°N/75.8°W	1.5	6.7
July 7, 1947	19.9°N/75.3°W	1.5	6.7

*Source parameters have been estimated (from Mocquet, 1984). Numerical values are approximate.

Seismic strain rates

Kostrov (1974) showed that the average irrotational strain in a volume caused by slips on different rupture planes is given by

$$\bar{\varepsilon}_{ij} = \frac{1}{2\mu\Delta V}\sum_{x=1}^{n} M_{ij}^{(x)} \quad (3)$$

where M_{ij} is the seismic moment components summed over n rupture planes and ΔV is the deformed volume. As before, we assumed a shear modulus μ of 3.3×10^{11} dyn/cm^2 and a 440-km-long × 40-km-wide × 15-km-deep deformed volume extending along the Oriente fault from 78°E to 74°E. Using the CMT source parameters (Table 3), we obtain the following strain tensor:

$$\bar{\varepsilon}_{ij} = \begin{pmatrix} 31.27 & 42.46 & -49.49 \\ 42.46 & -33.06 & 30.40 \\ -49.49 & 30.40 & 1.79 \end{pmatrix} \times 10^{-9}/\text{yr}, \quad (4)$$

with the rows and columns corresponding to the east, south, and vertical components, respectively. The average slip corresponding to this strain tensor is 13.9 mm/yr in a N84.5°E direction. Given the uncertainties associated with the determination of earthquake source parameters, this direction is in fair agreement with the NUVEL-1 prediction of N78°E. The slip rate agrees well with the value of 13.3 mm/yr obtained using Anderson's method (Table 2) and with the NUVEL-1 prediction of 12 ± 3 mm/yr.

This strain tensor represents the strain released during earthquakes in the deformed area averaged over 18 yr (data time span). It shows east-west extension and north-south contraction of about the same amount and with little vertical extension. Figure 15 shows that the strain tensor is almost isotropic in a vertical east-west–trending plane, as one should expect within a shear plane. This agrees with the Oriente fault being the main shear zone. In the horizontal plane, the strain tensor shows left-lateral shear (Fig. 15) consistent with strike-slip displacement along the plate boundary. In a vertical north-south–trending plane, the strain tensor shows right-lateral shear and strong contraction along an axis dipping 45° to the north, both consistent with southward thrusting and north-south shortening in the Santiago deformed belt.

DISCUSSION: SEISMICITY, ACTIVE STRUCTURES, AND PLATE MOTION

Marine geophysical data and the seismicity characteristics show that the tectonic regime along the southern Cuban margin is a combination of left-lateral shear and north-south compression along the whole margin. This transpressive regime is not restricted to the Santiago deformed belt, where surficial evidence of transpression is pronounced, but also extends into the Cabo Cruz basin, where the major tectonic structures indicate transtension. However, transtension in the Cabo Cruz pull-apart area is related to the local geometry of the main fault zone, and does not preclude the existence of a regional compressional component along the fault zone.

The seismic slip rate along the Oriente fault deduced from different methods and from data spanning up to 108 yr agrees well with the 12 ± 3 mm/yr NUVEL-1 prediction (Table 2). This result suggests that all the stress accumulated by the Caribbean–North America plate motion is released seismically along the southern Cuban margin. On the other hand, 97% of the total seismic energy released along the southern Cuban mar-

TABLE 2. SLIP AND MOMENT RATES ALONG THE SOUTHERN CUBAN MARGIN*

	Geology (NUVEL-1)	Cumulative Seismic Moment, 1887 to Present	Cumulative Seismic Moment, 1960 to Present	Anderson (1979)	Kostrov (1974)
Moment rate (10^{24} dyn.cm/yr)	26	34	14	29	30
Slip rate (mm/yr)	12.0 ± 3	15	6.3	13.3	13.8
Time span (years)	3×10^6	108	35	35	18

*Rates estimated from five different methods (explanations in text).

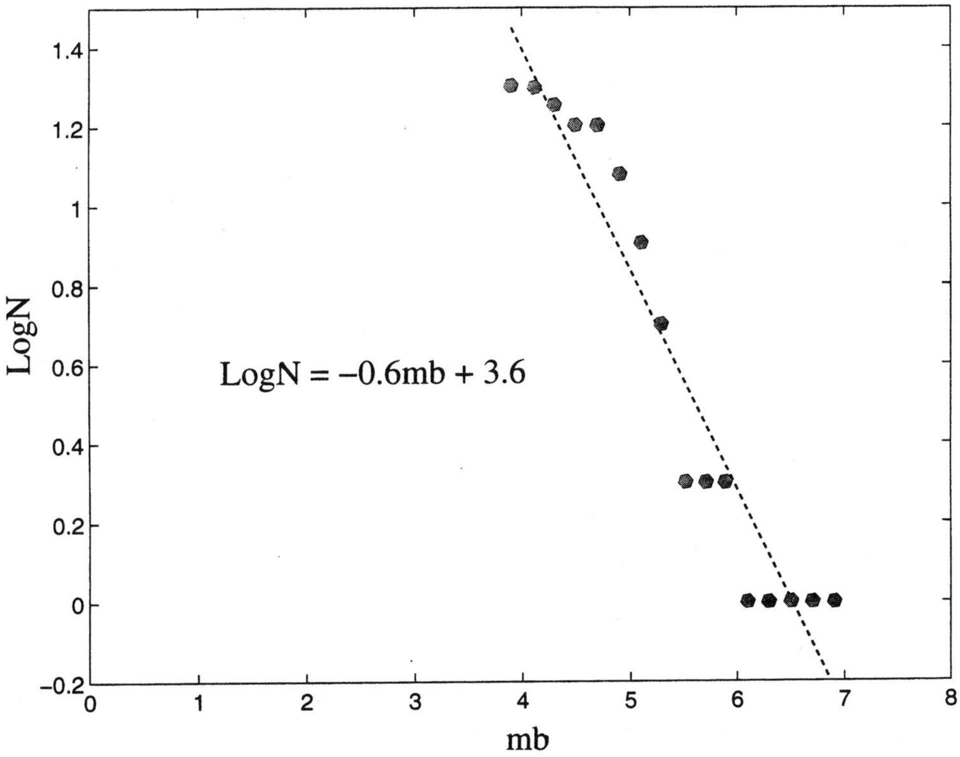

Figure 13. Cumulative frequency (*N*) versus magnitude (m_b) plot for the southern Cuban margin seismicity between 78°W and 74°W. The plotted regression line corresponds to the relation indicated in the figure.

TABLE 3. SOURCE PARAMETERS OF M_b >5 EARTHQUAKES ALONG SOUTHERN CUBAN MARGIN FROM 1970 TO PRESENT

Date	Time	Strike	Dip	Rake	m_b	M_s	M_o	M_{rr}*	M_{ss}*	M_{ee}*	M_{rs}*	M_{re}*	M_{se}*
11/13/78	07:43	330°	69°	73°	5.1	4.7	0.63	0.41	0.02	-0.43	-0.3	0.35	0.69
09/01/85	01:00	228°	53°	82°	5.1	4.5	0.44	0.44	-0.28	-0.17	-0.06	-0.11	-0.18
02/12/89	14:27	286°	29°	43°	5.2	4.6	0.75	0.44	-0.28	-0.16	0.11	-0.55	0.31
05/22/90	20:36	290°	43°	27°	5.0	4.9	0.13	0.57	-0.07	-0.50	-0.35	-0.82	0.74
08/26/90	07:53	039°	62°	57°	5.7	5.1	2.2	-1.4	-0.72	2.2	-0.69	-0.59	0.82
09/04/90	08:03	302°	21°	21°	5.2	4.5	0.73	0.24	-0.07	-0.17	-0.26	-0.63	0.23
05/25/92	16:55	248°	43°	01°	6.3	7.0	201	4	-85	81	81	-127	109

*M_{rr}, M_{ss}, M_{ee}, M_{rs}, M_{re}, and M_{se} are the moment tensor elements in 10^{24} dyn.cm expressed in an r-up, s-South, e-East coordinate system.

gin over the past 35 yr is due to the May 25, 1992, earthquake (Table 3). Also noticeable is the lack of moderate-size earthquakes ($5.0 < M_s < 6.0$) in this region, as shown by Figure 13. This tends to show that most of the seismic stress is released during a few strong earthquakes, such as the 1992 Cabo Cruz event, and is consistent with the historical seismicity that shows the occurrence of only one or two earthquakes with a magnitude greater than 7.0 per century.

These results single out this segment of the northern Caribbean strike-slip plate boundary from oceanic transform faults, where up to 95% of the total slip can occur aseismically (Trehu and Solomon, 1983; Steward and Okal, 1983; Dziak et al., 1991). In the case of oceanic transform faults, this process is usually related to an anomalously high heat flow that prevents seismic failure (Burr and Solomon, 1978; Kawasaki et al., 1985). Heat-flow measurements in the eastern part of the Cayman trough do not show any significant anomaly (Rosencrantz et al., 1988). Moreover, the Oriente fault along the southern Cuban margin crosses island-arc rocks, not oceanic material (Edgar et al., 1971; Mercier de Lépinay et al., 1989). We also show that the relative displacement of the Caribbean and North America plates is slightly oblique to the Oriente fault trend along the southern Cuban margin. This could participate in "locking" the fault zone by increasing the coupling between the Caribbean and North America lithospheres. These characteristics may explain the highly seismic nature of the northern Caribbean plate boundary along the southern Cuban margin.

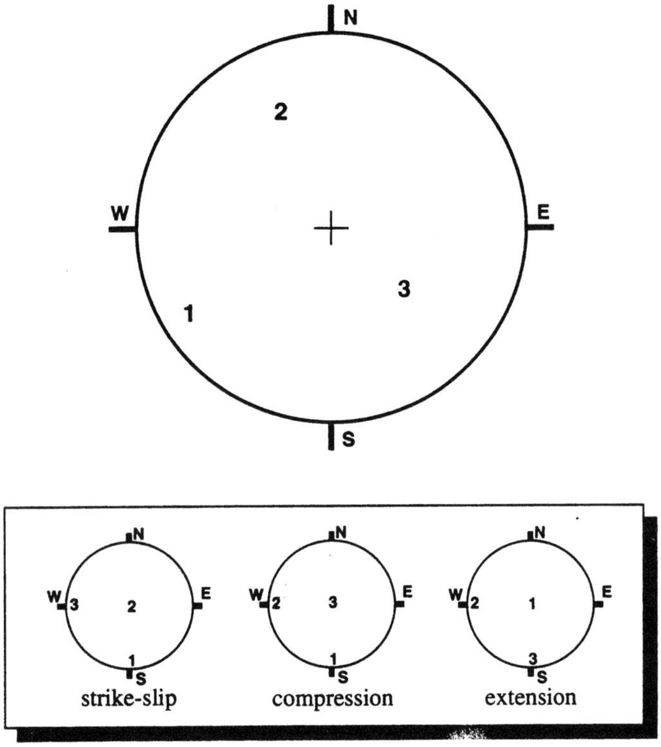

Figure 14. Principal axis of the stress tensor along the southern Cuban margin (Schmidt projection, lower hemisphere). Inset: theoretical stress tensor axis in the case of pure strike-slip, pure compression, and pure extension.

CONCLUSION

The analysis of the active tectonic structures along the northern Caribbean plate boundary from Cuba to Hispaniola reveals a narrow deformation zone dominated by left-lateral strike-slip faulting. The shear displacement is probably concentrated along the most prominent structural feature, the Oriente fault. This fault has a discontinuous trace with left-stepping offsets that generate tensional strain and cause the subsidence of pull-apart basins like the Cabo Cruz basin. However, the most striking tectonic structures display transpression (Santiago deformed belt) with active folding and reverse faulting. The earthquakes' distribution and their source parameters are well correlated with the active tectonic structures. Both types of data reveal the combination of north-south compression and left-lateral shear starting at about 78°W long. in the Cabo Cruz basin area and continuing all the way along the southern Cuban margin. We infer the existence of a component of convergence added to the shear displacement between the Caribbean and North America plates that would induce the transpressive strain regime and cause folding and reverse faulting along structures parallel to the main fault. We compute a seismic slip rate of 13–15mm/yr that agrees well with the 12 ± 3 mm/yr NUVEL-1 estimate along the northern Caribbean plate boundary. This result implies that all the slip along the southern Cuban margin occurs seismically and could suggest strong lithospheric coupling between the Caribbean and North America plates in that region.

ACKNOWLEDGMENTS

We are grateful to the ship crew and scientific members of the Seacarib II (R/V *Jean Charcot*) and Oriente (R/V *Le Suroit*) oceanographic campaigns during which most of the marine geophysical data presented here were collected. We thank Jim Dolan, William Dillon, and James Dewey for their helpful reviews. Géoscience Azur publication 167.

REFERENCES CITED

Anderson, J. G., 1979, Estimating the seismicity from geological structure for seismic-risk studies: Bulletin of the Seismological Society of America, v. 69, p. 135–158.

Angelier, J., and Mechler, P., 1979, A graphic method applied to the localization of principal stresses for fault tectonics and seismology; the right dihedral method: Bulletin de la Societé Géologique de France, v. 19, p. 1309–1318.

Brune J. N., 1968, Seismic moment, seismicity, and rate of slip along major fault zones: Journal of Geophysical Research, v. 73, p. 777–784.

Burr N. C., and Solomon, S. C., 1978, The relationship of source parameters of oceanic transform earthquakes to plate velocity and transform length: Journal of Geophysical Research, v. 83, p. 1193–1205.

Calais, E., and Mercier de Lépinay, B., 1990, A natural model of active transpressional tectonics: The en échelon structures of the Oriente Deep, along the northern Caribbean transcurrent plate boundary: Revue de l'Institute Francais du Pétrole et Annales des Combustible Liquides, v. 45, p. 147–160.

Calais, E., and Mercier de Lépinay, B., 1991, From transtension to transpression along the northern Caribbean plate boundary off Cuba: Implications for the recent motion of the Caribbean plate: Tectonophysics, v. 186, p. 329–350.

Calais, E., and Mercier de Lépinay, B., 1995, Strike-slip tectonic processes in the northern Caribbean between Cuba and Hispaniola (Windward Passage): Marine Geophysical Research, v. 17, p. 63–95.

Case, J. E., and Holcombe, T. L., 1980, Geologic-tectonic map of the Caribbean region: U.S. Geological Survey, Miscellaneous Investigations, Map I-1100, Scale 1:2,500,000.

Cobbold P. R., 1975, Fold propagation in single embedded layers: Tectonophysics, v. 27, p. 333–351.

DeMets, C., Gordon, R. G., Argus, D. F., and Stein, S., 1990, Current plate motions: Geophysical Journal of the Royal Astronomical Society, v. 101, p. 425–478.

Dziak R. P., Fox, C. G., and Embley, R. W., 1991, Relationship between the seismicity and geologic structure of the Blanco transform zone: Marine Geophysical Researches, v. 13, p. 203–208.

Dziewonski, A. M., Chou, T. A., and Woodhouse, J. H., 1981, Determination of earthquake source parameters from waveform data for studies of global and regional seismicity: Journal of Geophysical Research, v. 86, p. 2825–2852.

Edgar, N. T., Ewing, J. I., and Hennion, J., 1971, Seismic refraction and reflection in the Caribbean Sea: American Association of Petroleum Geology Bulletin, v. 55, p. 833–870.

Goreau, P. D. E., 1983, The tectonic evolution of the north-central Caribbean plate margin [Ph.D. thesis]: Cambridge, Massachusetts, Massachusetts Institute of Technology, 245 p.

Kawasaki, I., Yasutosi, K., Ikuko, T., and Kosugi, N., 1985, Mode of seismic

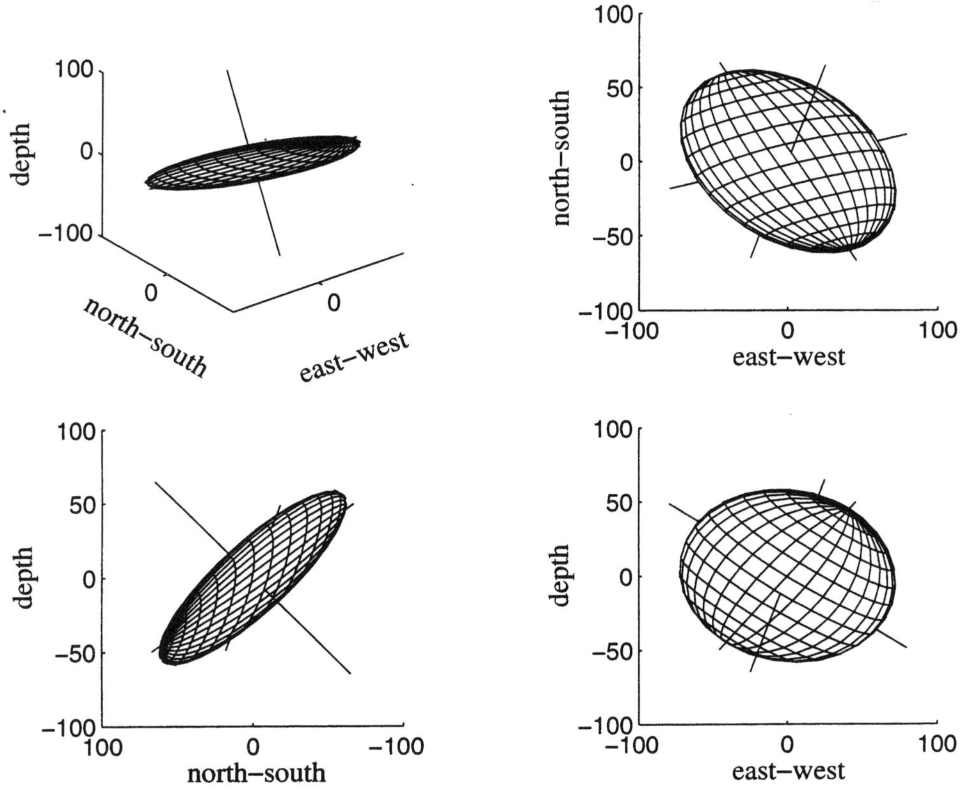

Figure 15. Ellipsoid representation of the strain tensor along the southern Cuban margin.

moment release at transform faults: Tectonophysics, v. 118, p. 313–327.

Kostrov, V. V., 1974, Seismic moment and energy of earthquakes, and seismic flow of rocks: Physics Solid Earth, v. 1, p. 13–21.

Mann, P., and Burke, K., 1984, Neotectonics of the Caribbean: Review of Geophysics and Space Physics, v. 22, p. 309–362.

Mascle, A., and Letouzey, A., 1990, Geological map of the Caribbean: Rueil-Malmaison, France, L'institute Français du Pétrole, Z sheets.

Mercier de Lépinay, B., Renard, V., Calais, E., Calmus T., Capet, X., Delteil, J., Gardner, O., Hernandez, G., Mauffret, A., Momplaisir, R., Quentric, I., Rodriguez, R., Ross, M., Tardy, M., and Vila, J. M., 1989, Morphology and structure along a major marine strike-slip fault zone: The northern Caribbean plate boundary between Cuba, Hispaniola and Jamaica, *in* Proceedings of the 28th International Geological Congress, July 9–19, 1989: Washington D.C.

Mocquet, A., 1984, Vitesses de déplacement discontinu le long d'une zone limite de plaques: Caraibes-Amérique du Nord (Discontinuous slip rates along the Caribbean–North America plate boundary) [unpublished M.A. thesis]: France, University of Rennes, 53 p.

Molnar, P., and Sykes, L. R., 1969, Tectonics of the Caribbean and Middle America regions from focal mechanisms and seismicity: Geological Society of America Bulletin, v. 80, p. 1639–1684.

Mount, V. S., and Suppe, J., 1987, State of stress near the San Andreas Fault; implications for wrench tectonics: Geology, v. 15, p. 1143–1146.

Odonne, F., and Vialon, P., 1983, Analogue models of folds above a wrench fault: Tectonophysics, v. 99, p. 31–46.

Perrot, J., Deschamps, A., Farra, V., and Virieux, J., 1994, Azimuthal distorsion of the seismic focal sphere: Application to earthquakes in subduction zone: Physics of the Earth and Planetary Interiors, v. 84, p. 240–270.

Perrot, J., Calais, E., and Mercier de Lépinay, B., 1997, Waveform simulation of the May 25th, 1992, Ms = 6.7 Cabo Cruz earthquake (Cuba): New constraints on the tectonic regime along the northern Caribbean plate boundary: PAGEOPH, v. 149, p. 475–487.

Richard, P., and Cobbold, P., 1989, Structures en fleur positives et décrochements crustaux: Modélisation analogique et interprétation dynamique (Positive flower structures and crustal strike-slip faults: analog models and dynamic interpretation): Comptes Rendus de l'Académie des Sciences des Paris, p. 553–560.

Rosencrantz, E., Ross, M. I., and Sclater, J. G., 1988, The age and spreading history of the Cayman Trough as determined from depth, heat flow and magnetic anomalies: Journal of Geophysical Research, v. 93, p. 2141–2157.

Salterain, P., 1883, Ligera resena de los tremblores de tierra ocurridos en la isla de Cuba (The earthquakes of the island of Cuba): Boletin de la Comision del Mapa Geologico de Espana, v. 10, p. 371–385.

Steward, L. M., and Okal, E. A., 1983, Seismicity and aseismic slip along the Eltanin fracture zone: Journal of Geophysical Research, v. 88, p. 10495–10507.

Taber, S., 1922, The seismic belt in the greater Antilles: Bulletin of the Seismological Society of America, v. 4, p. 199–219.

Thatcher W., Hileman, J. A., and Hanks, T. C., 1975, Seismic slip distribution along the San Jacinto fault zone, southern California, and its implications: Geological Society of America Bulletin, v. 86, p. 1140–1146.

Tréhu, A. M., and Solomon, S. C., 1983, Earthquakes in the Orozco transform zone: Seismicity, source mechanisms, and tectonics: Journal of Geophysical Research, v. 88, p. 8203–8225.

Wilcox, R. E., Harding, T. P., and Seely, D. R., 1973, Basic Wrench Tectonics: American Association of Petroleum Geologists Bulletin, v. 57, p. 74–96.

Manuscript Accepted by the Society January 9, 1998

Printed in U.S.A.

Geological Society of America
Special Paper 326
1998

The 1943–1953 north-central Caribbean earthquakes: Active tectonic setting, seismic hazards, and implications for Caribbean–North America plate motions

James F. Dolan
Department of Earth Sciences, University of Southern California, Los Angeles, California 90089
David J. Wald
U.S. Geological Survey, Pasadena, California 91125

ABSTRACT

The 1943 northern Mona Passage earthquake and the 1946 northeastern Hispaniola earthquake were two of the largest events to strike the north-central Caribbean region in more than 400 yr. Both were oblique, left-lateral thrust events generated by the rupture of large parts of the gently south-dipping, plate-interface fault that separated the obliquely subducting North America (Atlantic) slab from the overriding Hispaniola and Puerto Rico blocks of the northern Caribbean plate boundary zone (NCPBZ). These ruptures occurred at strongly coupled collisional asperities associated with active, collisional underthrusting of high-standing carbonate banks of the southeastern Bahamas. This is particularly well illustrated by the 1946 mainshock, where the 200-km-long aftershock zone exactly matches the limits of the active collision zone between northern Hispaniola and Silver and Navidad carbonate banks. A large aftershock of the 1946 earthquake in 1948 deepened the mainshock rupture plane, and another major aftershock in 1953 extended the rupture 15 km westward to the leading edge of the oblique, westward-propagating Bahamas collision.

Bathymetric data reveal a 7-km-deep pull-apart basin along a major left-lateral strike-slip fault in the Hispaniola slope between the 1943 and 1946 rupture planes, suggesting that this section of the plate-interface oblique thrust fault may not be undergoing active collision. We speculate that such noncollisional parts of the plate-interface fault may not be as strongly coupled as the collisional asperities. The occurrence of both the 1943 and 1946 events—the two largest thrust events known to have occurred on the plate-interface fault since at least 1787—suggests that in the northern Caribbean such large thrust earthquakes may typically nucleate only at strongly coupled collisional asperities.

Rakes derived from focal mechanisms that we calculated for the 1943 and 1946 mainshocks (030° and 034°, in map view, respectively) indicate that slip along the plate-interface fault is oblique to the 080–110° trend plate margin in the northeastern Hispaniola–northwestern Puerto Rico area. Much of the plate-boundary slip in this region, however, is partitioned onto the left-lateral, Septentrional fault onshore Hispaniola. This fault probably merges into the south-dipping plate interface along the top of the underthrust Atlantic slab at ~15 km depth beneath northern Hispaniola. The exact location of the downward termination of the Septentrional fault is

Dolan, J. F., and Wald, D. J, 1998, The 1943–1953 north-central Caribbean earthquakes: Active tectonic setting, seismic hazards, and implications for Caribbean–North America plate motions, *in* Dolan, J. F., and Mann, P., eds., Active Strike-Slip and Collisional Tectonics of the Northern Caribbean Plate Boundary Zone: Boulder, Colorado, Geological Society of America Special Paper 326.

unknown, but it probably occurs downdip from the 1943 and 1946 rupture planes. Thus, the rakes derived from the 1943 and 1946 mainshocks show that relative plate motion in the region must be oriented more easterly than ~030°. Furthermore, the oblique-thrust mechanism of the 1943 mainshock indicates that plate motion along the 080°-trending plate boundary off northwestern Puerto Rico must be oriented more northerly than 080°. This is supported by geological data that reveal evidence for a slight component of convergence across this part of the plate margin. Thus, the direction of relative motion between the obliquely subducting North America (Atlantic) plate and the Hispaniola and Puerto Rico blocks of the NCPBZ is between 030° and 080°. These data support several published models that predict east-northeast relative plate motion in the region. Conversely, the earthquake data rule out several models that predict that relative motion between the North America plate and the Puerto Rico block of the NCPBZ plate motion is oriented east-west. The transpressional regime revealed by the earthquake data also preclude models in which Puerto Rico has been proposed to be (1) actively rotating counterclockwise or (2) underlain by an active, south-dipping extensional decollement.

Evidence for a component of oblique left-lateral slip on the plate-interface thrust, coupled with published, preliminary Global Positioning System geodetic data, suggests that a significant portion of the relative motion between the North America plate and the Hispaniola and Puerto Rico blocks of the NCPBZ may be accommodated along the south-dipping plate-interface thrust fault. This fault thus represents a major seismic hazard to the densely populated northern parts of Hispaniola and Puerto Rico. Other major seismic sources in the region include the Septentrional fault, large normal faults in the Mona Passage between Hispaniola and Puerto Rico, and the left-lateral Enriquillo-Plantain Garden strike-slip fault and north-dipping Los Muertos plate-interface thrust fault along the southern edge of the NCPBZ.

The 1943–1953 earthquakes were the most recent major events in a cluster of large events that have occurred on faults along the northern margin of the NCPBZ since 1842. Available historical data suggest that this cluster of activity in the northern deformation belt is anti-correlated—both temporally and geographically—with a similar cluster of earthquakes that occurred between the early 1600s and 1860 on a parallel system of left-lateral strike-slip and thrust faults that stretches along the southern boundary of the NCPBZ. The cause of the intriguing pattern of "deformation switching" remains unknown but represents an important target of future research.

INTRODUCTION

Although the densely populated northern Caribbean islands of Hispaniola and Puerto Rico (Fig. 1) have the longest historical earthquake record in the Americas, dating back to the mid-16th century (Scherer, 1912; Kelleher et al., 1973; Chalas-Jimenez, 1989; W. R. McCann and L. Feldman, written communication, 1995), seismic hazards in the north-central Caribbean region remain relatively poorly understood. One of the main reasons for this is that, although the general tectonic setting of the northern Caribbean plate margin has been known for some time (e.g., Molnar and Sykes, 1969; Bracey and Vogt, 1970; Schell and Tarr, 1978), the major active faults in the plate boundary have not been studied in detail until recently. In many cases, basic faulting parameters, such as location, extent, geometry, kinematics, and earthquake history have not been known. For example, although the Septentrional fault is a major left-lateral strike-slip fault that stretches for several hundred kilometers across the length of northern Hispaniola, it was not fully identified as such until the early 1980s (Mann et al., 1984; Mann and Burke, 1984). The study of Mann et al. (this volume) is the first detailed study to be published on this major fault. The offshore part of the plate boundary in the Hispaniola region was, until recently, even less well studied than the onshore faults. Austin (1983) and Dillon et al. (1992) recognized active thrust faulting off northwestern Hispaniola, but a companion chapter by Dolan et al. (this volume) is the first detailed description of the active structures and tectonics of a 400-km length of the offshore plate boundary between north-central Hispaniola and northwestern Puerto Rico.

Major earthquakes in the region have also received relatively little study until recently. Prior to the 1943–1953 sequence discussed in this chapter, which was studied by Kelleher et al. (1973), only a few historic events have been documented in any detail (e.g., the 1842 northwest Hispaniola earthquake *in* Scherer, 1912; and the 1918 Mona Passage earthquake *in* Reid and Taber, 1919) despite the long historic record in the region.

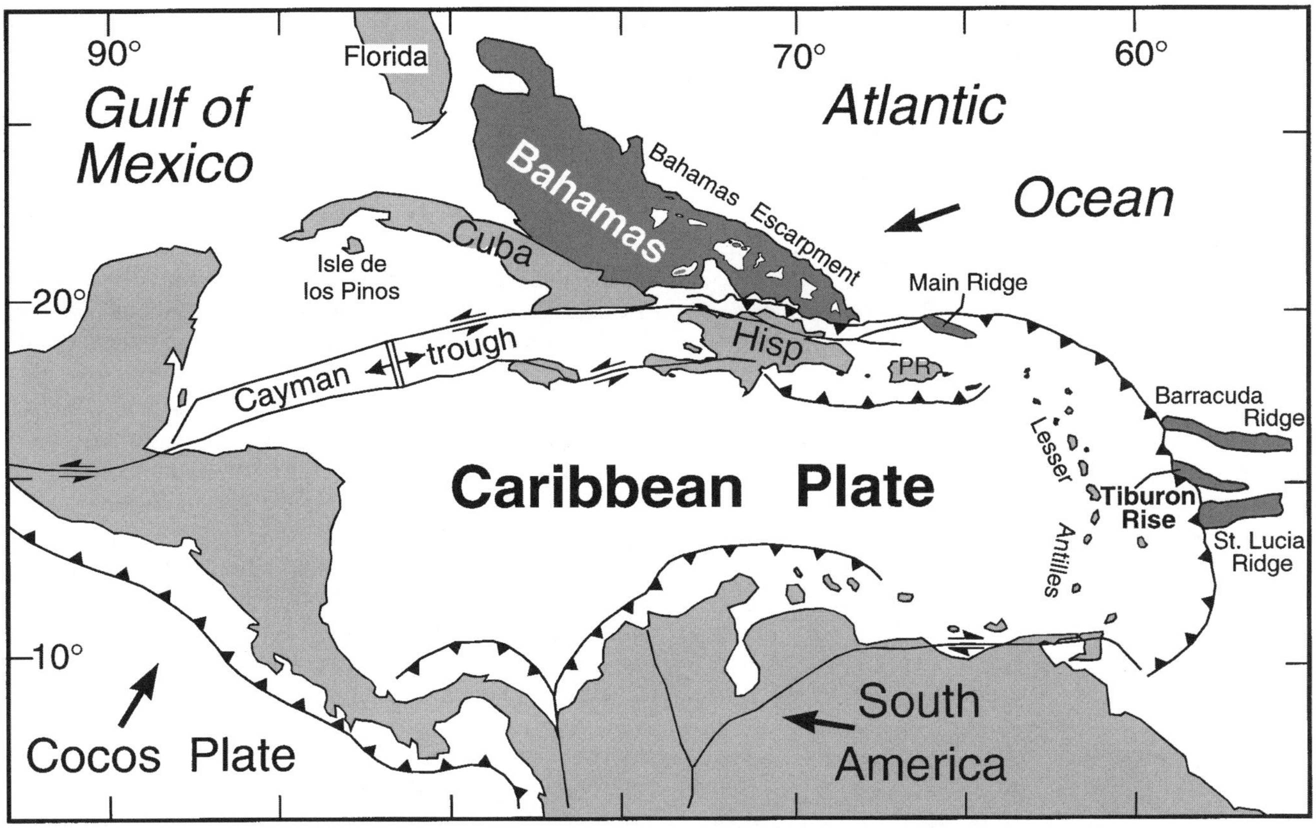

Figure 1. Location map of the Caribbean region (after Mann et al., 1984). Arrow in northeast denotes best estimate of northern Caribbean plate boundary zone (NCPBZ)–North America (Atlantic) plate relative motion vector determined in this study. Other arrows show plate motion directions from Mann et al. (1984). Westward motion of both the North America and South America plates relative to the Caribbean plate south of the NCPBZ results in predominantly left-lateral strike-slip motion along the northern boundary, and right-lateral strike-slip motion along the southern margins of the Caribbean plate, respectively; the Caribbean plate is stationary relative to a hot-spot reference frame. Dark shading denotes Bahamas carbonate province. PR is Puerto Rico; Hisp is Hispaniola.

Furthermore, paleoseismological studies have only just begun in the region. Prentice et al. (1993) published the first paleoseismologic data from the northern Caribbean based on their trench studies of the Septentrional fault in north-central Hispaniola (see also Mann et al., this volume).

Another major factor in our lack of knowledge about the seismotectonics and seismic hazards of the northern Caribbean is the fact that relatively few large earthquakes have occurred along the plate boundary during the instrumental era. This chapter focuses on the two largest instrumentally recorded earthquakes—the 1943 northern Mona Passage earthquake and the 1946 northeastern Hispaniola earthquake—that have occurred in the northern Caribbean region. In this study we (1) examine the relationship between these earthquakes and zones of active, collisional underthrusting along the northern Caribbean plate boundary zone (NCPBZ); (2) discuss the significance of these collisions for seismic hazard assessment in the region; (3) use the earthquake focal mechanisms that we have determined to address the long-standing disagreement about relative plate motions in the north-central Caribbean region; and (4) discuss the 1943 and 1946 earthquakes in the context of the temporal and spatial patterns of strain release along the NCPBZ.

GEOLOGIC SETTING

As a result of long-lived and ongoing interactions with the surrounding North America and South America plates, the margins of the Caribbean plate are extremely complex, both structurally and kinematically (Fig. 1). The north-central Caribbean islands of Hispaniola, Puerto Rico, and Cuba represent the dismembered remnants of one or more Cretaceous to Eocene island arcs that developed above a southwest-dipping subduction zone (Pindell and Barrett, 1991). Since middle Eocene time both the North and South America plates have moved westward relative to the Caribbean plate (in a hot-spot reference frame), with active arc volcanism occurring in the Lesser Antilles in response to west-dipping subduction of Atlantic oceanic lithosphere beneath the Caribbean plate (Pindell and Barrett, 1990). This relative motion results in predominantly left-lateral motion along the northern boundary of the Caribbean plate (Molnar and Sykes, 1969; Bracey and Vogt, 1970; Schell and Tarr, 1978; Mann et al., 1984, 1990; Pindell and Barrett, 1990). Despite a general consensus on this point, both the direction and velocity of Caribbean–North America relative plate motion have been the subject of long-standing debate, with a range of proposed directions from 090° to 065°, and a range of proposed rates from 11 to 37 mm/yr (Molnar and Sykes, 1969; Jordan, 1975; Minster and Jordan, 1978; Sykes et al., 1982; Stein et al., 1988; DeMets et al., 1990; DeMets, 1993; Deng and Sykes, 1995; Farina et al., 1995).

Global plate motion model NUVEL-1, which is relatively poorly constrained for the northern Caribbean region, yielded near east-west relative plate motion at an overall rate of ~11 ± 3 mm/yr in the Hispaniola region (De Mets et al., 1990; DeMets, 1993). Analysis of magnetic lineations in the Cayman trough suggests that the long-term average plate motion rate is ~15 mm/yr along the Caribbean–North America plate boundary (Rosencrantz et al., 1988). However, more recent analysis of only post-early Miocene to Holocene magnetic lineations yielded a slightly faster estimate of ~19–22 mm/yr (Rosencrantz, 1995). MacDonald and Holcombe (1978) obtained a similar estimate (20 ± 4 mm/yr) using magnetic anomaly data for the past 2.3 Ma. Sykes et al. (1982) and Rosencrantz and Mann (1991), however, argued that these values underestimate total Caribbean–North America motion because some plate-boundary slip bypasses the northern part of the Cayman trough and is accommodated along the left-lateral Enriquillo-Plantain Garden fault zone in Jamaica and southern Hispaniola (Fig. 1) (Mann et al., 1984). Sykes et al. (1982), using the configuration and thermal character of the subducted Atlantic slab and slip vectors from plate-boundary earthquakes in the eastern and northeastern Caribbean, estimated a rate of 37 ± 5 mm/year along an azimuth of 065°. Deng and Sykes (1995) argued that this rate is probably an overestimate, and they use their calculated Euler pole together with GPS geodetic data from Dixon (1993) to estimate a rate of 20–30 mm/yr, with a best estimate of 23 mm/yr, along an azimuth of 070° in the north-central Caribbean. Dewey and Suarez (1991) obtained a rate of 27 mm/yr by assuming that the Caribbean is fixed with respect to a hotspot reference frame. Preliminary GPS geodetic data from a network of northern Caribbean stations yielded an overall Caribbean–North America plate relative plate motion rate of ~24 mm/yr in the Hispaniola region (Farina et al., 1995; Dixon et al., 1998). Collectively, these estimates indicate that the relative motion rate between the Caribbean and North America plates is ~20–25 mm/yr.

In the north-central Caribbean the plate boundary forms a complex, 250-km-wide zone of deformation known as the northern Caribbean plate boundary zone (NCPBZ) (Fig. 2; Mann et al., 1984). The NCPBZ in this region is highlighted by two major zones of deformation that occur along the northern and southern boundaries of the islands of Puerto Rico and Hispaniola (Mann et al., 1984, 1990; Burke, 1988). The northern zone, which is the focus of this study, is associated with south-dipping underthrusting of Atlantic oceanic lithosphere of the North America plate (McCann and Sykes, 1984; McCann and Pennington, 1990; Deng and Sykes, 1995; Dolan et al., this volume), whereas the southern deformation zone involves north-dipping underthrusting of Caribbean lithosphere at the Los Muertos trench (Ladd and Watkins, 1978; Byrne et al., 1985; Ladd et al., 1990; Dillon et al., 1994; Dolan et al., this volume). The northern deformation zone exhibits well-developed strain partitioning in Hispaniola (Dolan et al., this volume), with much of the plate-boundary, left-lateral strike-slip motion accommodated on the Septentrional fault system (Mann et al., 1984, 1990, this volume; Calais et al., 1992; Prentice et al., 1993). The Septentrional fault system extends westward into the Oriente fault zone, which forms the northern boundary of the Cayman trough pull-apart basin (Fig. 1) (Calais et al., 1989, 1992, this volume; Edgar, 1991; Rosencrantz and Mann, 1991). East of Hispaniola, left-lateral strike-slip motion occurs on the eastern Septentrional and northern Puerto Rico slope fault zones north of Puerto Rico (Masson and Scanlon, 1991; Dolan et al., this volume). Development of a small accretionary prism offshore northern Hispaniola attests to an active component of convergence across the plate boundary in the Hispaniola region (Austin, 1983; Dillon et al., 1992; Dolan et al., this volume).

Hispaniola–Bahamas collision zone

One of the most notable features of the plate boundary in the Hispaniola area is a zone of active collision between Hispaniola and high-standing carbonate banks of the southeastern Bahamas. SeaMARC II side-scan sonar data collected offshore northern Hispaniola reveal that two Bahamas carbonate banks are actively colliding with Hispaniola over a front more than 200 km long (Dolan et al., this volume). The collision, which extends along the plate boundary between ~70° 00′W and 68° 30′W, involves the southern margins of Silver and Navidad banks, which rise 5.5 km above the surrounding Atlantic abyssal plain (Fig. 2). The collision is best illustrated offshore of north-central Hispaniola, between 70° 05′W and 69° 42′W, where Silver Spur, a bathymetrically high-standing ridge of Silver bank carbonate, impinges on the Hispaniola slope and projects beneath the Rio San Juan Peninsula of the Dominican Republic (Dolan et al., this volume). Throughout the rest of the collision zone, high-reflectivity carbonate bank material directly abuts lower reflectivity material of the Hispaniola slope.

To the west of the collision zone, the high-standing Mouchoir bank is separated from the deformation front at the base of the northern Hispaniola slope by the 20–30-km-wide Hispaniola basin (Fig. 3). Thus, this part of the Hispaniola margin is not experiencing active collision with the buoyant Bahamas block. In contrast, to the east of the Silver-Navidad bank collision zone, the Bahamas ridge has already been obliquely subducted beneath the Puerto Rico block of the NCPBZ (Dolan et al., this volume). These major differences in the structural style of the plate margin are due to the diachronous nature of the Bahamas collision—the

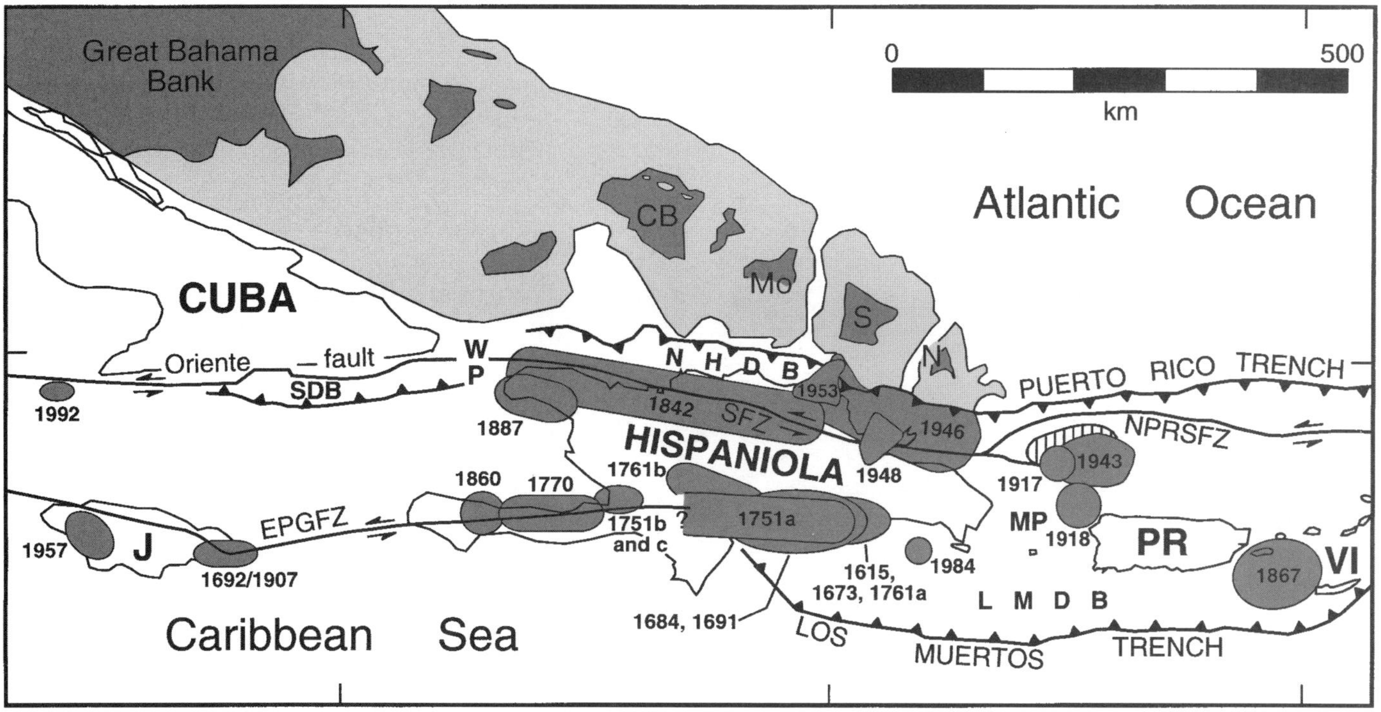

Figure 2. Neotectonic map of north-central Caribbean showing great width (250 km) and complexity of the northern Caribbean plate boundary zone (NCPBZ) as well as the location of major historical earthquakes (year of occurrence shown for each event; 1751a event occurred in October and 1751b earthquake occurred in November). Geologic data compiled from Dolan et al. (this volume), Mann et al. (1984, this volume), Masson and Scanlon (1991), Edgar (1991), Dillon et al. (1992), and Case and Holcombe (1980). Earthquake data compiled from Kelleher et al. (1973), Chalas-Jimenez (1989), McCann and Pennington (1990), and Mann et al. (this volume). CB is Caicos bank; EPGFZ is Enriquillo-Plantain Garden fault zone; J is Jamaica; LMDB is Los Muertos deformed belt; Mo is Mouchoir bank; MP is Mona Passage; N is Navidad bank; NHDB is northern Hispaniola deformed belt; NPRSFZ is northern Puerto Rico slope fault zone; PR is Puerto Rico; S is Silver bank; SDB is Santiago deformed belt; SFZ is Septentrional fault zone; VI is Virgin Islands; WP is Windward Passage.

generally westward motion and more northwesterly strike of the Bahamas ridge relative to the NCPBZ results in a westward-younging zone of active collision. Based on this reasoning, Dolan et al. (this volume) proposed that the Bahamas ridge collision has progressed much farther beneath northern Mona Passage than it has within the Silver-Navidad collision zone. They suggest that a zone of relatively shallow bathymetry in the northern part of Mona Passage (their Mona block) is the sea-floor manifestation of ongoing collisional underthrusting of an already subducted southeastern Bahamas bank, which they named Mona bank.

Dolan et al. (this volume) also used recent seismicity data to construct a structure-contour map on the top of the south- to south-southwest–dipping, underthrust Atlantic (North America) slab (Figs. 3 and 4). The strike and dip of the slab changes markedly at the proposed Mona bank collision. North of the Virgin Islands, at ~65° W, local network data reveal that the slab dips 45° south (Frankel, 1982; McCann and Sykes, 1984; Fischer and McCann, 1984). The dip of the slab shallows markedly westward toward the postulated Mona bank collision zone. At the locus of the collision at 67° 30′ the slab dips ~25° south (Fig. 3; Dolan et al., this volume). West of the collision the strike of the shallow part of the slab changes to west-northwest. Along the Silver-Navidad collision zone in northeastern Hispaniola the shallow part of the underthrust slab dips ~30° south-southwest.

THE 1943–1953 NORTH-CENTRAL CARIBBEAN EARTHQUAKE SEQUENCE

On July 29, 1943, a major earthquake occurred offshore of northwestern Puerto Rico (Fig. 5) (Kelleher et al., 1973). This event, which we refer to as the northern Mona Passage earthquake, was the first in a sequence of large earthquakes that struck the north-central Caribbean region during the decade 1943–1953. Three years after the northern Mona Passage earthquake, the region experienced an even larger event when, on August 4, 1946, a great earthquake occurred beneath the coast of northeastern Hispaniola, ~200 km west of the 1943 epicenter (Fig. 5) (Lynch and Bodle, 1948; Small, 1948; Kelleher et al., 1973).

In the following discussion of these earthquakes and their aftershocks we use the surface-wave magnitude (M_s) estimates of

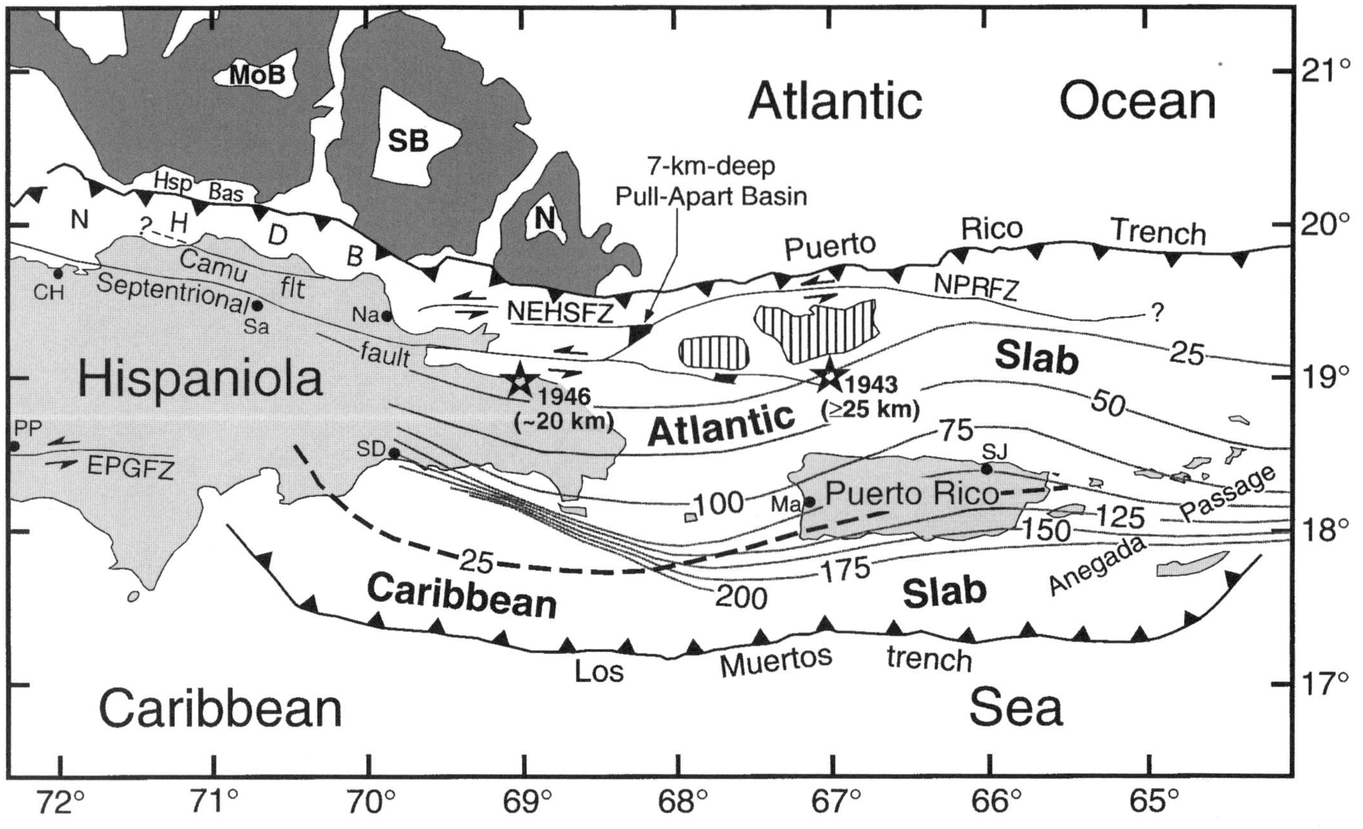

Figure 3. Structure contour map of underthrust Atlantic and Caribbean slabs in north-central Caribbean region. Depth contour interval is 25 km. Contours are from Dolan et al. (this volume) and are based on cross sections of recent seismicity that occurred between 1962 and 1994 and that were taken from the International Seismological Centre (ISC) Catalog of Earthquakes (1994). Slab contours shown as solid gray lines are constructed along the top of a south-dipping zone of seismicity interpreted to represent the top of the underthrust slab of Atlantic oceanic lithosphere of the North America plate. Slab contour shown by black dashed line is constructed along top of north-dipping upper limit of a north-dipping zone of seismicity thought to represent the top of underthrust Caribbean lithosphere. Note the epicenters and approximate focal depths of the 1943 northern Mona Passage and 1946 northeastern Hispaniola earthquakes, which are discussed in this chapter. The minimum focal depth shown for the 1943 event was determined in this study. The approximate focal depth for the 1946 event is from Russo and Villaseñor (1995). Dark shading denotes limits of Bahamas carbonate province (from Dolan et al., this volume). Black circles show locations of major cities. CH is Cap Haitien, Haiti; EPGFZ is Enriquillo-Plantain Garden fault zone from Mann et al. (1984); Hsp Bas is Hispaniola basin; Ma is Mayaguez, Puerto Rico; MoB is Mouchoir bank; N is Navidad bank; Na is Nagua, Dominican Republic; NEHSFZ is northeast Hispaniola slope fault zone; NHDB is northern Hispaniola deformed belt, which includes the northern Hispaniola accretionary prism; NPRFZ is northern Puerto Rico fault zone compiled from Masson and Scanlon (1991) and Dolan et al. (this volume); PP is Port-au-Prince, Haiti; Sa is Santiago, Dominican Republic; SB is Silver bank; SD is Santo Domingo, Dominican Republic; SJ is San Juan, Puerto Rico. Irregular black areas northeast of Hispaniola denote pull-apart basins at left-steps along the Septentrional fault system (from Dolan et al., this volume).

Kelleher et al. (1973), despite the fact that subsequent studies suggest that these estimates are systematically too large. For example, Kelleher et al. (1973) estimated an M_s of 8.1 for the 1946 mainshock, whereas Abe (1981) estimated M_s 8.0, and both Pacheco and Sykes (1992) and Russo and Villaseñor (1995) estimated M_s 7.8. Nevertheless, for the sake of consistency we use the Kelleher et al. (1973) estimates because only their study estimated magnitudes for all of the events we discuss below. Where available, we include more recent magnitude estimates in parentheses after the Kelleher et al. (1973) estimate. For example, the 1946 northeast Hispaniola earthquake is referred to as M_s 8.1 (Kelleher et al., 1973; M_s 8.0 in Abe, 1981; M_s 7.8 in both Pacheco and Sykes, 1992, and Russo and Villaseñor, 1995); the 1943 northern Mona Passage earthquake is referred to as M_s 7.8 (Kelleher et al., 1973; M_s 7.7 in Abe, 1981; M_s 7.5 in Pacheco and Sykes, 1992).

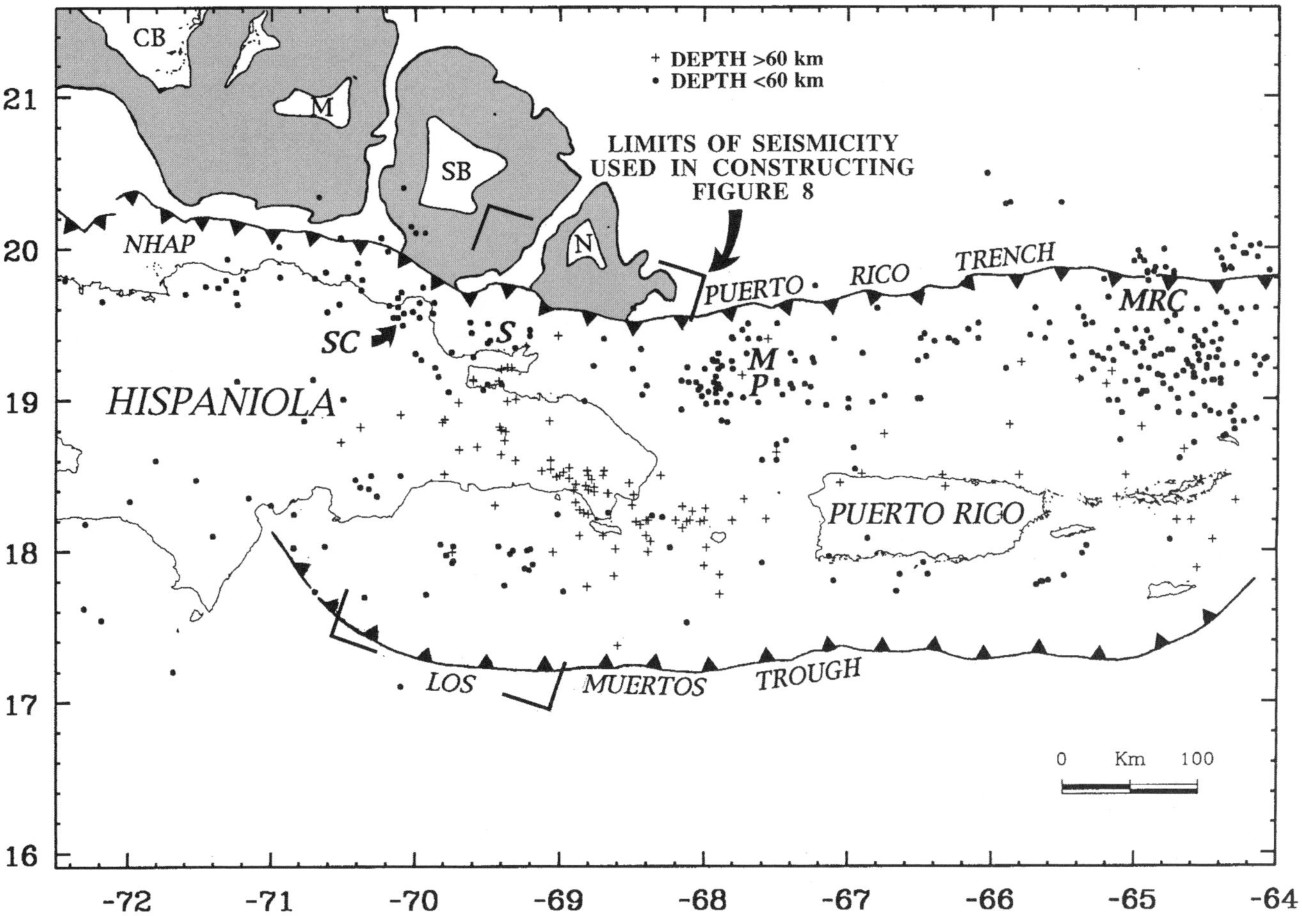

Figure 4. Map of recent seismicity (1962–1992) from ISC catalog (slightly modified from Dolan et al., this volume). Gray shading denotes limits of southeastern Bahamas carbonate province. Box denotes limits of seismicity used to construct cross section shown in Figure 8. CB is Caicos bank; M is Mouchoir bank; MP is Mona Passage seismicity cluster; MRC is Main Ridge seismicity cluster; N is Navidad bank; NHAP is northern Hispaniola accretionary prism; S is Samana seismicity cluster; SB is Silver bank; SC is Sosua seismicity cluster. See Dolan et al. (this volume) for detailed discussion of seismicity distribution.

The north-central Caribbean region was first settled permanently during Columbus's second voyage in 1494, and consequently the area has the longest historical earthquake record in the New World, dating back to the mid-16th century (Scherer, 1912; Kelleher et al., 1973; Chalas-Jimenez, 1989; W. R. McCann and L. Feldman, written communication, 1995). Based on this historical record, the 1943 and 1946 mainshocks appear to be the largest earthquakes that have struck the northeastern Hispaniola–Mona Passage–northwestern Puerto Rico area since at least November 2, 1564, when a large earthquake destroyed the northern Hispaniola cities of Santiago and La Vega (Scherer, 1912; McCann and Pennington, 1990; W. R. McCann and L. Feldman, written communication, 1995).

Despite the large magnitudes of the 1943 and 1946 earthquakes, they have been studied little; prior to this study and the recent study by Russo and Villaseñor (1995), neither their focal mechanisms or focal depths had been determined. To better understand the seismotectonics and seismic hazards of the area, we have determined focal mechanisms for the six largest earthquakes from the 1943–1953 sequence, including both the 1943 and 1946 mainshocks, and the four largest aftershocks of the 1946 event (August 8, 1946; October 4, 1946; April 21, 1948; and May 31, 1953) (Figs. 6 and 7). We initially determined the range of possible first motion mechanisms for each event (Fig. 6). For all but one of the six events (the M_s 7.6 event on August 8, 1946, discussed in a following section), one nodal plane is well constrained to strike west-northwest and dip ~70–75° north-northeast. The best-fitting conjugate planes are consistent with shallowly dipping thrust faulting with fault dips ranging from south-southeast to due west. To resolve the strike of the shallowly dipping planes, we used the well-constrained, steeply dipping nodal planes as a constraint in body-waveform inversions to determine the conjugate plane most consistent with available P and S bodywaves for each earthquake. The detailed results of our waveform inversions will be reported in detail in a later paper by Wald and Dolan. Because the number of high-qual-

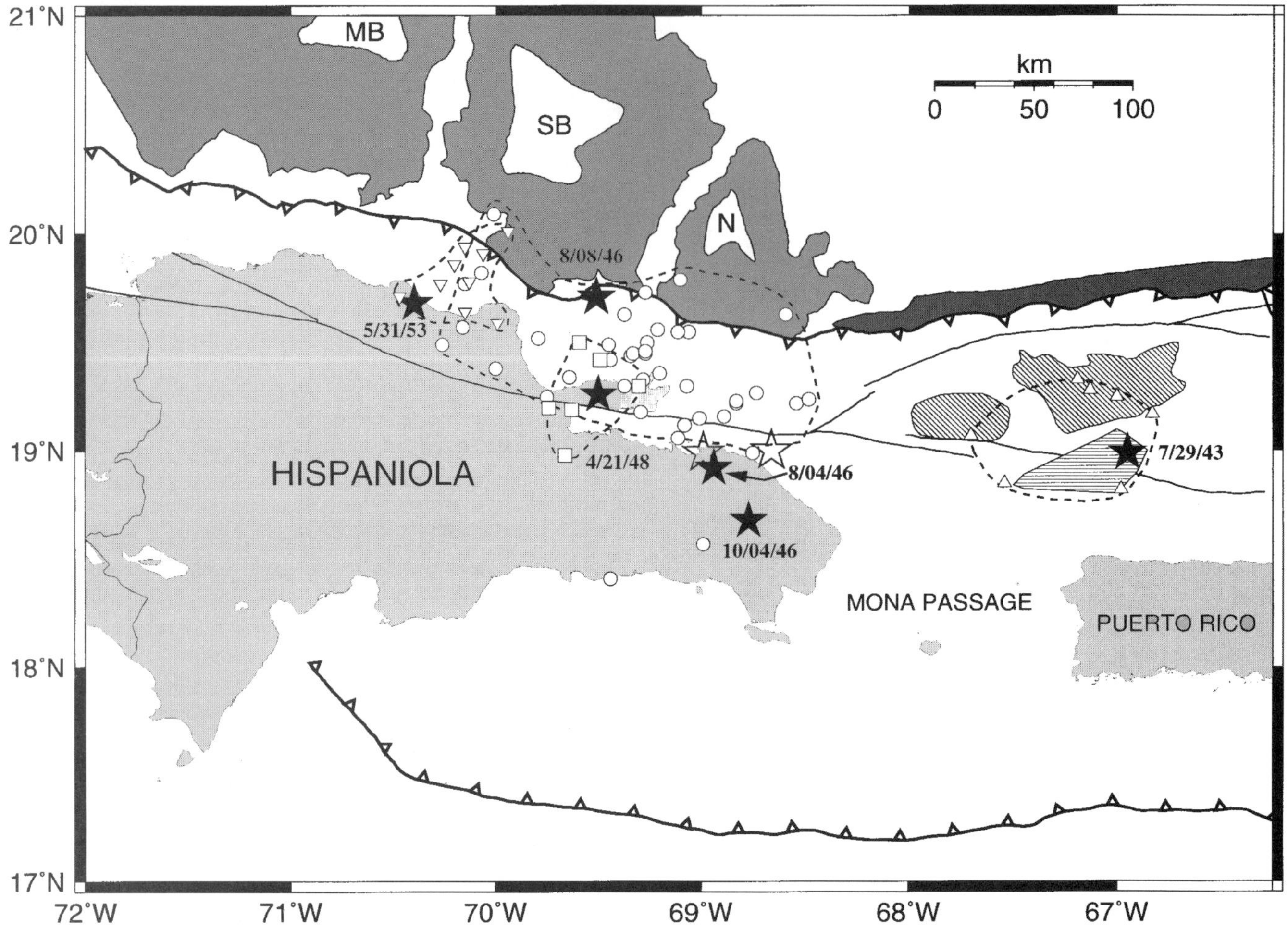

Figure 5. Relocated mainshocks and aftershocks of the 1943 northern Mona Passage earthquake and the 1946 northeastern Hispaniola earthquake from Kelleher et al. (1973). Mainshocks and large aftershocks, including the four largest aftershocks of the 1946 event (August 8, 1946; October 4, 1946; April 21, 1948; and May 31, 1953), are denoted by black stars. Smaller aftershocks denoted by various open symbols. Open stars show relocations of the 1946 mainshock and October 4, 1946, aftershock by Russo and Villaseñor (1995). Russo and Villaseñor's (1995) relocation of the 1946 mainshock epicenter does not change significantly from that of Kelleher et al. (1973). Russo and Villaseñor's (1995) relocation of the October 4, 1946, aftershock does change significantly, however, shifting 20 km to the north-northeast of the Kelleher et al. (1973) location. Note coincidence of 1946 aftershock zone with the 200-km-long Silver-Navidad bank collision zone. Solid lines with teeth are deformation fronts of the Puerto Rico trench–northern Hispaniola accretionary prism system (north of Hispaniola and Puerto Rico) and the Los Muertos trench accretionary prism (south of the islands). Dark shading denotes limits of southeast Bahamas carbonate banks (Dolan et al., this volume). The two irregular zones of diagonal hatching north-northwest of Puerto Rico represent areas of anomalously shallow bathymetry ("anomalous block" of McCann and Sykes, 1984). Dolan et al. (this volume) interpret these shallow regions as the surficial manifestation of collisional underthrusting of a southeastern Bahama bank, which they named Mona bank. The triangular zone of horizontal hatching north of Puerto Rico denotes the rupture plane of the 1943 northern Mona Passage earthquake, as determined in this study. Note the proximity of the 1943 rupture zone to the proposed Mona bank collision. Solid lines without teeth are major, left-lateral strike-slip faults of the Septentrional fault system (SFS). Note the bifurcation of the SFS in the northern Mona Passage and the divergence of two eastern strands around the Mona bank zone of anomalously shallow bathymetry. Dark shading represents basinal sediments in the Puerto Rico trench. Geologic data compiled from Dolan et al. (this volume), Mann et al. (1984), Dillon et al. (1991), Edgar (1991), Masson and Scanlon (1990), and Case and Holcombe (1980). MB is Mouchoir bank; N is Navidad bank; SB is Silver bank.

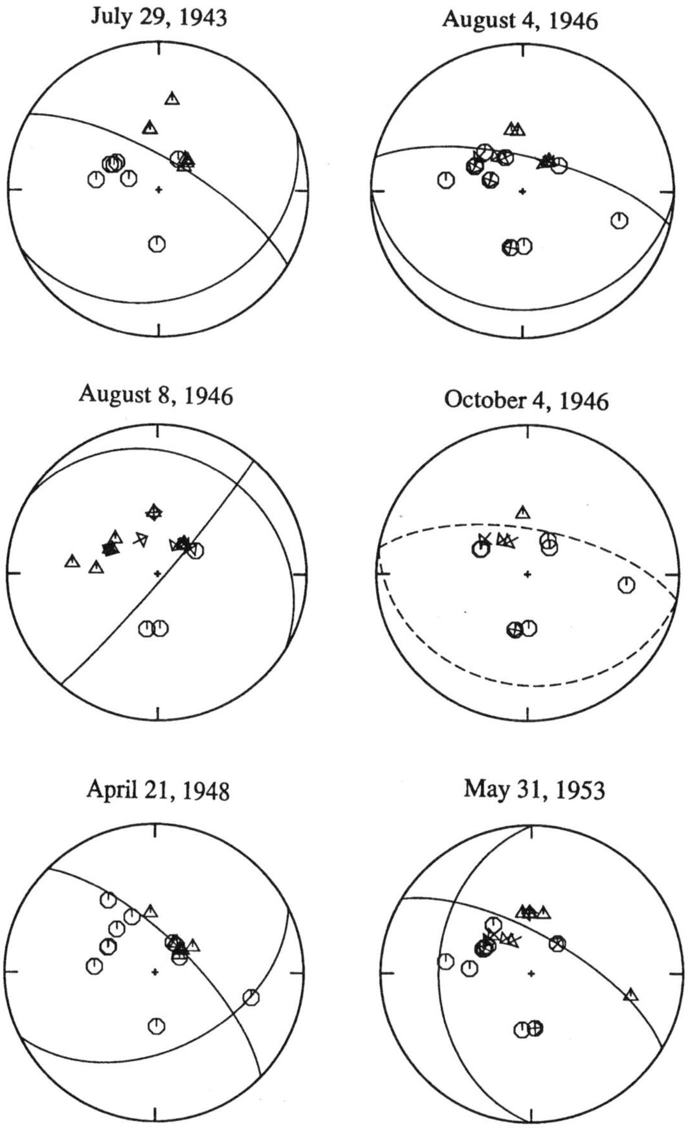

Figure 6. Focal mechanisms determined in this study for the six largest earthquakes in the 1943–1953 north-central Caribbean sequence (lower hemisphere projections). Mechanisms for July 29, 1943, northern Mona Passage mainshock; the August 4, 1946, northeastern Hispaniola mainshock; and the May 31, 1953, Sosua earthquake (an aftershock of the 1946 mainshock) show the results of our body-waveform inversions. The mechanisms for the other three events are P-wave first-motion focal mechanisms. Circles, triangles, and plus symbols denote dilational, nodal, and compressional arrivals, respectively.

ity waveforms is limited for this time period, constraining one nodal plane based on the more numerous polarity observations (for the steeply dipping plane) is beneficial in adding stability to the waveform inversion. We used the methodology of Kikuchi and Kanamori (1991), which we modified to accommodate both the nodal plane constraint and the varied mechanical and electromagnetic instrument types available during this time period.

Our body-waveform inversions were most successful for three events—the 1943 and 1946 mainshocks and the 1953 Silver Spur aftershock (discussed in following sections)—and these are shown in Figures 6 and 7. The focal mechanisms shown for the other three events—the two large 1946 aftershocks and the April 21, 1948, Samana earthquake—are first-motion mechanisms only. In general it proved difficult to accurately constrain the exact strike of the shallowly dipping thrust plane for these events; rotating the strike of this plane by as much as 60°, from near east-west to north-northwest, does not result in significantly different waveform results. For these events we assumed that the strike of the shallowly dipping plane, which we interpret as the faulting plane, was parallel to the west-northwest–trending long axis of the 1946 aftershock zone (Fig. 5). Given the parallelism of the aftershock zone with the ~275–290° orientation of all major active faults in this part of the plate boundary, including the Septentrional strike-slip fault and the offshore deformation front, we feel that this is an appropriate assumption. In the following section we discuss each of the six earthquakes we studied in detail.

1946 northeastern Hispaniola earthquake

The tectonic setting of the 1946–1953 northeastern Hispaniola earthquake sequence is somewhat better defined than that for the 1943 northern Mona Passage earthquake, and we therefore discuss the northeastern Hispaniola earthquake and its aftershocks first. The M_s 8.1 mainshock (Kelleher et al., 1973; M_s 8.0 in Abe, 1981; M_s 7.8 in both Pacheco and Sykes, 1992, and Russo and Villaseñor, 1995) occurred on August 4, 1946, and was followed by numerous small to moderate aftershocks. Kelleher et al. (1973) relocated 53 of the largest aftershocks. Their relocated aftershocks define a 195-km × 95-km zone that extends west-northwest along the northern slope of Hispaniola, basically between the offshore deformation front and the coast of Hispaniola; several early aftershocks (i.e., before the first large aftershock on August 8, 1946) extend northward almost to the deformation front (Fig. 5). Russo and Villaseñor (1995) also relocated 61 aftershocks of the 1946 mainshock. Their distribution is generally similar to that shown by Kelleher et al. (1973), although the spatial clustering of their relocated events is somewhat tighter than that shown by Kelleher et al. (1973). The aftershock zone (as defined both by Kelleher et al., 1973, and Russo and Villaseñor, 1995) exactly matches the limits of the Hispaniola–Bahamas collision zone as defined by SeaMARC II side-scan sonar and seismic reflection data, from the southeast edge of Navidad bank to the western margin of Silver Spur (Fig. 5; Dolan et al., this volume). Our body-waveform inversions yielded a focal mechanism for the 1946 mainshock that shows that it was a thrust fault event that occurred on either a shallowly south-dipping plane that strikes 085°, or a steeply northeast-dipping plane that strikes 110° (Fig. 6).

As noted above, the strike of the shallowly dipping plane is rather poorly constrained. Several lines of evidence, however, show that the event occurred on the shallowly south-dipping plane. The great width of the 1946 aftershock zone relative to its length is incompatible with a steeply dipping fault plane and

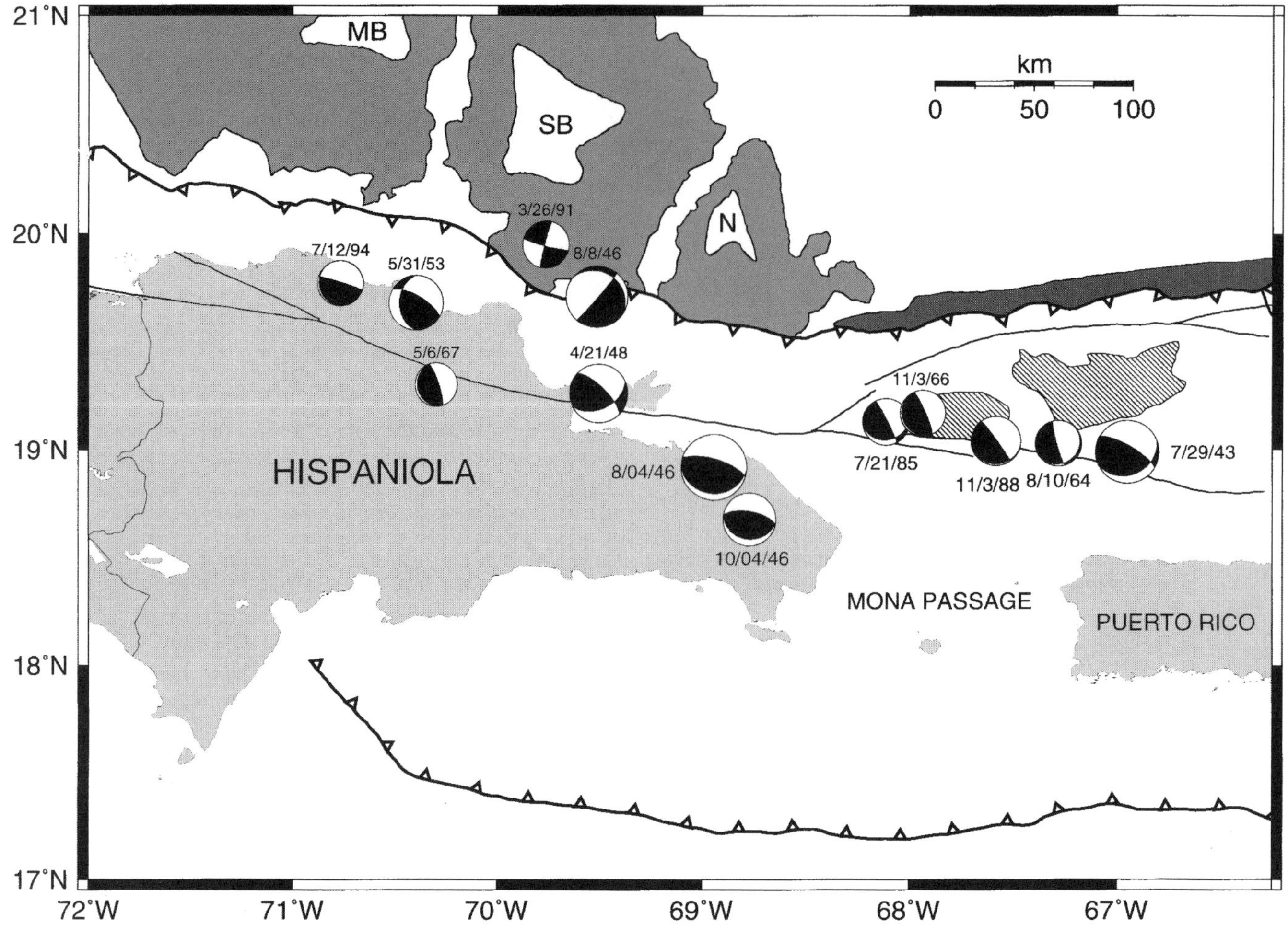

Figure 7. Map of the six focal mechanisms determined in this study (large mechanisms) as well as seven other instrumentally recorded events (small mechanisms) compiled from Molnar and Sykes (1969); Kafka and Weidner (1979); and Harvard Centroid Moment Tensor Catalog (1994). Focal mechanisms are lower hemisphere projections; black areas denote compression, white areas show dilation. MB is Mouchoir bank; N is Navidad bank; SB is Silver bank.

shows that the earthquake occurred along the shallowly dipping nodal plane (Fig. 5). Furthermore, the coincidence of the northern edge of the aftershock zone with the collisional deformation front indicates that the rupture plane dipped south beneath Hispaniola. Finally, Russo and Villaseñor's (1995) relocated aftershocks define a gently south- to southwest-dipping plane that extends from the deformation front toward the mainshock focus (Fig. 8). In our body-waveform inversion, the depth of the mainshock focus is relatively poorly constrained but was between 15 and 40 km. Russo and Villaseñor (1995) estimate a depth of ~20 km. The dip of the shallowly south-dipping nodal plane on our focal mechanism is 23° (Fig. 6). If we fit a plane extending from the deformation front at the sea floor, through the south-dipping cluster of aftershocks, to the mainshock focus at 20-km depth, we calculate a dip of the fault plane of ~20°, nearly identical to that derived from our focal mechanism (Fig. 8). The mainshock focus occurred near the downdip (southern) edge of the rupture plane, near the southeastern edge of the aftershock zone. These data suggest that the 1946 mainshock rupture propagated updip and to the west-northwest.

In addition to the shallow, gently south-dipping cluster of aftershocks extending downdip to the mainshock focus, Russo and Villaseñor (1995) also relocated 22 deeper 1946 events that they interpret as aftershocks (all events relocated by Russo and Villaseñor, 1995, shown by triangles in Fig. 8). These deep 1946 events occurred in a diffuse zone that encompassed most of the thickness of the underthrust Atlantic slab. As can be seen from Figure 8, the deep 1946 events are part of a long-lived sequence of intraplate seismicity associated with the underthrust Atlantic slab, probably related to bending stresses within the slab (and possibly also deep extension as the slab sinks, as suggested by McCann and Sykes, 1984). In contrast, the shallow, south-dipping zone of aftershocks was temporally anomalous, and relatively few events have occurred in this region during the past

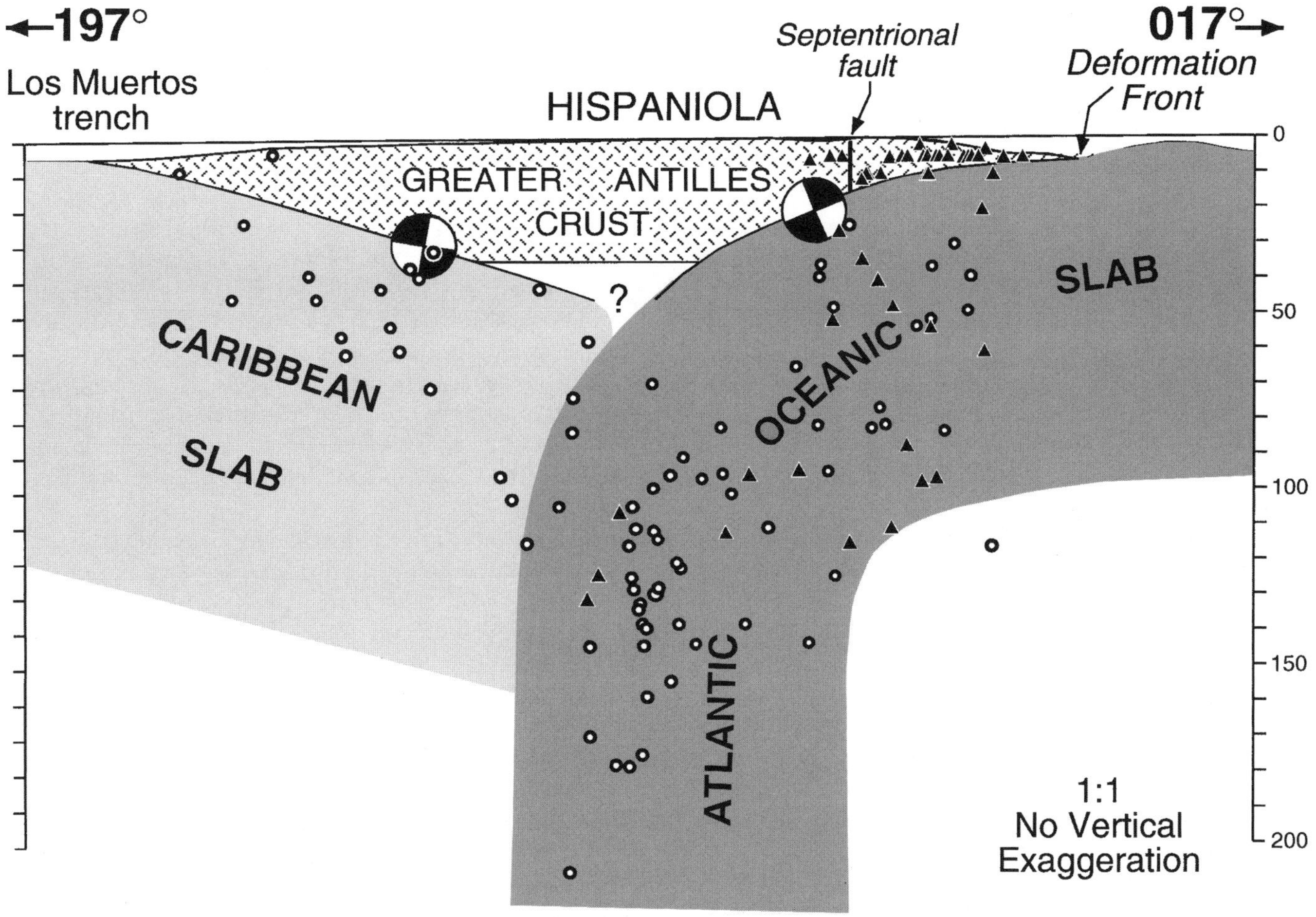

Figure 8. Cross section oriented 017° showing recent seismicity in the region of the 1946 northern Hispaniola earthquake (modified from Dolan et al., this volume). Heavy line at top of figure represents bathymetry-topography along center line of cross section. No vertical exaggeration. Note that although profile is oriented approximately perpendicular to regional strike, the profile is oblique to some tectonic and topographic features. Shading denotes approximate thickness of different lithospheric plates, based on assumptions that the Atlantic plate consists of normal oceanic lithosphere ~90–100 km thick and that the thicker crust of the Caribbean plate (Edgar et al., 1971; Case et al., 1990) results in a thicker lithosphere. All seismicity data from National Earthquake Information Center Catalog of Earthquakes (1994) 1962–1992 except as noted. Only earthquakes with well-defined depths have been used; earthquakes assigned default depths of 10 km and 33 km were not used in constructing these sections. Triangles denote 1946 aftershocks relocated by Russo and Villaseñor (1995). Note the well-defined, shallowly southwest-dipping zone of 1946 aftershocks. Also note the diffuse, southward-dipping zone of aftershocks associated with underthrusting of Atlantic oceanic crust and Bahamas platform carbonates. These shallow aftershocks extend almost to the deformation front defined by the SeaMARC II data, suggesting that the 1946 mainshock may have locally ruptured all the way to the sea floor in this collisional region of enhanced mechanical coupling. Note the near-total absence in the recent ISC data of well-located earthquakes associated with the south-dipping plate interface between underthrust Atlantic oceanic crust and Hispaniola that ruptured during the 1946 northeastern Hispaniola earthquake. In contrast, the relocated 1946 aftershocks define a shallowly southwest-dipping plane approximately along our proposed 1946 mainshock rupture surface. Note that the deeper relocated 1946 aftershocks occur within a long-lived zone of deep background seismicity defined by the past 30 yr of ISC catalog data. The gentle southwestward dip of the shallow 1946 aftershock zone, coupled with its location north of the Septentrional fault, confirms that the 1946 mainshock occurred along the shallowly southwest-dipping plate interface, rather than along a strand of the Septentrional fault system as has been suggested by Russo and Villaseñor (1995). Focal mechanisms are shown for the 1946 northeast Hispaniola mainshock (from this study) and for the M_S 6.7 earthquake of June 24, 1984, which occurred along the north-dipping Los Muertos trench zone of underthrusting (from Byrne et al., 1985).

30 yr. These observations indicate that the shallow aftershocks were directly related to afterslip on or near the rupture plane of the 1946 mainshock.

Russo and Villaseñor (1997) note that the deep aftershock activity increased 15-fold in the year after the 1946 mainshock and suggest that this indicates that the earthquake occurred on a steeply northeast-dipping plane. There are two major problems with this interpretation. First, it does not explain the shallow, south-dipping aftershock zone. Second, it merely suggests mechanical continuity of the slab, not necessarily that the deep slab ruptured in the earthquake. Such mechanical continuity of the slab is entirely consistent with our model, and in fact would be predicted by it. We would expect that as the Atlantic slab underwent 5–10 m of coseismic underthrusting during the 1946 mainshock, the slab would respond to the consequent change in internal stresses by producing the intraslab earthquakes observed as Russo and Villaseñor's deep "aftershocks." We would expect these to be most frequent right after the mainshock, as has been shown to be the case by Russo and Villaseñor (1995). In summary, while the deep 1946 events relocated by Russo and Villaseñor (1995) were technically aftershocks of the 1946 event (having occurred within one rupture length of the mainshock), they do not define the rupture plane, as suggested by Russo and Villaseñor. Rather, the 1946 rupture plane is defined by the temporally anomalous, gently south-dipping zone of shallow aftershocks.

1946 tsunamis. Immediately after the August 4, 1946, mainshock, a tsunami struck ~30 km off the eastern coastline of the Rio San Juan Peninsula, drowning nearly 100 people in the village of Mantanzas (then known as Julia Molina), just south of the present city of Nagua (Fig. 3) (Lynch and Bodle, 1948; Small, 1948). Although the tsunami height was only 2.5 m at Mantanzas, the low beach relief in the area allowed the tsunami to propagate inland for several kilometers, accounting for the high loss of life (Lynch and Bodle, 1948). Farther north, at the mouth of Rio Boba, higher beach relief overtopped by the tsunami required tsunami heights of 4–5 m (Lynch and Bodle, 1948; Small, 1948). Anecdotal accounts noted that the tsunami propagated from north to south down the coast, indicating a source to the north. The location of the tsunami implies that it was probably caused by either: (1) sea-floor uplift associated with propagation of the mainshock rupture to the sea floor at the collisional deformation front; or (2) widespread, earthquake-induced slumping of the Hispaniola slope or the southern slopes of Silver or Navidad banks. The southern slopes of the carbonate banks exhibit numerous slump scars and avalanche chutes, some of which are very large (Mullins et al., 1991, 1992; Dolan et al., this volume). Similarly, the Hispaniola slope within the collision zone is highlighted by widespread, recent slumping (Dolan et al., this volume). The location of the most widespread zone of major slumps on the Hispaniola slope 100 km east of the tsunami landfall, however, suggests that the 1946 tsunami was probably not directly related to the widespread slumping of the northeastern Hispaniola slope. In addition to the large tsunami immediately after the August 4 mainshock, smaller tsunamis continued for several days after the mainshock (Seismological Notes, 1946). The post-mainshock occurrence of these tsunamis suggests that they may have been caused by submarine slumps and slides triggered by aftershocks.

Large 1946 aftershocks

The M_s 7.6 earthquake of August 8, 1946 (Kelleher et al., 1973; M_s 7.3 in Russo and Villaseñor, 1995), represents the largest aftershock of the 1946 northeast Hispaniola earthquake. Our preliminary analysis of the waveforms for the August 8 event indicates that it was a complex rupture, probably involving several subevents. Because of this complexity, the first-motion-only focal mechanism we show for the August 8 earthquake is poorly constrained and should be considered uncertain; we plan to continue our attempts to model the waveforms of this complex event. Despite our lack of confidence in the focal mechanism shown in Figure 6, one aspect of the August 8 earthquake is quite clear: The preliminary focal mechanism is markedly different from all the other events in the 1943–1953 sequence. Unlike the other events, all of which show thrust faulting on shallowly south- or southwest-dipping planes, the August 8, 1946, earthquake exhibits either right-lateral motion on a near-vertical plane striking 040°, or left-lateral motion on a moderately north-dipping plane that strikes 126° (Fig. 6).

Kelleher et al. (1973) calculate an M_s of 7.0 for a second major aftershock on October 4, 1946. Their results show this event as having occurred at the extreme southeastern edge of the 1946–1953 earthquake sequence, more than 25 km southeast of the 1946 mainshock (Fig. 5). In contrast, Russo and Villaseñor (1995) calculated an M_s of 6.1 for this event. Furthermore, their relocation shows that the earthquake actually occurred much closer to the mainshock, 20 km north-northeast of the position calculated by Kelleher et al. (1973) (Fig. 5). Although we have calculated only a first-motion mechanism for this earthquake, it appears to be remarkably similar to the mechanism that we determined for the 1946 mainshock, with thrust faulting on a plane that dips 17° south (Fig. 6).

1948 Samana earthquake

The 1946 earthquakes were followed by two other large events within the collision zone. On April 21, 1948, an M_s 7.3 event occurred beneath the Samana Peninsula, within the central part of the 1946 aftershock zone (Fig. 5). Relocated aftershocks for this event (Kelleher et al., 1973) indicate a rather narrow aftershock zone that extended only ~50 km along strike, but ~65 km perpendicular to strike (Fig. 5). Our first-motion focal mechanism shows that this earthquake was similar to the 1946 mainshock, with oblique left-lateral–thrust motion occurring on a shallowly south-dipping plane (Fig. 5). The relatively large strike–perpendicular width of the aftershock sequence of this event strengthens our interpretation of a low-angle, south-dipping 1946 rupture plane (Fig. 6). Aftershocks of the 1948 earthquake extend ~15 km south of the aftershock zone of the 1946 earthquakes, suggesting that the 1948 event may have extended the 1946 rupture plane downdip in the central part of the collision zone.

1953 Silver Spur earthquake

On May 31, 1953, an M_s 7.0 earthquake occurred beneath northern Hispaniola, just south of the coast, along the southwestern projection of the western edge of the Silver Spur collision (Fig. 2) (International Seismological Summary of 1953, 1961); Sykes and Ewing, 1965). Our body-waveform inversion yielded a focal mechanism for this earthquake that is similar to the 1946 and 1948 ruptures, revealing thrust faulting on a shallowly south-southwest–dipping plane (Figs. 6 and 7). Aftershocks for this earthquake define an elongated triangle that extends northeastward from the mainshock along the western edge of the Silver Spur collision (Fig. 5). Although the 1953 mainshock occurred ~15 km west of the western edge of the 1946 aftershock zone, the eastern half of the 1953 aftershock zone overlaps with the westernmost part of the 1946 aftershock zone (Fig. 5). The focal depth of the Silver Spur earthquake was ~15–25 km.

1943 northern Mona Passage earthquake

The July 29, 1943, M_s 7.8 northern Mona Passage earthquake (Kelleher et al., 1973; M_s 7.7 in Abe, 1981; M_s 7.5 in Pacheco and Sykes, 1992) ruptured a large segment of the plate boundary north of the Mona Passage between Hispaniola and Puerto Rico (Fig. 5). Our body-waveform inversion yielded a focal mechanism that is generally similar to the 1946, 1948, and 1953 northeastern Hispaniola earthquakes in that it shows thrust faulting on either a shallowly south-dipping plane, or a steeply west-northwest–dipping plane (Fig. 6). However, the focal mechanism indicates that the 1943 event exhibited slightly more oblique, left-lateral motion than the 1946 mainshock (Fig. 6).

We have used the waveform inversion to constrain the dimensions of the main rupture area for the 1943 mainshock. These data reveal a roughly triangular asperity that extends westward from the mainshock focus for ~50 km; almost no strain release occurred east of the mainshock (Fig. 5). The rupture plane is ~40 km wide (viewed perpendicular to the shallowly dipping faulting plane) along its eastern edge and narrows to a point at its western end. Seven aftershocks of the 1943 earthquake (relocated by Kelleher et al., 1973) define a 90-km × 65-km "halo" that extends well beyond the rupture plane (Fig. 5). The aftershock zone closely corresponds to the zone of anomalously shallow bathymetry at the north end of Mona Passage, which Dolan et al. (this volume) interpret as the sea-floor manifestation of the Mona bank collision. The rupture zone of the earthquake is parallel to, and just south of the zone of shallow bathymetry (Fig. 5). Unlike the 1946 mainshock, the 1943 northern Mona Passage event does not appear to have ruptured to the sea floor at the deformation front. Rather, the waveform analysis indicates that slip occurred no shallower than ~25 km depth. The absence of shallow rupture probably explains the lack of any reported tsunamis for this earthquake. A 25–30-km-deep focus for the 1943 mainshock places the event near the top of a long-lived, south-dipping cloud of small to moderate earthquakes identified by Dolan et al. (this volume) as the top of the underthrust Atlantic (North America) slab (Fig. 3; see also Dolan et al., this volume).

Furthermore, although the 1943 mainshock epicenter was located close to the sea-floor trace of the southeastern strand of the left-lateral Septentrional strike-slip fault defined by SeaMARC II side-scan sonar data (Dolan et al., this volume), the large strike–perpendicular width of the 1943 aftershock zone relative to its along-strike length, with the occurrence of aftershocks almost 40 km north of the surface trace of the fault as well as 25 km to the south, suggests that the earthquake occurred on a shallowly south-dipping plane. This observation, coupled with the occurrence of the earthquake along the inferred top of the underthrust slab and the similarity of the 1943 focal mechanism to the shallowly south- to southwest-dipping thrust earthquakes of the 1946–1953 sequence to the west, indicates that the 1943 earthquake did not occur on the Septentrional fault. Rather, as with the 1946–1953 earthquakes beneath Hispaniola, the focal mechanism and aftershock data indicate that the 1943 earthquake ruptured the shallowly south-dipping, interplate thrust interface along the top of the underthrust Atlantic slab (Fig. 6).

Focal mechanisms for recent (1962–1994) moderate earthquakes

In addition to our focal mechanisms for the 1943–1953 earthquakes, focal mechanisms of seven moderate earthquakes that have occurred in the region since 1962 have been published (Molnar and Sykes, 1969; Kafka and Weidner, 1979; Harvard Centroid Moment Tensor Catalog, 1994). Six of the seven published focal mechanisms are similar to the 1943–1953 earthquakes, showing predominantly thrust motion along shallowly south- and southwest-dipping planes (Fig. 7). As with the larger 1943–1953 events, rake angles of the six recent moderate thrust-fault events indicate that motion was oblique to the strike of the plate margin. Obliquity of slip during these earthquakes to the plate margin ranges from ~30° to near 90°, indicating a range of motion from mixed thrust–left-lateral displacement to left-lateral strike-slip offset nearly parallel to the strike of the underthrust Atlantic slab (Fig. 7). In general, the focal mechanisms for the smaller events show more highly oblique motion than the larger 1943–1953 earthquakes. The seventh moderate earthquake, which occurred just north of the deformation front near 69° 45′W, exhibited near-pure strike-slip motion on vertical planes, either left-lateral slip on a 100°-striking plane, or right-lateral slip on a 015°-striking plane.

DISCUSSION

The most striking characteristic of the 1943 and 1946 earthquakes is their spatial coincidence with active collision zones involving buoyant, high-standing carbonate banks of the southeastern Bahamas. This relationship is particularly clear in the case of the 1946–1953 north Hispaniola sequence. When compared with the structure contour map of the underthrust Atlantic slab beneath Hispaniola, the focal depth, the shallow southward dip, and the west- to west-northwest strike of the 1946–1953 faulting planes are all consistent with the earthquakes having occurred on the plate interface along the top of the underthrust North America (Atlantic) slab (Fig. 7). This observation, coupled with the spatial coincidence of the aftershock zones of the 1946–1953 earthquakes with the Silver-Navidad bank collision zone defined by Dolan et al. (this volume), suggests that the earthquake ruptured the strongly coupled interface between Hispaniola and the underthrust portions of the colliding Bahama banks. The fact that the shallow, gently south-dipping zone of 1946 aftershocks extends to the deformation front further implies that the earthquake may have ruptured updip from the focus at ~20 km all the way to the sea-floor deformation front.

The tectonic significance of the August 8, 1946, M_s 7.6 (Kelleher et al., 1973; M_s 7.3 in Russo and Villaseñor, 1995) aftershock that struck offshore Hispaniola just south of the collisional deformation front is still somewhat unclear because of the uncertain nature of the focal mechanism and the absence of a well-constrained focal depth. One possibility is that the earthquake represents right-lateral tear faulting within the underthrust Bahama bank as part of collisional readjustment of the colliding block to slip during the northeastern Hispaniola earthquake.

The focal mechanism and location of the October 4, 1946, M_s 7.0 (Kelleher et al., 1973; M_s 6.1 in Russo and Villaseñor, 1995) aftershock suggest that this earthquake may have re-ruptured the extreme downdip (southern) limit of the 1946 mainshock rupture plane. We speculate that the October 4 earthquake may have ruptured down to the extreme downdip edge of underthrust, buoyant Bahamas material.

The location of the 1948 Samana earthquake suggests that this event re-ruptured, and deepened, the central part of the shallowly south-dipping 1946 mainshock rupture plane (Fig. 5). Aftershocks of the 1946 mainshock and recent moderate earthquakes are both anomalously sparse in the eastern part of the 1946 rupture zone to the east of the 1948 rupture (Figs. 4 and 5). These data suggest that the 1948 earthquake may have ruptured a zone of incomplete strain release during the 1946 mainshock. The Samana earthquake also occurred directly downdip (southward) of the projection of Navidad Channel—the deep, linear, north-northeast–trending channel that separates Silver and Navidad banks (Dolan et al., this volume)—suggesting the possibility that the 1948 rupture occurred along a zone of weakness associated with the underthrust portion of Navidad Channel. The 1948 earthquake occurred beneath the Samana Peninsula, which exhibits much more rugged topography than areas to the east or west (de Zoeten et al., 1991; Dolan et al., this volume; their fig. 11A). This observation suggests a third possibility, that the Samana earthquake may have been localized at a collisional asperity beneath the Samana Peninsula. We speculate that the anomalously high topography of the peninsula may be related to underthrusting of a particularly shallow portion of the underthrust Bahama banks, in much the same way that we propose that the zone of anomalously shallow bathymetry in the northern part of Mona Passage is caused by underthrusting of Mona bank.

The coincidence of the western limit of 1953 aftershock zone with the leading (western) edge of the Silver Spur collision suggests that this earthquake ruptured the area of strong coupling between Hispaniola and the underthrust portion of Silver Spur. The 1953 aftershock zone extended ~10–15 km west of the western limit of the 1946 aftershocks (Fig. 5), indicating that the 1953 earthquake may have extended the 1946 rupture to the extreme western edge of colliding carbonate basement. Molnar and Sykes (1969) have determined a focal mechanism for a 1967 M 5.3 event that occurred at a depth of 39 km, ~15 km southeast of the 1953 event (Fig. 7). Their mechanism is consistent with west-southwestward underthrusting along a shallowly west-southwest–dipping fault plane. Comparison of the depths of the 1953 and 1967 earthquakes with the structure contour map of the underthrust Atlantic slab shows that both earthquakes were generated by rupture of the plate-boundary thrust interface along the top of the underthrust slab (Fig. 7).

On the basis of recent seismicity, the 25- and 50-km depth contours of the top of the underthrust Atlantic slab can only be confidently traced northwestward to ~69° 45′W (Fig. 3; Dolan et al., this volume). However, based on the similarity of the 1953 and 1967 focal mechanisms with those of the 1943, 1946a, 1946c, and 1948 earthquakes, we suggest that the shallowly south-dipping North America (Atlantic) slab extends at least as far west as 70° 15′W, the western edge of the 1953 rupture. Furthermore, the consistent southerly dip of structures in the accretionary prism of the northern Hispaniola deformed belt west of the Silver Spur collision, and the shallow southward dip of underthrust carbonate basement beneath the deformation front along the southern edge of the Hispaniola basin (Dolan et al., this volume), imply that the underthrust Atlantic slab may exist west of the collision. The slab is not seismically resolvable in north-central and northwestern Hispaniola, however, and the deep crustal–upper mantle structure west of 70° 15′W remains poorly understood.

As with the 1946 northeastern Hispaniola earthquake, the 1943 northern Mona Passage earthquake appears to have been localized along an active collision. As discussed above, Dolan et al. (this volume) have proposed that the anomalously shallow bathymetry of the northern Mona Passage area has developed in response to underthrusting of a buoyant Bahama bank, which they named Mona bank. Since the Bahamas ridge strikes more northwesterly than the plate boundary deformation front, collisional underthrusting is diachronous, younging from east to west (Dolan et al., this volume). Thus, the collision has progressed farther in the Mona Passage area than to the west in the region of the 1946 earthquake. Based on this reasoning, Dolan et al. (this volume) suggest that the Mona bank has already been thrust beneath the insular slope, which has been uplifted in response to the collision.

Comparison of the 1943 mainshock focal mechanism data with the structure contour map of the underthrust Atlantic slab indicates that the 1943 rupture occurred along the shallowly south-dipping interface between the underthrust Atlantic slab and the overriding Caribbean margin (Figs. 3 and 7). The focal mechanisms and focal depths of several moderate, recent earthquakes from beneath the northwest part of Mona Passage also support this interpretation (Dolan et al., this volume). All three recent events were shallowly dipping thrust-faulting earthquakes that occurred at depths that plot along the top of the slab interface (Fig. 7).

One of the most puzzling aspects of the 1943 mainshock rupture was the absence of widespread damage from the event in northwestern Puerto Rico, which is only ~50 km south of the epicenter. We suggest that the answer lies in the rupture characteristics of the earthquake. The mainshock focus is located at the extreme eastern edge of the rupture plane, and our body-waveform inversion suggests that several subevents propagated westward along the rupture plane. These data indicate a strong westward directivity during this event. If so, then most of the energy from the earthquake probably propagated west, away from Puerto Rico, and dissipated along the northern, offshore slope of Hispaniola. Furthermore, the relatively great depth of the earthquake (>25 km) precluded generation of any significant tsunami, thus sparing low-lying coastal areas in northwestern Puerto Rico and northeastern Hispaniola from inundation.

Implications of the 1943–1953 earthquake sequence for seismic versus aseismic strain release along the plate-interface thrust fault

What do the 1943 and 1946 earthquakes tell us about the degree of seismic coupling along the plate boundary? We have proposed that the 1943 and 1946 earthquakes were generated by rupture of strongly coupled, collisional asperities along the plate-interface thrust fault. We suspect that elastic strain energy along these collisional parts of the plate-interface is released predominantly during coseismic slip. What can we say about the degree of coupling along the noncollisional parts of the plate boundary?

The absence of rupture in the 200-km-wide section of the plate boundary between the 1943 and 1946 earthquakes is an interesting anomaly. As noted above, our waveform inversions suggest that the 1943 earthquake ruptured unilaterally westward, which probably resulted in a large dynamic strain pulse that may have loaded the section of the plate interface to the west of the 1943 earthquake toward failure. The fact that the western segment did not rupture could be because of either: (1) the recent occurrence of a moderately large earthquake that relieved the strain along the fault between the 1943 and 1946 ruptures or (2) a long-term, slower strain accumulation rate along the unruptured part of the plate-interface thrust fault.

If the absence of rupture on the fault between the 1943 and 1946 earthquakes was due to the recent release of strain in a large, pre-1943 earthquake, then that earthquake was either not felt in Puerto Rico or Hispaniola, or it was not recorded historically. The only large historical event that might have occurred on the plate-interface thrust fault in this area is the M_s 7.0 event of 1917 (Kelleher et al., 1973; McCann and Sykes, 1984). Rupture of the entire unruptured part of the fault could have produced an M ~7.5 earthquake comparable to that in 1943, but even farther from Puerto Rico than the 1943 event. Thus, it was possible that such an event would have not been felt strongly in Puerto Rico. This would place the event closer to Hispaniola, however, making it more likely that the postulated event would have been felt and recorded there. The directivity of the event could also play an important role. As noted above, the 1943 mainshock, despite its proximity to coastal Puerto Rico, produced only a maximum Modified Mercali Intensity of V in northwestern Puerto Rico (W. R. McCann, personal communication, 1995). It would be especially easy for an event not to be recorded the farther back in time it occurred. This, however, would allow a longer period of strain accumulation on the plate-interface thrust fault.

Alternatively, the portion of the plate-interface thrust fault plane between the 1943 and 1946 ruptures may not store elastic strain energy as efficiently as the collisional asperities along the fault that ruptured in those two earthquakes. The bathymetry of the Hispaniola slope provides important constraints on the location of active collisional underthrusting (Dolan et al., this volume). Although the eastern part of the unruptured zone overlaps with the western part of the zone of shallow bathymetry in the northern Mona Passage, most of the unruptured zone lies in an area of complicated bathymetry associated with a major bifurcation of the Septentrional strike-slip fault system at ~68° 25′W (Fig. 3). Centered in the middle of the unruptured zone is a 7-km-deep, 35-km-long pull-apart basin at a major left-step in the left-lateral strike-slip fault system (Dolan et al., this volume). The presence of this deep basin contrasts markedly with the slopes to the east and west, and seems incompatible with collisional underthrusting of a bathymetrically high-standing block beneath the unruptured part of the plate boundary. Based on this reasoning, we suggest that the Mona bank and Silver-Navidad bank collision zones are separated by a section of the plate boundary that is not undergoing collisional underthrusting of buoyant Bahamas material. If so, then the degree of seismic coupling between the plates in the unruptured segment of the plate interface may be less than that along the collisional asperities. Thus, the elastic strain necessary to generate slip in earthquakes may be stored more slowly than in the surrounding, more strongly coupled parts of the plate-interface thrust fault, and a significant portion of the interplate slip may be accommodated aseismically along the noncollisional parts of the plate-interface thrust fault. This leads us to speculate that the noncollisional parts of the fault may participate in large thrust earthquakes less frequently than more strongly coupled portions of the fault. This inference is supported by the fact that the two largest instrumentally recorded events to strike the region—the 1943 and 1946 earthquakes—both occurred at sites of presumed collisional underthrusting and consequent strong interplate coupling. The exact degree to which aseismic slip contributes to interplate slip along the thrust interface, however, remains unknown.

Based on the observations discussed above, we speculate that large interplate thrust earthquakes most commonly nucleate at collisional asperities. This does not mean that large thrust ruptures cannot nucleate outside the zones of active collisional underthrusting, only that the inferred higher degree of interplate coupling in the collision zones results in faster, more effective strain accumulation necessary to generate large earthquakes. Furthermore, although large earthquakes may commonly nucleate at collisional asperities, the earthquake ruptures may propagate beyond those asperities into less strongly coupled, noncollisional parts of the plate-interface thrust fault.

Extent of underthrust, buoyant Bahamas province in the north-central Caribbean

If large thrust earthquakes along the northern margin of the NCPBZ nucleate more effectively at collisional asperities, then it is of obvious importance to understand the three-dimensional distribution of underthrust, buoyant Bahamas material. Furthermore, the extent of buoyant Bahamas material along the plate boundary represents one of the key elements that must be understood before attempting any reconstruction of the tectonic and mechanical evolution of the NCPBZ. The southern margin of the Bahamas province, however, has already been subducted beneath the NCPBZ. We must therefore use indirect means to infer the subducted extent of the Bahamas province.

First, and most basically, the Bahamas province is thought to have developed above a northwest-striking oceanic fracture zone associated with opening of the Atlantic Ocean (Mullins and Lynts, 1977). The linear, fracture-zone–parallel nature of the Bahamas is reflected most prominently in the Bahamas escarpment, the steep, northeastern wall of the Bahamas (Fig. 1). As discussed above, the northwest strike of the Bahamas, coupled with the overall near–east-west relative North America (Atlantic) plate–NCPBZ motion, indicates that the Bahamas collision is diachronous, younging from the west. Thus, the collision has progressed farther in the east than in the west. Dolan et al. (this volume) propose that Mona bank, which they hypothesize has been subducted beneath the northern Mona Passage, represents the easternmost extent of underthrust, buoyant Bahamas.

As noted above, the bathymetry of the Hispaniola slope also places important constraints on the distribution of underthrust Bahamas material. Specifically, the presence of the 7-km-deep pull-apart basin above the middle of the unruptured section of the thrust fault suggests that this part of the plate boundary is not undergoing active collisional underthrusting. We suggest that this zone of noncollision probably extends from the eastern edge of the 1946 rupture plane at ~68° 30′W to the western edge of the northern Mona Passage zone of anomalously shallow bathymetry at ~68° 00′W (Fig. 5).

If, as we propose, 1943 and 1946 mainshocks nucleated at collisional asperities, then we can use the rupture planes of the 1946 and (with slightly less confidence) the 1943 earthquakes as proxies for the presence of underthrust buoyant Bahamas material. The clear spatial correlation between the rupture plane of the 1946 northeastern Hispaniola earthquake and the active Silver-Navidad bank zone of collisional underthrusting provides strong support for this inference.

Finally, side-scan sonar data show that collisional underthrusting terminates at the western edge of the Silver Bank collision zone (Dolan et al., this volume). West of the collision, the southern edges of high-standing Mouchoir and Caicos banks lie to the north of the deformation front (Fig. 2). For example, seismic-reflection data from the Hispaniola basin west of Silver Spur show a sharply defined southern limit to high-standing Mouchoir bank 8–25 km north of the deformation front; carbonate material underthrust beneath the Hispaniola slope consists of gently south-dipping slope deposits that lie more than 4.5 km below the crest of Mouchoir bank (Dolan et al., this volume). In other words, because of the highly irregular shape of the southern margin of the Bahamas province, and the diachronous, westward-younging nature of the oblique collision, collisional underthrusting of high-standing carbonate bank material has not yet occurred west of the Silver Spur collision, with a few minor exceptions (e.g., Mouchoir Spur; Dolan et al., this volume).

Based on these observations, we have constructed a map of the probable approximate downdip, southern limit of underthrust Bahama carbonates (Fig. 9). Note that although we have drawn the southern edge of postulated Mona bank along the southern edge of the 1943 rupture plane, the actual downdip limit of underthrust Mona bank basement in this area is unknown.

Relative plate motions in the north-central Caribbean based on earthquake focal mechanisms

To understand the mechanics and evolution of the NCPBZ, we must know the direction of relative plate motion between the North America plate (Bahamas province) and the Hispaniola and Puerto Rico blocks, which lie within the NCPBZ; the true Caribbean plate lies to the south of the NCPBZ. The direction of motion along the NCPBZ has been the subject of much debate during recent years. Regional plate models, based primarily on analysis of magnetic lineations formed during opening of the Cayman trough, the presence of left-lateral strike-slip faults along the northern plate boundary and right-lateral faults along the southern boundary, and active volcanism in the Lesser Antilles island arc indicate that both the North and South America plates have moved generally westward with respect to the Caribbean (relative to the hot-spot reference frame) since Eocene time (Pindell and Barrett, 1990; Mann et al., 1990). However, current absolute and relative plate motions in the north-central Caribbean region remain relatively poorly constrained. One of the major reasons for this is that, in marked contrast to most other plate boundaries worldwide, attempts to model Caribbean–North America plate motion vectors have largely proved unsuccessful (e.g., DeMets et al., 1990; DeMets, 1993). These results indicate that the Caribbean plate cannot be modeled as a rigid plane with discrete boundaries (DeMets, 1993). Rather, the Caribbean plate must be undergoing significant internal deformation. The validity of this concept is demonstrated by numerous geological and geophysical studies

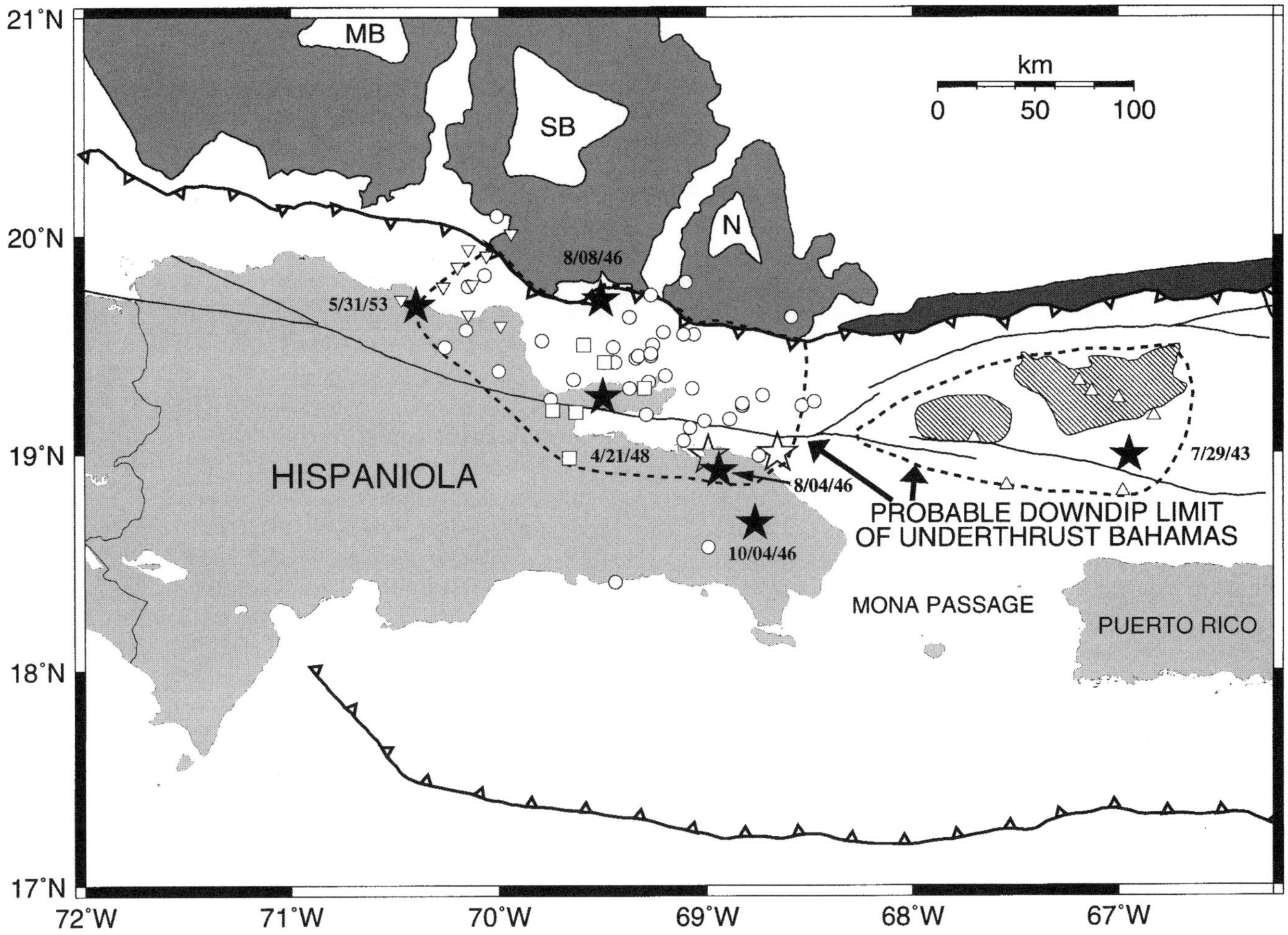

Figure 9. Map of proposed extent of underthrust Bahama bank material based on earthquake locations, limits of Silver-Navidad bank collision zone, and extent of Mona bank zone of anomalously shallow bathymetry. MB is Mouchoir bank; N is Navidad bank; SB is Silver bank.

that show that the margins of the Caribbean plate are characterized by wide, extremely complex zones of deformation (see overview in Mann et al., 1990). In the north-central Caribbean, for example, the 250-km-wide plate boundary zone is characterized by relative motions between several microplates and numerous smaller structural blocks (Mann et al., 1984, 1995; Byrne et al., 1985; Masson and Scanlon, 1991; Dolan et al., this volume).

As noted above, global plate motion models do not adequately constrain Caribbean–North America plate motions, and consequently there is disagreement about the exact relative plate motion along the NCPBZ (e.g., Sykes et al., 1982; McCann and Sykes, 1984; Stein et al., 1988; DeMets et al., 1990; DeMets, 1993). There is general agreement that both the North and South America plates are moving predominantly westward with respect to the Caribbean plate, resulting in a significant component of left-lateral strike-slip motion across the NCPBZ. Differences arise, however, in terms of the amount of oblique motion occurring along the margin, and even in the direction of any oblique motion.

Most models indicate either east-west relative motion (e.g., DeMets, 1993; Farina et al., 1995) or east-northeast motion (Bracey and Vogt, 1970; Schell and Tarr, 1978; Sykes et al., 1982; Deng and Sykes, 1995). All of these models predict predominantly left-lateral motion along the NCPBZ, with varying degrees of convergence. Masson and Scanlon (1991), on the basis of their analysis of the neotectonics of the Puerto Rico area, proposed an alternative scenario in which Puerto Rico behaves as a distinct platelet that is actively rotating counterclockwise between the Puerto Rico trench and Los Muertos trench within the overall left-lateral NCPBZ. Still other models suggest active transtensional and extensional deformation of the Hispaniola–Puerto Rico region (Larue and Ryan, 1990; Larue et al., 1990; Speed and Larue, 1991).

In the absence of high-resolution geodetic data for the northern Caribbean region, which are just becoming available (Farina et al., 1995; Dixon et al., 1998), the focal mechanisms that we have calculated, when coupled with geologic data, provide one of the most accurate means for placing constraints on the directions of relative plate motions in the region. The 1943 and 1946 earthquakes are among the largest (and the 1946 mainshock may, in fact, be the largest) events to have struck the northeast Hispaniola–northwest Puerto Rico area in at least 430 yr, and they record most of the strain released seismically along the plate interface during that time. It is therefore likely that they represent a significant portion of the long-term pattern of strain release along the margin. This inference is supported by the internal similarity of the focal mechanisms calculated for the individual earthquakes in the 1943–1953 sequence, as well as by the overall similarity of those focal mechanisms to mechanisms calculated for recent, moderate earthquakes that have occurred along the plate interface.

The rake that we calculated for the 1943 earthquake (030° in map view) indicates that during the earthquake motion along the east-northeast–trending plate interface (080° trend at the sea floor; ~065° trend at 25 km depth; Fig. 3) was highly oblique, with approximately equal amounts of convergence perpendicular to the margin and left-lateral motion parallel to it. The rake for the 1946 event (034° in map view) shows that motion along the plate interface during that earthquake was also oblique, but with only about one-half as much left-lateral motion parallel to the 100°-trending margin as thrust motion perpendicular to the margin (margin-parallel:margin-perpendicular ratio ~1:2). The rakes calculated for the other earthquakes from the 1946–1953 sequence give similar results (Fig. 6).

The earthquake data are supported by Dolan et al.'s (this volume) analysis of the active tectonics of the north-central Caribbean, which shows evidence of a convergent component of deformation along the entire northern boundary of the NCPBZ, from north of the Virgin Islands to north-central Hispaniola. Similarly, marine geophysical data from northwestern Hispaniola (Dillon et al., 1992) and southeastern Cuba (Calais et al., 1989, 1992, this volume) indicate that the contractional component of deformation continues as far west as the eastern end of the Cayman trough (Fig. 2). Even in a regime in which pure left-lateral motion would occur on east-west planes (e.g., NUVEL-1; DeMets et al., 1990), such contractional structures would be expected along the west-northwest–trending plate margin in the Hispaniola restraining bend (Mann et al., 1984). The occurrence of active contractional structures along the 080–085°-trending plate margin off northern Puerto Rico, however, requires that the relative motion vector between the North America plate and the Hispaniola and Puerto Rico blocks of the NCPBZ must be oriented more northerly than 080° to the north of Puerto Rico, given that the eastern strand of the Septentrional fault and the northern Puerto Rico slope fault zone—the only major strike-slip faults in the region—strike approximately east-west and northeast, respectively.

The relative motion data discussed above are based on the mechanisms of earthquakes that occurred at depths of 19–49 km along the south-dipping plate interface. Dolan et al. (this volume) have shown that the Hispaniola–Puerto Rico margin exhibits well-developed strain partitioning, with contractional deformation largely decoupled from left-lateral strike-slip motion. In northern Hispaniola, most transcurrent motion is accommodated along the Septentrional fault system (Mann et al., 1984, this volume; Prentice et al., 1993). East of 68° 15′W strike-slip motion appears to be accommodated along both the Septentrional system and the northeastern Hispaniola slope–northern Puerto Rico slope fault systems (Fig. 2) (Masson and Scanlon, 1991; Dolan et al., this volume). In such a strain-partitioned regime, motions along the decoupled contractional and transcurrent structures sum to the overall relative plate motion. These partitioned structures must merge at depth, with the deep portions of the transpressional plate boundary moving in the overall (oblique) relative plate motion direction (i.e., east-northeast). In the north-central Caribbean region, the partitioned strike-slip and contractional faults probably merge along a shallowly south-dipping decollement at the top of the underthrust North America (Atlantic) slab (Dolan et al., this volume).

For the rake of an earthquake to represent the true relative plate motion, partitioning of the overall oblique slip onto parallel strike-slip and contractional faults must occur *updip* of the rupture plane of the plate-interface earthquake. In other words, if a strike-slip fault merges with the decollement deeper than the rupture zone of a plate-interface earthquake, then the earthquake will exhibit a focal mechanism that is more contractional than the overall relative plate motion. Overall relative plate motion in such cases will depend on the relative slip directions and slip rates along the plate-interface thrust fault and the parallel strike-slip fault. Thus, the exact nature and location of the mechanical interaction between the strike-slip faults and the plate-interface decollement are critical for correctly interpreting the earthquake rake data.

Although we do not have definitive data on the location at which the partitioned strike-slip faults of the north-central Caribbean merge with the decollement, the marine geophysical data discussed in Dolan et al. (this volume) provide a general idea of where such interactions must take place. The mainshock focus of the 1943 northern Mona Passage earthquake was located almost directly beneath the eastern strand of the left-lateral Septentrional fault system, well to the south of the northern Puerto Rico slope fault zone (NPRSFZ) (Fig. 3). Thus, strike-slip on the NPRSFZ is partitioned off the south-dipping plate-interface oblique-thrust fault *updip* from the 1943 rupture plane. The rake from the 1943 earthquake can therefore be used to define the relative motion direction between the Puerto Rico block and the North America (Atlantic) slab for the area to the north (i.e., updip) of the 1943 rupture plane. In contrast, the rupture plane lies beneath, and locally slightly to the south of, the sea-floor trace of the Septentrional fault (Fig. 5). Although we do not known the exact nature and location of the interaction of the eastern strand of the Septentrional fault with the decollement, it would appear from the location of the surface trace of the fault that at least some left-lateral strike-slip motion may be partitioned off the decollement *downdip* from the 1943 earthquake plane. If so, then the rake derived from the 1943 earthquake focal mechanism would indicate more of a compressional component of deformation than the overall relative plate motion that is occurring downdip of the zone of strain partitioning. In other words, overall relative plate motion between the North America plate and the Puerto Rico block is probably more easterly than the 030° direction implied by the 1943 focal mechanism.

The data from the 1946 northeastern Hispaniola earthquake are similar. The sea-floor trace of the northeastern Hispaniola slope fault zone (NEHSFZ), a left-lateral strike-slip splay of the Septentrional fault system, is located only ~10–20 south of the deformation front, more than 40 km north of the epicenter of the 1946 mainshock (Fig. 3). Thus, assuming that the NEHSFZ dips steeply or vertically, most strike-slip motion on this fault is partitioned off the south-dipping plate interface updip from the major part of the 1946 rupture plane. In contrast, the 1946 epicenter was located just south of the surface trace of the Septentrional fault. Most of the aftershocks of the 1946 earthquake that were associated with the rupture plane, as well as the 1953 earthquake, occurred to the north of the Septentrional fault, indicating that the rupture plane extends well to the north of the Septentrional fault. The location of the Septentrional fault therefore suggests that a large percentage of the left-lateral motion is partitioned off the plate-interface fault to the south (i.e., downdip) of the 1946 rupture plane. As with the 1943 event, the rake determined for the 1946 earthquake therefore exhibits motion that is more contractional than the overall relative plate motion.

The data for the 1943 and 1946 earthquakes show that overall relative motion between the underthrust North America (Atlantic) plate and the Hispaniola and Puerto Rico blocks of the NCPBZ cannot be oriented more northerly than ~030°. Furthermore, the oblique-thrust focal mechanism for the 1943 event, together with geologic evidence showing some convergence along the margin north of Puerto Rico, requires that overall relative motion cannot be oriented more easterly than ~080° north of Puerto Rico. Thus, the relative plate motion direction along the northern deformation belt of the NCPBZ in the Puerto Rico–Hispaniola region lies between ~030° and 080°. These data cast doubt on models in which the direction of relative plate motion along the northern edge of the NCPBZ is oriented east-west (e.g., DeMets, 1993; Farina et al., 1995).

The exact relative plate motion vector depends on the ratio of slip rates on the left-lateral Septentrional fault system and the plate-interface oblique-thrust fault. The higher the slip rate on the Septentrional fault system, the closer the overall relative plate motion vector will be to 080°. Conversely, the higher the slip rate along the plate-interface thrust fault, the closer the overall relative motion vector will be to 030°. No well-constrained slip-rate data are currently available for motion along either the plate interface or the left-lateral strike-slip faults of the Septentrional fault system, although Mann et al. (this volume) show two independent estimates of the maximum possible slip rate for the Septentrional fault of 17 ± 6 and 19 ± 5 mm/yr.

Recent GPS geodetic data suggest that the Septentrional fault may accommodate only about 40% to 50% of the total motion along the NCPBZ (Dixon et al., 1998). Furthermore, these data imply that significant left-lateral motion occurs to the north of the Septentrional fault, offshore of Hispaniola. Because the oblique-left-lateral–thrust fault along the south-dipping plate interface is the dominant structure offshore northern Hispaniola, we suggest that it is likely that much of this off-Septentrional fault motion probably occurs during oblique slip on the south-dipping plate-interface fault. If so, then the offshore oblique thrust fault may have a relatively fast slip rate. Alternatively, some of the offshore left-lateral motion suggested by the GPS data may be accommodated on the northeastern Hispaniola slope fault and/or the Camu fault. Determination of the true slip rates of the major faults within the northern deformation belt of the NCPBZ will require more occupations of the recently established northern Caribbean geodetic network (Dixon et al., 1998), as well as the use of more sophisticated models that incorporate realistic fault geometries and the rakes derived from our focal mechanisms derived from events on the offshore plate-interface thrust fault.

The lack of detailed slip-rate data, coupled with our inability to determine exactly where strike-slip motion is partitioned off the plate interface, precludes an exact assessment of relative plate motion in north-central Caribbean, beyond placing maximum limits on the orientation of the relative motion vector between 030° and 080°. Even with detailed GPS geodetic data, care must be taken to model the effects of interseismic strain accumulation on the offshore oblique-thrust fault, as well as the Septentrional strike-slip fault. Other possible complications include the strain fields generated by the northeastern Hispaniola slope and Camu left-lateral strike-slip faults.

The similarity of the focal mechanisms of the 1943 and 1946–1953 earthquakes argues against active rotation of Puerto Rico as a separate microplate, as proposed by Masson and Scanlon (1991). The Masson and Scanlon (1991) model predicts transtensional deformation northwest of Puerto Rico, in the region of the 1943 earthquake. Our focal mechanism for that event, however, reveals oblique convergence along this part of the plate boundary. As noted above, marine geophysical data offshore northern Puerto Rico also show transpressional deformation across the entire margin (Larue and Ryan, 1998; Dolan et al., this volume). We believe that the post–10.9 Ma clockwise rotation of Puerto Rico observed in paleomagnetic data (Reid et al., 1991) record an earlier, since-terminated tectonic event.

The data discussed above also rule out models in which the NCPBZ is currently undergoing transtensional deformation (e.g., Larue and Ryan, 1990; Larue et al., 1990; Speed and Larue, 1991). We do not mean to imply by this that normal faults do not exist along the plate margin. Masson and Scanlon (1991) and Dolan et al. (this volume) describe large, oblique-normal faults along the northern margin of the Puerto Rico trench. Minor normal faults also occur along the southern edges of Navidad and Silver banks. Furthermore, it is possible that some predominantly strike-slip faults that Dolan et al. (this volume) describe from the northeastern Hispaniola slope also exhibit local transtensional motion (e.g., bounding faults of the two large pull-apart basins northeast of Hispaniola). However, the earthquake data discussed above indicate that motion along the NCPBZ is fundamentally transpressional, not transtensional, in the Hispaniola–Puerto Rico region.

In summary, the earthquake data presented in this chapter, coupled with geological evidence for a minor component of contractional deformation along the 080–085°-trending plate boundary north of Puerto Rico (Dolan et al., this volume), indicate that overall relative plate motion between the North America (Atlantic) plate and the Hispaniola and Puerto Rico blocks of the NCPBZ is oriented between ~030° and 080°. These data support earlier models in which relative plate motion is east-northeast in the region (e.g., Bracey and Vogt, 1970; Schell and Tarr, 1978; Sykes et al., 1982; Deng and Sykes, 1995).

Dolan et al. (this volume) point out that some north-south shortening also occurs along north-dipping thrust faults within the Los Muertos deformed belt (Fig. 2). This observation indicates that the relative motion vector between the true Caribbean plate, which lies south of the NCPBZ, and the North America plate, which lies north of the NCPBZ, must be oriented more northeasterly than the relative motion vector between the North America plate and the Hispaniola and Puerto Rico blocks. The true orientation of the plate motion vector between the Caribbean and North America plates depends on the slip rates of the plate-boundary–parallel, left-lateral Septentrional–northern Puerto Rico slope and Enriquillo-Plantain Garden fault systems, relative to the slip rates of the oblique-contractional and contractional fault systems of the northern Hispaniola and Los Muertos deformed belts—the higher the contractional slip rates relative to the strike-slip rates, the more northerly the true plate motion vector will be oriented; the higher the left-lateral strike-slip rates, the more easterly the relative motion vector will be oriented. Recent GPS geodetic data, however, suggest that rates of north-south shortening across the northern Caribbean plate boundary are probably only a few mm/yr (Dixon and Mao, 1997). Their data show that north-south convergence between the North America and South America plates is ~3.5 mm/yr at 65° W, increasing steadily to ~6 mm/yr at 75° W. Because some of this north-south motion must be accommodated along contractional structures along the southern Caribbean plate boundary zone (see Mann et al., 1990, for an overview of southern Caribbean tectonics), north-south shortening across the Los Muertos deformed belt is probably slow.

SEISMIC HAZARDS OF THE NORTHERN CARIBBEAN REGION

In this section we discuss the seismic hazards associated with major faults within the northern Caribbean plate boundary zone in the Hispaniola–Puerto Rico region. Historic earthquakes define two major belts of active faulting along the northern and southern boundaries of the NCPBZ (Fig. 2). The northern deformation belt comprises the offshore, plate-interface, oblique-thrust fault as well as the Septentrional, northern Puerto Rico slope, and northeastern Hispaniola slope left-lateral strike-slip faults. The southern deformation belt comprises the north-dipping Los Muertos plate-interface thrust fault and the Enriquillo-Plantain Garden left-lateral strike-slip fault.

Northern deformation zone of the NCPBZ

Earthquakes generated by slip on the plate-interface oblique thrust fault. In this chapter, we have documented the faulting parameters and tectonic setting of the 1943 and 1946 earthquakes, which we interpret to have occurred on the south-dipping, oblique-thrust fault separating the downgoing North America (Atlantic) slab from the overriding Hispaniola and Puerto Rico–Virgin Island blocks of the NCPBZ. The interplate thrust fault can certainly generate M ~8 earthquakes, such as the 1946 event, that are capable of producing extensive damage (Lynch and Bodle, 1948; Small, 1948). The gentle southward dip of this fault results in much greater potential fault-plane area relative to fault length than for the more steeply dipping strike-slip faults of the Septentrional system. Balancing this, however, is the fact that the subduction thrust fault lies well offshore of both Hispaniola and Puerto Rico. Even with the gentle dip of this fault, it lies ~25–30 km deep beneath the population centers of northern Hispaniola, and 75–100 km deep beneath the north coastal region of Puerto Rico (Fig. 3). Furthermore, as with most subduction-zone thrust-fault earthquakes worldwide, rupture in both the 1943 and 1946 mainshocks initiated near the base of the fault plane and propagated updip toward the deformation front—away from populated areas. In the case of the 1943 mainshock, rupture also appears to have propagated unilaterally to the west, directly away from the population centers of coastal Puerto Rico. Source directivity is not always so beneficial, however, as in the case of the 1946 mainshock, where probable unidirectional westward propagation—inferred from the location of the earthquake focus at the far eastern end of the rupture plane—probably directed much of the energy toward the major cities of the Cibao Valley and northern coast of Hispaniola.

In addition to the 1943 and 1946 thrust earthquakes, a poorly reported earthquake in 1787 caused strong ground shaking (Modified Mercalli Intensity [MMI] VIII) over a 150-km length of the northern coastal region of Puerto Rico (McCann and Sykes, 1984). This is the only potential great earthquake to have affected northern Puerto Rico during the >400-yr-long historic period (McCann and Sykes, 1984). The widespread nature of these reports of strong ground shaking led McCann and Sykes (1984) to propose that the 1787 earthquake was generated by rupture of a large part of plate-interface thrust fault (Fig. 2). The poorly defined location of the 1787 event suggests that it, like the 1943 and 1946 events, may have nucleated at a collisional asperity. We speculate that the 1787 earthquake may have nucleated at a collision involving Main Ridge, a west-northwest–trending ridge located 125 km north-northeast of Puerto Rico (Fig. 1). McCann and Sykes (1984) suggest that the Main Ridge is an aseismic ridge that, like the Bahamas, developed along a fracture zone associated with Mesozoic opening of the Atlantic Ocean. The proposed Main Ridge collision is the site of a persistent cluster of seismicity (McCann and Sykes, 1984; Dolan et al., this volume). Historical accounts of the 1787 event must be better documented to determine whether this event was, or was not, a great thrust earthquake.

The damage patterns reported for the 1787 event could also have been generated by a large, offshore strike-slip earthquake on the eastern continuation of the Septentrional fault system.

Accurate estimates of the seismic hazard posed by the south-dipping plate-interface thrust to the population centers of northern Hispaniola and Puerto Rico are complicated by the absence of a well-constrained slip rate for the fault (see discussion above), as well as by our lack of knowledge about the ratio of seismic to aseismic slip that occurs along the plate-interface thrust outside the collisional asperities. Determination of this parameter is of obvious importance for seismic hazard assessments for the region. If the plate-interface thrust fault is strongly coupled, even in noncollisional areas, then the seismic hazard to the Virgin Islands, Puerto Rico, and Hispaniola is much higher than if the noncollisional parts of the plate boundary are slipping largely aseismically.

Septentrional fault system. Slip on the Septentrional fault probably generated the large northwestern Hispaniola earthquakes of 1842 and 1887 (Fig. 2) (McCann and Pennington, 1990; Mann et al., this volume). The great length of the zone of high shaking intensities during the 1842 event suggests that it may have been a great earthquake (Scherer, 1912; Mann et al., 1984, this volume). Other evidence for the potential of the Septentrional fault system to produce large earthquakes comes from paleoseismological trenching along the central reach of the fault in Hispaniola. There, Prentice et al. (1993) and Mann et al. (this volume) have shown that the most recent surface rupture on the fault produced several meters of slip. These data indicate that this earthquake was almost certainly in excess of M 7. The recurrence of such a large event today in the rapidly developing northern part of Hispaniola would have devastating consequences given the high population density and frequently poor quality of construction in the region. The city of Santiago, Dominican Republic, in particular, with a population of ~750,000, lies within a few kilometers of the fault (Fig. 3) (Prentice et al., 1993; Mann et al., this volume). Old Santiago was destroyed in the earthquake of November 2, 1564, although it remains unclear what fault generated that event (Prentice et al., 1993; Mann et al., this volume).

Strike-slip faults offshore northern Puerto Rico Dolan et al. (this volume) show that the Septentrional fault system (SFS) extends eastward offshore Hispaniola. The fault bifurcates at ~68° 30′W, with one strand extending eastward toward the northern slope of Puerto Rico, and another strand extending northeastward to connect with the northern Puerto Rico slope fault zone (NPRSFZ) (Fig. 3) (Masson and Scanlon, 1991; Dolan et al., this volume). The presence of at least one active strike-slip fault of the eastern SFS beneath the northern Puerto Rico slope, within ~60 km of the northwestern tip of the island, represents a major potential hazard to the densely populated northern coastal region of Puerto Rico. In contrast, the NPRSFZ probably does not represent a major, independent seismic hazard of concern for northern Puerto Rico. The fault is located more than 110 km north of the Puerto Rico coast (Fig. 2). Furthermore, the location of the fault within a few kilometers of the deformation front of the south-dipping subduction thrust indicates that the NPRSFZ must merge downward into the subduction thrust at a depth of only a few kilometers (Dolan et al., this volume). Thus, the NPRSFZ is probably not capable of generating damaging earthquakes by itself, independently of major earthquakes generated by slip on the plate-interface, oblique-subduction thrust fault.

Normal fault earthquakes in Mona Passage. Mona Canyon, offshore of northwestern Puerto Rico, is controlled by large normal faults (Dolan et al., this volume). The steep eastern wall of the canyon, in particular, appears to be controlled by a single normal fault. These faults must be considered a potential seismic source, particularly in light of the fact that they lie within ~25 km of the northwestern tip of Puerto Rico. The location of the 1918 M_s ~7.5 Mona Passage earthquake suggests that it may have been generated by slip on a normal fault within Mona Passage (McCann and Sykes, 1984). The 1918 event resulted in extensive damage and loss of life in cities along the western coast of Puerto Rico (Reid and Taber, 1919, 1920).

Normal fault earthquakes in the underthrust Atlantic slab. Dolan et al. (this volume) ascribe the normal faults observed along the northern edge of the Puerto Rico trench to bending of the Atlantic slab during highly oblique underthrusting. On December 25, 1969, an M_s 7.5 normal fault earthquake occurred in the Atlantic oceanic slab being subducted beneath the Lesser Antilles island arc (Stein et al., 1982, 1983). Similar normal fault earthquakes along the northern margin of the Caribbean plate may pose a significant seismic hazard to the region. The direction of underthrusting in the northern Caribbean is highly oblique, however, and consequently the rate of slab bending is much slower than along the Lesser Antilles subduction zone. Thus, normal fault earthquakes associated with slab bending along the northern plate boundary will be less frequent and/or smaller than those along the Lesser Antilles margin where subduction is nearly perpendicular to the plate boundary. Nevertheless, the increase in deep, intraslab seismicity following the 1946 northeast Hispaniola mainshock shows that the oblique thrust component of motion associated with oblique underthrusting does cause readjustment of the underthrust slab after major slab-interface thrust earthquakes.

Tsunami hazards. Earthquakes on the offshore subduction thrust fault can cause destructive tsunamis (e.g., numerous small- to moderate-sized tsunamis generated by the 1946 northeastern Hispaniola earthquake) that are a cause for concern for the many low-lying coastal communities in Hispaniola and Puerto Rico. Such tsunamis can be generated either by displacement of the sea floor associated with updip rupture of the thrust fault to the sea floor, as well as by underwater landsliding. Because strong ground motions can trigger these underwater landslides, earthquakes on offshore strike-slip faults (e.g., eastern Septentrional fault system; northeastern Hispaniola slope fault; northern Puerto Rico slope fault), although they may not directly cause major elevation changes in the sea floor through fault slip, must also be considered potential triggers for destructive tsunamis. The numerous tsunamis that occurred for several days after the 1946 northeastern Hispaniola mainshock were probably generated by landsliding triggered by aftershocks.

Southern deformation zone of the NCPBZ

Los Muertos deformed belt. The Los Muertos deformed belt (LMDB) forms the southern edge of the NCPBZ for more than 650 km, from the Anegada Passage in the east to the San Pedro basin south of central Hispaniola in the west (Fig. 3). Active accretionary prism development within the LMDB is characterized by offscraping of Caribbean sediments and the development of north-dipping thrust faults and folds above a gently north-dipping thrust fault. This thrust fault defines the boundary between the overriding Puerto Rico–Virgin Islands and Hispaniola blocks of the NCPBZ and the true Caribbean plate, which lies to the south of the NCPBZ (Ladd and Watkins, 1978; Masson and Scanlon, 1991). The 1984 M_s 6.7 San Pedro basin thrust earthquake, the largest instrumentally recorded event to occur within the LMDB, demonstrates that the subduction thrust along the top of the underthrust Caribbean slab is seismogenic (Byrne et al., 1985). Little is known about this major seismic source, save its basic geometry as defined by side-scan sonar and seismic-reflection surveys and the dip of the Wadati-Benioff zone (Ladd and Watkins, 1978; Byrne et al., 1985; Ladd et al., 1990; Masson and Scanlon, 1991; Dolan et al., this volume). No known large historical earthquakes can be reliably associated with the Los Muertos subduction thrust, despite the >400-yr-long historic record in Puerto Rico.

Anegada Passage. The Anegada Passage, which extends northeastward across the Virgin Islands, connects the Los Muertos deformed belt with the Puerto Rico trench at ~63° W (Fig. 3). The Anegada Passage appears to encompass a zone of northeast-trending strike-slip faults and several pull-apart basins (Murphy and McCann, 1979; Jany et al., 1990). Isoseismals and maps of seismic seawaves generated by the large earthquake of 1867 indicate that the event occurred within the Anegada Passage (Reid and Taber, 1920). McCann and Sykes estimated a magnitude of ~7.75 for this event.

Enriquillo-Plantain Garden fault. First identified by Mann et al. (1984), the Enriquillo-Plantain Garden fault zone is a major left-lateral strike-slip fault that extends from near the western termination of the Los Muertos deformed belt in south-central Hispaniola westward for >800 km through the southern peninsula of Haiti and offshore through Jamaica to the southern edge of the Cayman trough (Fig. 2). No paleoseismological information is available for this fault, but it may have generated many of the major earthquakes that have struck southern Hispaniola (Fig. 2) (Mann et al., 1984, this volume; Chalas-Jimenez, 1989; McCann and Pennington, 1990). By analogy to the strain-partitioned system of faults beneath the northern zone of deformation of the NCPBZ, earthquakes generated by slip on thrust faults oriented parallel to the Enriquillo-Plantain Garden fault zone must also be considered a potential hazard to the southern Hispaniola and Jamaica regions (Mann et al., this volume).

Temporal occurrence of major earthquakes in the NCPBZ

Earthquakes in the NCPBZ in the Hispaniola area have clustered in two major temporal and geographic sequences—a period of numerous large events along the southern deformation belt during the 17th and 18th centuries, and a period of elevated activity along the northern belt of the NCPBZ during the 19th and 20th centuries (Figs. 2 and 10) (Scherer, 1912; Reid and Taber, 1919, 1920; Lynch and Bodle, 1948; Kelleher et al., 1973; Chalas-Jimenez, 1989). We refer to this alternation of major earthquake activity between the two deformation belts as "deformation switching."

Since 1842, the northern deformation zone of the NCPBZ has been in an active period of relatively frequent earthquakes (Fig. 10). These earthquakes include large events in 1842 and 1884 that probably occurred on the Septentrional fault system in northwestern Hispaniola (Mann et al., 1984, this volume), the 1917 and 1918 Mona Passage earthquakes (which may have occurred on either normal faults within Mona Passage [McCann and Sykes, 1984], the eastern Septentrional fault system, or the plate-interface thrust), and the 1943 northern Mona Passage and 1946 northeastern Hispaniola earthquakes, which occurred on the plate-interface oblique-thrust fault (this study) (Fig. 2). The 1953 Silver Spur earthquake was the most recent major event in this active period.

The active period along the southern deformation belt of the NCPBZ began in the early 1600s and included large earthquakes in October and November 1751, 1770, and 1860 (Fig. 2) (Scherer, 1912). These four events define a westward-propagating sequence that may have occurred along the Enriquillo-Plantain Garden fault system in southern Hispaniola (Fig. 2) (Mann et al., this volume). The occurrence of a tsunami during the October 1751 event, however, led McCann and Sykes (1984) to suggest the alternative possibility that this event may have occurred along the westernmost part of the north-dipping, plate-interface thrust fault beneath the Los Muertos deformed belt. The occurrence of a tsunami during an earthquake, however, does not provide definitive proof of an offshore source, as strong ground shaking generated by an earthquake on a nearby onshore fault could also trigger submarine landslides. The most likely possibility is that the 1751–1860 southern Hispaniola earthquake sequence was a domino-like, east-to-west "cascade" of events similar to the 1939–1967 sequence of earthquakes along the North Anatolian fault in Turkey, in which ~1,000 km of that large strike-slip fault broke in a westward-propagating series of large to great earthquakes (Ambraseys, 1970; Barka, 1992).

Although the first major event in the northern active sequence (1842) occurred before the final major event in the southern active period (1860), during most of the 250-yr-long southern active period, no large earthquakes appear to have occurred within the northern zone of deformation. Conversely, no large events—aside from the moderate-sized (M_s 6.7) 1984 San Pedro basin earthquake beneath the Los Muertos deformed belt—have occurred along the southern deformation belt since 1860 (Fig. 10). The southern active period was preceded by the 1564 earthquake, which devastated the city of Santiago in the Cibao Valley of northern Hispaniola (Scherer, 1912; Prentice et al., 1993; Mann et al., this volume; L. Feldman and W. R. McCann, written communication, 1995). This is apparently the earliest well-recorded event in

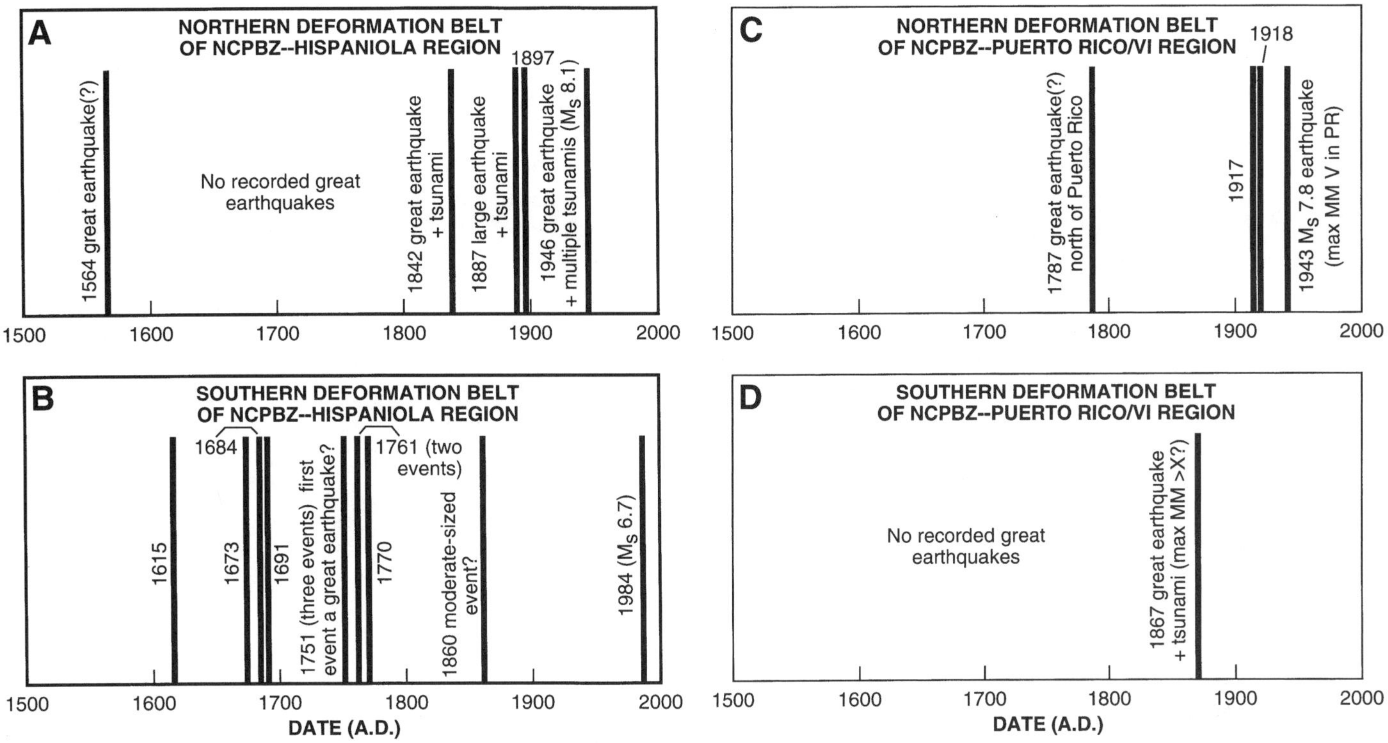

Figure 10. Plots of temporal and geographic distribution of major historical earthquakes in the Hispaniola–Puerto Rico region of the northern Caribbean plate boundary zone (NCPBZ) (modified from Chalas-Jimenez, 1989). Catalog is probably complete for large to great earthquakes since ca. 1800. Although some large offshore events may be missing during this period, it is unlikely that any great earthquakes were not recorded during the past 200 yr (Kelleher et al., 1973). Catalog prior to 1800 is of variable quality and may be incomplete, especially for offshore earthquakes. Events listed here had maximum Modified Mercali Intensities of >VIII, and many exhibited MM Intensities of X and XI. (A) Occurrence of large recorded earthquakes in the northern belt of deformation of the NCPBZ in the Hispaniola region (includes offshore, south-dipping interplate thrust fault and the Septentrional fault system). Note cluster of events beginning in 1842 and apparent absence of great earthquakes between 1564 and 1842. (B) Occurrence of large earthquakes in the southern deformation belt of the NCPBZ in the Hispaniola region (includes the left-lateral Enriquillo-Plantain Garden strike-slip fault system [Mann et al., 1984, this volume] and offshore thrust faults of the western Los Muertos deformed belt). Note cluster of large to great earthquakes during 17th and 18th centuries, as well as absence of large earthquakes since 1860. This cluster appears to be anticorrelated with the 19th and 20th century cluster of events in the northern deformation belt. (C) Occurrence of large recorded earthquakes along the northern deformation belt of the NCPBZ in the Puerto Rico–Virgin Islands region (includes offshore, south-dipping interplate thrust fault and left-lateral strike-slip faults of the eastern Septentrional and northern Puerto Rico slope fault systems). In general, Puerto Rico appears to have experienced far fewer large to great earthquakes than Hispaniola (Kelleher et al., 1973). This does not appear to be a result of better record keeping in Hispaniola than Puerto Rico, as the historical record of earthquakes in Puerto is actually better preserved than that for Hispaniola (L. Feldman, personal communication, 1995). Rather, it may be a function of the fact that all of the major active plate-boundary faults are offshore in the Puerto Rico region, in contrast to Hispaniola, where the major strike-slip faults are onshore for most of their lengths. In addition to a great earthquake in 1787, the northern deformation zone off Puerto Rico produced a cluster of earthquakes in the first half of the 20th century. The largest of these shocks was the 1943 Mona Passage earthquake. This cluster is broadly contemporaneous with the recent cluster in northern Hispaniola. (D) Occurrence of large earthquakes in the southern deformation belt of the NCPBZ in the Puerto Rico–Virgin Islands area. The only well-recorded, large event to strike this region during the historic period was the great 1867 Anegada Passage earthquake.

the Western Hemisphere (L. Feldman and W. R. McCann, written communication, 1995). Thus, it remains unknown whether it records the final large event in an earlier northern active period.

Far fewer large earthquakes have been recorded in the Puerto Rico–Virgin Islands areas than in the Hispaniola region, despite comparable historical records in both areas (Kelleher et al., 1973; McCann and Pennington, 1990). In the 450-yr-long historic record from the Puerto Rico area, only the poorly recorded 1787 event (McCann and Sykes, 1984), the Mona Passage earthquakes of 1917, 1918, and 1943, and the Anegada Passage earthquake of 1867 have ruptured large parts of the plate-boundary faults (Figs. 2 and 10). Unlike Hispaniola, however, all of the major plate-boundary faults in the Puerto Rico region (northern Puerto Rico slope fault zone, eastern Septentrional fault zone, Mona Passage normal faults, south- and north-dipping plate-interface thrusts north and south of the island) are located offshore. Offshore earthquakes are less likely to be recorded than onshore events, and the catalog for Puerto Rico may be missing large events. Nevertheless, with the possible exception of the 1787 event north of Puerto Rico, no large earthquakes appear to have occurred near San Juan, Puerto Rico, since the mid-16th century (Kelleher et al., 1973).

Because of the rarity of recorded events and the possibility of a significantly incomplete catalog, we cannot determine whether the Puerto Rico–Virgin Islands region has experienced alternating periods of activity on the northern and southern deformation belts of the NCPBZ. We can say, however, that the 1917–1943 sequence of earthquakes in the northern Mona Passage occurred within the active period along the northern deformation zone in Hispaniola (Fig. 10). As the Mona Passage is located adjacent to Hispaniola, this is perhaps not a surprise. It does, however, suggest that whatever mechanism is controlling the temporal and geographic clusters of major earthquakes operates over an area encompassing most of the northern belt of deformation, from northwestern Hispaniola to the Mona Passage, a distance of 750 km (Fig. 2).

Although it is difficult to generalize from only one cycle of "deformation switching," the striking temporal and geographic separation of the "active" and "quiet" earthquake periods suggests that active deformation alternates between fault systems along the northern and southern edges of the NCPBZ for time periods on the order of hundreds of years. We suggest that it is likely that within each of the deformation zones, an active sequence merely records a stress perturbation that propagates throughout large parts of the fault network. In other words, all but the first earthquake in each active sequence may be triggered by static stress changes (an increase in Coulomb failure stress) within the fault network caused by an earlier earthquake in that deformation belt. Although the mechanics of this stress triggering are not yet fully understood, this idea has recently been successfully used to explain much of the earthquake distribution in California during the past 185 yr (Harris and Simpson, 1992, 1996; Stein et al., 1992, 1994; Harris et al., 1995; Deng and Sykes, 1997) and along the North Anatolian fault during the 1939–1967 rupture sequence (Stein et al., 1997). Alternatively, time-dependent loading caused by viscous relaxation of the lower crust after a large earthquake may provide another mechanism for generating clusters of earthquakes (Ben-Zion et al., 1993).

Although the idea of static stress triggering provides a reasonable, and intuitively understandable, explanation for the clustered earthquake behavior within each of the deformation belts, it does not readily explain the apparent limitation of active deformation to one belt of deformation at a time. This pattern of earthquake occurrence raises a number of interesting questions: Why does deformation in one belt shut off when deformation begins in the other belt? Or does the locus of active deformation truly switch from one edge of the plate boundary zone to the other? Is the "quiet" edge of the plate boundary zone only quiet with respect to earthquakes? In other words, do aseismic processes dominate during the quiet intervals? More detailed geodetic measurements should be able to resolve the question of whether aseismic deformation is an important contributor to strain accommodation, especially on onshore faults in Hispaniola. The other questions, however, particularly the question of what triggered the 19th and 20th century active period in the northern belt of deformation, can probably best be addressed through numerical modeling, both of static stress changes related to earlier earthquakes and of the effects of viscous relaxation of the lower crust.

It is possible, of course, that the apparent switching of active deformation from the southern to the northern edge of the NCPBZ is merely a temporal and geographic coincidence—an artifact of a historical catalog that is not long enough. It is also possible that the "deformation switching" is an artifact of an incomplete historical record. However, although the possibility of an incomplete catalog cannot be ruled out, particularly for the earlier events, it seems unlikely that any large earthquakes are missing from the catalog since the mid-18th century, at least for onshore events. The possibility of an incomplete catalog needs to be investigated further, with a more complete review of original historical documents.

CONCLUSIONS

Our focal mechanisms show that the 1943 northern Mona Passage earthquake and the 1946 northeastern Hispaniola earthquake were both oblique-thrust events that occurred on the gently south-dipping plate-interface between the underthrust North America (Atlantic) plate and the Hispaniola and Puerto Rico blocks of the northern Caribbean plate boundary zone (NCPBZ). Furthermore, a comparison of the aftershock zones of these two events with marine geophysical data reveals that the earthquakes ruptured strongly coupled asperities on the plate-interface thrust associated with zones of active collisional underthrusting of high-standing carbonate banks of the southeastern Bahamas. These two earthquakes represent two of the largest events to strike the northeastern Hispaniola–northwestern Puerto Rico region during the past 400 yr, leading us to suggest that such large thrust events may typically nucleate only at strongly coupled collisional asperities on the plate interface. The rakes derived from our focal mechanisms for these events reveal a relatively constant, 030–035° direction of slip (in map view) along the plate interface. The northern deformation belt of the NCPBZ, however, exhibits well-developed strain partitioning, with much of the left-lateral slip along the plate boundary partitioned off the south-dipping plate interface and onto the steeply dipping Septentrional (and related) strike-slip faults. The depth at which the strike-slip fault merges with the south-dipping plate interface is poorly constrained, but it probably lies downdip of the 1943 and 1946 rupture planes. This indicates that the rakes derived from those events will show a plate motion direction that is more contractional than the true vector. Thus, the 030–035° orientation of slip along the south-dipping oblique-thrust plate-interface fault represents the most northerly possible orientation of the relative motion vector. Furthermore, the fact that the oblique-thrust 1943 event occurred along an 080°-trending margin requires that relative motion along the northern edge of the NCPBZ be oriented more northerly than 080°, given that the only major strike-slip faults in the region strike approximately east-west to east-northeast.

This estimate is in sharp contrast to published, preliminary GPS data (Farina et al., 1995), which show the overall motion vector oriented ~090°, as well as earlier studies of global plate models (e.g., NUVEL-1; DeMets et al., 1990; DeMets, 1993), which show predominantly east-west relative plate motion. More recent GPS data suggest that significant left-lateral motion may occur offshore northern Hispaniola (Dixon et al., 1998). The marine geophysical and seismicity data discussed by Dolan et al. (this volume) show that the south-dipping, plate-interface thrust fault is the dominant structure north of the Septentrional fault. Our focal mechanisms show that oblique-slip along this interface accommodates significant left-lateral motion. Thus, we suggest that it is likely that the slip rate on the offshore oblique thrust fault may be higher than previously assumed in models suggesting near-pure, east-west relative plate motion. The only other possible fault that could accommodate left-lateral slip north of the Septentrional fault in Hispaniola is the Camu fault. This fault, however, does not appear to be a continuous feature across the Silver-Navidad bank collision zone, and it may no longer be a major active fault. Any future modeling of GPS data must incorporate realistic slip vectors and geometries for the offshore plate-interface fault.

The south-dipping, plate-interface oblique-thrust fault and the Septentrional strike-slip fault—the two major faults within the northern deformation belt of the NCPBZ—represent only two of the seismic hazards affecting the north-central Caribbean region. Other major seismic sources that must be considered in any comprehensive seismic hazard assessment of the region include the two major faults of the southern deformation belt—the left-lateral Enriquillo-Plantain Garden strike-slip fault and the north-dipping Los Muertos plate-interface thrust fault—as well as normal faults in the Mona Passage and large, offshore normal-fault earthquakes within the underthrust Atlantic slab. Direct displacement of the sea floor during offshore thrust and normal-fault events represents a major hazard to the densely populated coastal regions of Hispaniola and Puerto Rico, as many cities are built close to sea level. Tsunamis may also be generated by underwater landslides triggered by strong shaking from nearby ruptures. Because such tsunamis do not require direct tectonic displacement of the sea floor, they can be generated by ruptures on onshore faults. The seismic risk is tremendously increased by the relatively poor construction techniques common to many parts of the region.

The 1943 and 1946 mainshocks were the most recent large earthquakes in a series of major events that occurred on faults in the northern part of the NCPBZ between 1842 and 1953. The historical record suggests that this earthquake cluster was preceded by more than 250 yr of relative quiescence on faults along the northern edge of the NCPBZ. Ruptures on faults along the southern edge of the NCPBZ (Enriquillo-Plantain Garden fault, Los Muertos thrust fault) exhibit exactly the opposite temporal sequence. An ongoing period of quiescence since 1860 was preceded by a 250-yr-long period of intense activity beginning in the early 1600s. Within each of the two belts of deformation, such on-off clustering of events may be triggered by either static stress changes (e.g., Harris and Simpson, 1992, 1996; Stein et al., 1992, 1994, 1997; Harris et al., 1995; Deng and Sykes, 1997) or by time-dependent loading related to viscous relaxation of the ductile lower crust (Ben-Zion et al., 1993). These mechanisms, however, fail to explain in any obvious way why earthquakes should occur in only one of the belts of deformation at a time. This pattern could, of course, be merely coincidental—the function of an incomplete or too-short historical record. Nevertheless, it represents a potentially fruitful subject for both static stress modeling and models of long-distance, long-term, visco-elastic interactions between widely separated fault networks.

ACKNOWLEDGMENTS

Dolan was supported by a grant from the National Science Foundation (OCE 8812082) and Wald was supported as a postdoctoral research fellow by the U.S. Geological Survey and the National Research Council. We would like to thank Paul Mann, Bill McCann, Eric Calais, and Carol Prentice for helpful discussions and thoughtful reviews. Larry Feldman and Bill McCann kindly shared information from their unpublished compilation of historical earthquakes in the northern Caribbean region.

REFERENCES CITED

Abe, K., 1981, Magnitudes of large shallow earthquakes from 1904–1980: Physics of the Earth and Planetary Interiors, v. 27, p. 72–92.

Ambraseys, N. N., 1970, Some characteristic features of the North Anatolian fault zone: Tectonophysics, v. 9, p. 143–165.

Austin, J. A., 1983, Overthrusting in a deep-water carbonate terrane, *in* Balley, A. W., ed., Seismic expression of structural styles: American Association of Petroleum Geologists Studies in Geology 15, p. 3.4.2-167–3.4.2-172.

Barka, A., 1992, The North Anatolian fault zone: Annales Tectonicae, Special Issue—Supplement to v. VI, p. 164–195.

Ben-Zion, Y., Rice, J. R., and Dmowska, R., 1993, Interaction of the San Andreas fault creeping segment with adjacent great rupture zones and earthquake recurrence at Parkfield: Journal of Geophysical Research, v. 98, p. 2135–2144.

Bracey, D. W., and Vogt, P. R., 1970, Plate tectonics in the Hispaniola area: Geological Society of America Bulletin, v. 81, p. 2855–2860.

Burke, K., 1988, Tectonic evolution of the Caribbean: Annual Review of Earth and Planetary Sciences, v. 16, p. 201–230.

Byrne, D. B., Suarez, G., and McCann, W. R., 1985, Muertos Trough subduction—Microplate tectonics in the northern Caribbean: Nature, v. 317, p. 420–421.

Calais, E., Mercier de Lépinay, B., Renard, V., and Tardy, M., 1989, Geometrie et regime tectonique le long d'une limite de plaques en coulissage: La frontiere nord-Caraibe de Cuba a Hispaniola, Grandes Antilles, (Geometry and tectonic regime along a strike-slip plate boundary: The northern Caribbean plate boundary from Cuba to Hispaniola, Greater Antilles): Comptes Rendus de l'Académie des Sciences Paris, v. 308, p. 131–135.

Calais, E., Bethoux, N., and Mercier de Lépinay, B., 1992, From transcurrent faulting to frontal subduction: A seismotectonic study of the northern Caribbean plate boundary from Cuba to Puerto Rico: Tectonics, v. 11, p. 114–123.

Case, J., and Holcombe, T., 1980, Geologic-tectonic map of the Caribbean region: U.S. Geological Survey Miscellaneous Investigations Map I-1100, scale 1:2,500,000.

Case, J. E., MacDonald, W. D., and Fox, P. J., 1990, Caribbean crustal provinces; seismic and gravity evidence, *in* Dengo, G., and Case, J. E., eds., The Caribbean region: Boulder, Colorado, Geological Society of America, The Geology of North America, v. H, p. 15–36.

Chalas-Jimenez, J. A., 1989, Probabilidad de un sismo catastrofico en la Republica Dominicana: Historia sismica, prediccion, y magnitudes esperadas: Revista Geofisica (Instituto de Geografia e Historia), v. 31, p. 185–193.

DeMets, C., 1993, Earthquake slip vectors and present-day plate motions: Journal of Geophysical Research, v. 98, p. 6703–6714.

DeMets, C., Gordon, R. G., Argus, D. F., and Stein, S., 1990, Current Plate Motions: Geophysical Journal International, v. 101, p. 425–478.

Deng, J., and Sykes, L. R., 1995, Determination of Euler pole for contemporary relative motion of Caribbean and North American plates using slip vectors of interplate earthquakes: Tectonics, v. 14, p. 39–53.

Deng, J., and Sykes, L. R., 1997, Evolution of the stress field in southern California and triggering of moderate-sized earthquakes: A 200-year perspective: Journal of Geophysical Research, v. 102, p. 9859–9886.

Dewey, J. W., and Suarez, G., 1991, Seismotectonics of Middle America, *in* Slemmons, D. B., Engdahl, E. R., Zoback, M. D., and Blackwell, D., eds., Neotectonics of North America: Boulder, Colorado, Geological Society of America Decade Map Volume 1, p. 309–321.

de Zoeten, R., Draper, G., and Mann, P., 1991, Geologic map of the northern Dominican Republic, scale 1:100,000, *in* Geologic and Tectonic development of the North America–Caribbean plate boundary in Hispaniola, Mann, P., Draper, G., and Lewis, J., eds., Geological Society of America Special Paper 262, Plate 1.

Dillon, W. P., Austin, J. A., Scanlon, K. M., Edgar, N. T., and Parson, L. M., 1992, Structure and development of the insular margin north of western Hispaniola: A tectonic accretionary wedge on the Caribbean plate boundary: Marine and Petroleum Geology, v. 9, p. 70–88.

Dillon, W. P., Edgar, N. T., Scanlon, K. M., and Coleman, D. F., 1994, A review of the tectonic problems of the strike-slip northern boundary of the Caribbean Plate and examination by GLORIA, *in* Gardner, J. V., Field, M. E., and Twichell, D. C., eds., Geology of the United States' Seafloor: The View from Gloria: United Kingdom, Cambridge University Press, p. 135–164.

Dixon, T. H., 1993, GPS measurement of relative motion of the Cocos and Caribbean plates and strain accumulation across the Middle American Trench: Geophysical Research Letters, v. 20, p. 2167–2170.

Dixon, T. H., and Mao, A., 1997, A GPS estimate of relative motion between North and South America: Geophysical Research Letters, v. 24, p. 535–538.

Dixon, T., Farina, F., DeMets, C., Jansma, P., Mann, P., and Calais, E., 1998, Relative motion between the Caribbean and North American plates and associated boundary zone deformation based on a decade of GPS observations: Journal of Geophysical Research (in press).

Edgar, N. T., 1991, Structure and geologic development of the Cibao Valley, *in* Mann, P., Draper, G., and Lewis, J. F., eds., Geologic and tectonic development of the North America–Caribbean plate boundary in Hispaniola: Geological Society of America Special Paper 262, p. 281–299.

Edgar, N. T., Ewing, J. I., and Hennion, J., 1971, Seismic refraction and reflection in the Caribbean Sea: American Association of Petroleum Geologists Bulletin, v. 55, p. 833–870.

Farina, F., Calais, E., DeMets, C., Dixon, T., Jansma, P., and Mann, P., 1995, GPS measurements across the northern Caribbean plate boundary zone: Preliminary results: Eos (Transactions of the American Geophysical Union), v. 76, p. S94.

Fischer, K. M., and McCann, W. R., 1984, Velocity modeling and earthquake relocation in the North East Caribbean: Seismological Society of America Bulletin, v. 74, p. 1249–1262.

Frankel, A., 1982, A composite focal mechanism for microearthquakes along the northeastern border of the Caribbean plate: Geophysical Research Letters, v. 9, p. 511–514.

Harris, R., A., and Simpson, R. W., 1992, Changes in static stress on southern California faults after the 1992 Landers earthquake: Nature, v. 360, p. 251–254.

Harris, R. A., and Simpson, R. W., 1996, In the shadow of 1857—Effect of the great Ft. Tejon earthquake on subsequent earthquakes in southern California: Geophysical Research Letters, v. 23, p. 229–232.

Harris, R. A., Simpson, R. W., and Reasonberg, P. A., 1995, Influence of static stress changes on earthquake locations in southern California: Nature, v. 375, p. 221–224.

Harvard Centroid Moment Tensor Catalog, 1993 (http://www.seismology.harvard.edu): Harvard University, Department of Earth and Planetary Sciences, Cambridge, Massachusetts.

International Seismological Centre Catalog of Earthquakes, 1994 (http://www.isc.ac.uk): Cambridge, U.K., International Seismological Centre.

International Seismological Summary of 1953, 1961, prepared and edited by the International Union of Geodesy and Geophysics, The British Association Seismological Committee: Richmond, Surrey, U.K., Kew Observatory, 823 p.

Jany, I., Scanlon, K. M., and Mauffret, A., 1990, Geological interpretation of combined SEABEAM, GLORIA, and seismic data from Anegada Passage (Virgin Islands, North Caribbean): Marine Geophysical Research, v. 12, p. 173–196.

Jordan, T. H., 1975, The present-day motions of the Caribbean plate: Journal of Geophysical Research, v. 80, p. 4433–4449.

Kafka, A. L., and Weidner, T. J., 1979, The focal mechanisms and depths of small earthquakes as determined from Rayleigh wave radiation patterns: Bulletin of the Seismological Society of America, v. 69, p. 1379–1390.

Kelleher, J., Sykes, L., and Oliver, J., 1973, Possible criteria for predicting earthquake locations and their application for major plate boundaries of the Pacific and the Caribbean: Journal of Geophysical Research, v. 78, p. 2547–2585.

Kikuchi, M., and Kanamori, H., 1991, Inversion of complex Body Waves-III: Bulletin of the Seismological Society of America, v. 81, p. 2335–2350.

Ladd, J. W., and Watkins, J. S., 1978, Tectonic development of trench-arc complexes on the northern and southern margins of the Venezuela Basin, *in* Watkins, J. S., Montadert, L., and Dickerson, P., eds., Geophysical and geological invetigations of continental margins: American Association of Petroleum Geologists Memoir 29, p. 363–372.

Ladd, J. W., Holcombe, T. L., Westbrook, G. K., and Edgar, N. T., 1990, Caribbean marine geology; Active margins of the plate boundary, *in* Dengo, G., and Case, J. E., eds., The Caribbean region: Boulder, Colorado, Geological Society of America, The Geology of North America, v. H., p. 261–290.

Larue, D., and Ryan, H., 1990, Extensional tectonism in the Mona Passage, Puerto Rico and Hispaniola: A preliminary study: Transactions of the 12th Caribbean Geological Conference, St. Croix, Virgin Islands: Miami,

Florida, Miami Geological Society, p. 223–230.

Larue, D. K., and Ryan, H. F., 1998, Seismic reflection profiles of the Puerto Rico Trench: Shortening between the North American and Caribbean plates, *in* Lidiak, E. G., and Larue, D. K., eds.: Boulder, Colorado, Geological Society of America Special Paper 322, p. 193–210.

Larue, D. K., Joyce, J., and Ryan, H. F., 1990, Neotectonics of the Puerto Rico Trench: Extensional tectonism and forearc subsidence: Transactions of the 12th Caribbean Geological Conference, St. Croix, Virgin Islands: Miami, Florida, Miami Geological Society, p. 231–247.

Lynch, J. J., and Bodle, R. R., 1948, The Dominican earthquakes of August, 1946: Bulletin of the Seismological Society of America, v. 38, p. 1–17.

MacDonald, K. C., and Holcombe, T. L., 1978, Inversion of magnetic anomalies and sea-floor spreading in the Cayman Trough: Earth and Planetary Science Letters, v. 40, p. 407–414.

Mann, P., and Burke, K., 1984, Neotectonics of the Caribbean: Reviews of Geophysics and Space Physics, v. 22, p. 309–362.

Mann, P., Burke, K., and Matumoto, T., 1984, Neotectonics of Hispaniola: Plate motion, sedimentation, and seismicity at a restraining bend: Earth and Planetary Science Letters, v. 70, p. 311–324.

Mann, P., Schubert, C., and Burke, K., 1990, Review of Caribbean neotectonics, *in* Dengo, G., and Case, J. E., eds., The Caribbean region: Boulder, Colorado, Geological Society of America, The Geology of North America, v. H., p. 405–432.

Mann, P., Taylor, F., Edwards, R. L., and Ku, T., 1995, Actively evolving microplate formation by oblique collision and sideways motion along strike-slip faults: An example from the northeastern margin of the Caribbean plate: Tectonophysics, v. 246, p. 1–69.

Masson, D. G., and Scanlon, K. M., 1991, The neotectonic setting of Puerto Rico: Geological Society of America Bulletin, v. 103, p. 144–154.

McCann, W. R., and Pennington, W. D., 1990, Seismicity, large earthquakes, and the margin of the Caribbean Plate, *in* Dengo, G., and Case, J. E., eds., The Caribbean Region: Boulder, Colorado, Geological Society of America, The Geology of North America, v. H, p. 291–306.

McCann, W. R., and Sykes, L. R., 1984, Subduction of aseismic ridges beneath the Caribbean Plate: Implications for the tectonics and seismic potential of the northeastern Caribbean: Journal of Geophysical Research, v. 89, p. 4493–4519.

Minster, J. B., and Jordan, T. H., 1978, Present-day plate motions: Journal of Geophysical Research, v. 83, p. 5331–5354.

Molnar, P., and Sykes, L. R., 1969, Tectonics of the Caribbean and Middle America regions from focal mechanisms and seismicity: Geological Society of America Bulletin, v. 80, p. 1639–1684.

Mullins, H. T., and Lynts, G. W., 1977, Origin of northwestern Bahama Platform: Review and reinterpretation: Geological Society of America Bulletin, v. 88, p. 1447–1461.

Mullins, H. T., Dolan, J., Breen, N., Andersen, B., Gaylord, M., Petruccione, J. L., Wellner, R. W., Melillo, A. J., and Jurgens, A. D., 1991, Retreat of carbonate platforms: Response to tectonic processes: Geology, v. 19, p. 1089–1092.

Mullins, H. T., Breen, N., Dolan, J., Wellner, R. W., Petruccione, J. L., Gaylord, M., Anderses, B., Melillo, A. J., Jurgens, A. D., and Orange, D., 1992, Carbonate platforms along the southeast Bahamas-Hispaniola collision zone: Marine Geology, v. 105, p. 169–209.

Murphy, A. J., and McCann, W. R., 1979, Preliminary results from a new seismic network in the northeastern Caribbean: Bulletin of the Seismological Society of America, v. 69, p. 1497–1513.

National Earthquake Center Catalog of Earthquakes, 1994 (http://wwwneic.cr.usgs.gov): Golden, Colorado, U.S. Geological Survey.

Pacheco, J. F., and Sykes, L. R., 1992, Seismic moment catalog of large, shallow earthquakes, 1900–1989: Bulletin of the Seismological Society of America, v. 82, p. 1306–1349.

Pindell, J. L., and Barrett, S. F., 1990, Geologic evolution of the Caribbean region; a plate-tectonic perspective, *in* Dengo, G., and Case, J. E., eds., The Caribbean region: Boulder, Colorado, Geological Society of America, The Geology of North America, v. H., p. 405–432.

Prentice, C. S., Mann, P., Taylor, F. W., Burr, G., and Valastro, S., 1993, Paleoseismicity of the North American–Caribbean plate boundary (Septentrional fault), Dominican Republic: Geology, v. 21, p. 49–52.

Reid, H., and Taber, S., 1919, The Puerto Rico earthquakes of October–November 1918: Bulletin of the Seismological Society of America, v. 9, p. 95–127.

Reid, H., and Taber, S., 1920, The Virgin Islands earthquakes of 1867–1868: Bulletin of the Seismological Society of America, v. 10, p. 9–30.

Reid, J. A., Plumley, P. W., and Schellekens, J. H., 1991, Paleomagnetic evidence for late Miocene counterclockwise rotation of north coast carbonate sequence, Puerto Rico: Geophysical Research Letters, v. 18, p. 565–568.

Rosencrantz, E., 1995, Opening of the Cayman Trough and evolution of the northern Caribbean plate boundary: Geological Society of America Abstracts with Programs, v. 27, no. 6, p. A–153.

Rosencrantz, E., and Mann, P., 1991, SeaMARC II mapping of transform faults in the Cayman Trough, Caribbean Sea: Geology, v. 19, p. 690–693.

Rosencrantz, E., Ross, M., and Sclater, J. G., 1988, Age and spreading history of the Cayman Trough as determined from depth, heat flow, and magnetic anomalies: Journal of Geophysical Research, v. 93, p. 2141–2157.

Russo, R. M., and Villaseñor, A., 1995, The 1946 Hispaniola earthquake and the tectonics of the North America–Caribbean plate boundary zone, northeastern Hispaniola: Journal of Geophysical Research, v. 100, p. 6265–6280.

Russo, R. M., and Villaseñor, A., 1997, The 1946 Hispaniola earthquake and the tectonics of the North America–Caribbean plate boundary zone, northeastern Hispaniola: Reply: Journal of Geophysical Research, v. 102, p. 793–802.

Schell, B. A., and Tarr, A. C., 1978, Plate tectonics of the northeastern Caribbean Sea region: Geologie En Mijnbouw, v. 57, p. 319–324.

Scherer, J., 1912, Great earthquakes in the island of Haiti: Bulletin of the Seismological Society of America, v. 2, p. 161–180.

Seismological Notes, 1946, Bulletin of the Seismological Society of America, v. 36, p. 373–383.

Small, W. M., 1948, A short description of the general geology of the Dominican Republic, with notes on the earthquake of August 4, 1946: Bulletin of the Seismological Society of America, v. 38, p. 19–32.

Speed, R. C., and Larue, D. K., 1991, Extension and transtension in the plate boundary zone of the northeastern Caribbean: Geophysical Research Letters, v. 18, p. 573–576.

Stein, R. S., King, G. C. P., and Lin, J., 1992, Change in failure stress on the southern San Andreas fault system caused by the 1992 Magnitude M=7.4 Landers earthquake: Science, v. 258, p. 1328–1332.

Stein, R. S., King, G. C. P., and Lin, J., 1994, Stress triggering of the 1994 M=6.7 Northridge, California, earthquake by its predecessors: Science, v. 265, p. 1432–1435.

Stein, R. S., Barka, A., and Dieterich, J. H., 1997, Progressive failure on the North Anatolian fault since 1939 by earthquake stress triggering: Geophysical Journal International, v. 128, p. 594–604.

Stein, S., Englen, J., Wiens, D., Speed, R., and Fujita, K., 1982, Subduction seismicity and tectonics in the Lesser Antilles arc: Journal of Geophysical Research, v. 87, p. 8642–8664.

Stein, S., Engeln, J. F., Wiens, D. A., Speed, R. C., and Fujita, K., 1983, Slow subduction of old lithosphere in the Lesser Antilles: Tectonophysics, v. 99, p. 139–148.

Stein, S., DeMets, C., Gordon, R. G., Brodholt, J., Argus, D., Engeln, J. F., Lundgren, P., Stein, C., Wiens, D. A., and Woods, D. F., 1988, A test of alternative Caribbean Plate relative motion models: Journal of Geophysical Research, v. 93, p. 3041–3050.

Sykes, L. R., and Ewing, M., 1965, The seismicity of the Caribbean region: Journal of Geophysical Research, v. 70, p. 5065–5074.

Sykes, L. R., and McCann, W. R., and Kafka, A. L., 1982, Motion of the Caribbean plate during the last 7 million years and implications for earlier Cenozoic movements: Journal of Geophysical Research, v. 87, p. 10656–10676.

Manuscript Accepted by the Society January 9, 1998

Printed in the U.S.A.

Index

[Italic page numbers indicate major references]

Printed in U.S.A.